PLANT CHROMOSOMES

PLANT CHROMOSOMES

Analysis, Manipulation and Engineering

Arun Kumar Sharma

and

Archana Sharma

Centre of Advanced Study (Cell and Chromosome Research)
Department of Botany, University of Calcutta
35, Ballygunge Circular Road
Calcutta 700 019
India

CRC Press
Taylor & Francis Group
Boca Raton London New York

CRC Press is an imprint of the
Taylor & Francis Group, an **informa** business

First published 1999 by Harwood Academic Publishers

Published 2019 by CRC Press
Taylor & Francis Group
6000 Broken Sound Parkway NW, Suite 300
Boca Raton, FL 33487-2742

© 1999 by Taylor & Francis Group, LLC
CRC Press is an imprint of Taylor & Francis Group, an Informa business

First issued in paperback 2019

No claim to original U.S. Government works

ISBN 13: 978-0-367-45565-1 (pbk)
ISBN 13: 978-90-5702-387-3 (hbk)

Visit the Taylor & Francis Web site at
http://www.taylorandfrancis.com

and the CRC Press Web site at
http://www.crcpress.com

British Library Cataloguing in Publication Data

A catalogue record for this book is available from the British Library.

CONTENTS

CONTENTS

CONTENTS

CONTENTS

CONTENTS

PREFACE

During the past two decades, there has been remarkable progress in research on chromosomes of the plant system. The analysis of physical and chemical details of chromosomes has been able to resolve cryptic genotypic differences at the population level, thus widening the scope of analysis of biodiversity. The chromosome structure has been clarified in great detail, enabling identification of gene sequences at the microscopic level. A solid foundation has been laid for the analysis of biodiversity at genetic and chromosomal levels. The importance of chromosome study in genetic improvement of crop species is well established. It provides a basic idea of the status, affinities and relationship of taxa which serve as prerequisites for undertaking any breeding programme.

Simultaneously, there have been significant developments in the study of gene structure, analysis of its DNA component and mapping of gene loci, utilizing extracted DNA and chromosomes *in situ*. The combination of molecular and cellular methods is best exemplified by the preparation of probes and their application to chromosomes for gene identification. Modern biotechnology is providing very effective methods of tracing the origin and identification of biodiversity.

The manipulation and engineering of chromosomes have been aided by novel methods, combining conventional and modern techniques of biotechnology. For those workers interested in the study of genetics, biodiversity, and biotechnology for gene transfer, experience in the methods of analysis, manipulation and engineering in the plant system is essential. The plant system, because of totipotency, with each cell capable of regeneration, requires a specialized treatment for manipulation and transfer, as compared to animals. This book is planned to fulfil the needs of botanists and plant geneticists, who are engaged in the study of evolution, biodiversity, chromosome manipulation and transgenesis. The treatise attempts to cover all the latest methods with the principles and scope, ranging from the most simple to most sophisticated ones. The earlier book by the authors on *Chromosome Techniques—a Manual* (1994) dealt mainly with the then available protocols for chromosome analysis of plants, animals and man. The wide coverage of all groups of living organisms prevented a special emphasis on analysis of the plant system, as well as manipulation and engineering. Moreover, since the book was written in 1993, there have been numerous publications on the plant system alone, principally in the areas of localization and manipulation of chromosome segments as well as transgenesis. These developments and the need for a special treatment for plant systems alone, stressed the importance of a book on techniques designed to serve the needs of plant scientists, and hence this publication.

For the convenience of the reader, the subject matter and the methods are presented in five sections. With an introduction on the general scenario in Section I, the textual part starts with Section II. This section deals principally with chromosome analysis—at structural, ultrastructural, chemical and molecular levels. These four aspects have been covered in seven chapters. There are three chapters on structural aspects, two on chemical aspects following extraction and *in situ* estimation, and one dealing with the molecular analysis *in situ*. The chapter on microscopy covers ultrastructural analysis as well using electron microscopy. It is followed by a chapter on cytochemical analysis, also involving microscopy. The *in situ* molecular analysis deals with all levels of hybridization, including chromosome painting.

Section III on molecular pattern analysis and mapping of genes on chromosomes is divided into six chapters encompassing sequence and length polymorphism, cloning and library, mapping through chromosome walking, isolation and mapping mediated through transposons and finally DNA fingerprinting. This aspect is essential for the documentation of biodiversity and study of phylogeny.

Chromosome manipulation and engineering are covered in Section IV in six chapters. The first chapter deals with the use of physical and chemical agents on chromosomes and cell cycle, which form the initial steps in manipulation. Since these findings are basic to monitoring of genotoxic effects of environmental pollutants, this facet has been covered in the next chapter. The methods for chromosome manipulation start with the conventional techniques; utilizing hybridization in Chapter IV.3, followed by tissue and protoplast culture in Chapter IV.4. Genome and chromosome transfer are covered in Chapter IV.5. Finally gene transfer, development and identification of transgenics have been dealt with in Chapter IV.6.

The last Section V, on microdissection and engineering, outlines the latest methods of chromosome dissection and transfer at the microscopic level, utilizing laser beam and micromanipulator.

In the preparation of this book, in addition to the earlier treatises of C.D. Darlington and L.F. La Cour, D.H. Johansen and E. Gurr, some of the comparatively recent publications including, *Methods in Plant Molecular Biology and Biotechnology* (edited by B.R. Glick and D.E. Thompson) CRC Press, 1993; *Chromosome Analysis Protocols* (edited by J.R. Gosden) Humana Press, 1994; *Studies on Plant Molecular Biology—a Laboratory Course Manual* (edited by Maliga P. *et al.*), Cold Spring Harbor, 1995; and also by Dr J. Sambrook and associates (1989) have been freely consulted. In the section on analysis, there have been a few unavoidable duplications with our previous book. In outlining some of the standard protocols, the names of the scientists, already cited in our earlier books, have been omitted for the sake of brevity.

The authors are grateful to several scientists who have been of considerable help by supplying their reprints, material, and recipes. We are indeed grateful to Professor M. D. Bennett (Jodrell Laboratory), Professor I. Schubert and Professor J. Fuchs of Plant Genetic Institute, Gatersleben and Professor K. Fukui, Hokuriku National Agricultural Experimental Station, Niigata, Y. Mukai, (Kyoiku University, Osaka, Japan) for their invaluable help. Dr U.C. Lavania, Assistant Director, CIMAP, Lucknow, one of our erstwhile associates, helped us considerably, supplying literature from time to time. Lastly, the patience of Mr S. K. Sur in deciphering the scripts and Dr Anusri Banerjee, Director of Neo Dimension and her associate, Mr K. Singha in preparing the final manuscript is gratefully acknowledged.

The book is intended to serve the needs of students and research workers in plant sciences, genetics, and those scientists interested in biotechnology and the analysis of biodiversity. The authors would consider their efforts a success if these objectives are fulfilled.

SECTION—I

CHROMOSOME RESEARCH IN BIOTECHNOLOGY, BIODIVERSITY AND GENETICS

The study of chromosome structure has become completely a synthetic discipline depending to a large measure on biophysical and biochemical tools for clarification of the genetic skeleton. At the molecular level, it represents a giant complex molecule made up of smaller but less complex molecules, the genes, which are arranged in a linear order in the nucleoprotein skeleton. The study has gradually been shifted to a chemical and molecular level from a purely cytogenetic standpoint. Such an analysis has given one an understanding of the pattern of organization at the submicroscopic stage. Techniques of structural study, absorption analysis, high resolution autoradiography, electron microscopy are responsible for resolving the chromosome structure as made up of fibrils 20–30 Å in thickness, folded several times to yield a fibre 100 Å approx. thick. It is a long chain nucleoprotein fibre in which condensed and decondensed segments alternate, the condensed segments representing an aggregate of beads of nucleosomes made up of histone octamers and a surrounding DNA fibre. Pulse labelling autoradiography has further revealed that chromosome structure is made up of many replicons or replicating units, which divide as discrete units, instead of the division starting from one end and moving to the other. The structure is linear, multirepliconic and uninemic. The backbone of the chromosome is a core of protein fibre, surrounded by DNA loops, forming a scaffold, where non-histone proteins are involved in forming DNA loop domain. The methods have also helped in resolving the sequence pattern of genes in higher organisms, where the genes are split, made up of essential and non-essential sequences.

Wide genetic diversity between the different groups does not evidently permit a similar response to methods of analysis and hardly any technique can be considered to be of universal application for critical analysis. Consequently, a series of methods utilizing a common technology, such as electron microscopy or fluorescence, have been devised to deal with diverse groups. Such differential response of plant species to physical and chemical methods is also an index of genotypic control, *vis-à-vis* biodiversity.

Estimates of nuclear DNA in higher organisms have revealed the presence of multiple copies of similar sequences forming the bulk of the DNA. In the linear structure, such highly homogeneous repeats are located at one locus, minor and moderate repeats being interspersed throughout in the unique sequences, between and within the genes. It is indeed ironical, that more than

75% to 90% of DNA represent the excess amount and only a small fraction is responsible for the entire gene complement in higher organisms, including man. The evolutionary strategy which has kept this amount of DNA in chromosome, without recognizable function, is still not clear. The basis of their conservation against rigours of selection, is still a challenging problem in biology.

In addition to resolution of chromosome structure, further refinements in methods have led to identification of gene loci and mapping of genes at the chromosome level. The use of probes, in identification of chromosome segments and gene loci in chromosomes, was indeed a remarkable development. It is based on the principle of molecular hybridization and involves strand separation of the double helix by heat denaturation and immobilization, followed by hybridization using the selected probe tagged with the isotopic or nonisotopic element. The selected probe hybridizes with the complementary strand in the chromosome which can be detected through autoradiography. The radioactive spot indicates the site of the gene in the chromosome. Through this method, coupled with banding technique, a number of genes have now been mapped in the chromosomes, which has enabled further location of a large number of genes. Conventional methods would have involved breeding followed by study of linkage and crossing-over in subsequent progenies. Moreover, breeding technique permits localization of relative distance and not visual localization.

The process of molecular hybridization has been much refined in recent years, substituting the isotopic probe with nonisotopic ones, tagged with fluorescent compounds. The detection of fluorescence at the chromosome site, marks the site of the gene at the chromosome level. Later, the technique has even been further modified with the use of multiple probes tagged with compounds of different colours, so that a number of genes can be simultaneously located in the chromosomes. They can be detected through differential fluorescence at different chromosome loci. The entire procedure of locating a large number of genes simultaneously in the chromosome, with the aid of different fluorescent colours, is otherwise termed as "Chromosome Painting" at the molecular level. This method permits even detection of single copy sequences of genes in chromosomes. It is at present possible not only to locate the gene but even to trace the foreign genome in case of a hybrid.

The genes, once localized in the chromosomes, need to be isolated for analysis and manipulation. However, chromosome manipulation and transfer of chromosomes and segments had also been in vogue earlier utilizing refined breeding methods. Modern developments have enriched this potential by extending it to horizontal transfer, through technology permitting quicker isolation and manipulation.

Isolation and analysis normally involve the application of restriction enzymes to isolated DNA for cutting into pieces and ligation to a vector for cloning on a bacterium. The cloning is needed to secure an adequate amount of gene sequences of DNA through bacterial multiplication for analysis. The process of restriction cut, ligation and cloning, is rather complex.

At the chromosome level, the isolation and separation of chromosome fragments have been much simplified by laser dissection of chromosomes under the microscope with a micromanipulator. The manipulation of minute gene sequences at the molecular level under the microscope has become a reality because of laser surgery.

For further analysis and convenience in manipulation of the DNA, the use of polymerase chain reaction or PCR is indeed a landmark development. It enables a billion-fold amplification of a short sequence within a few hours. The entire reaction is carried out in a programmed system in which the DNA double helix is denatured into two strands through high temperature followed by replication through specific primers and enzymes, resulting in duplication. The total process of denaturation and annealing of duplex is repeated in a programmed fashion in which the thermostable enzyme Taq polymerase plays the crucial role. The later developments include polymerization of even minute sequences at the chromosome level. Thus the use of specific probes for identification, laser beam for dissection, and PCR for amplification has indeed revolutionized the chromosome research at the molecular level. The visual manipulation of genes under the microscope permits chromosome manipulation and engineering with predictable accuracy.

Such landmark developments in biotechnology for analysis, manipulation and engineering are proving to be of immense value in genome analysis and horizontal gene transfer for genetic improvement. The identification of finer segments of chromosomes and their topography are essential for comparative analysis, study of evolution and biodiversity. The determination of functional sequences and location at the chromosome level has aided gene mapping at the site of the chromosome. The understanding of chromosome structure, the differentiated segments, their function and behaviour in somatic and germinal line, are prerequisites for every programme on genetic improvement.

The analysis of biodiversity at the genetic level has been aided through these refinements in methods. The molecular documentation too for all groups of plants through fingerprinting is within the reach of the investigator.

The analysis, manipulation of chromosomes or chromosome segments and direct transfer of genes, have widened the scope of genetics and plant improvement in a spectacular fashion. The production of wide hybrids, and transgenic individuals with desired medicinal properties, or resistance to biotic factors or genes of industrial importance are all realities. The success of these ventures owes principally to refinements in methods which have been outlined in the subsequent chapters. In outlining the methods, the detailed protocol and their principles, along with advances achieved in general, have been presented.

SECTION—II

CHROMOSOME ANALYSIS—SCOPE AND ADVANCES

The methods for the study of structure and chemical components of chromosomes have undergone gradual refinements aided through cytogenetic, biochemical and biophysical tools. Such methods have led to the resolution of chromosome in its finer details, localization of gene loci and analysis of its components. The unravelling of the structural, functional and behavioural components of chromosomes are of prime importance in the understanding of biodiversity, evolutionary pathways and programmes for improvement.

The role of numerical and structural differences of chromosomes and karyotype diversity in different taxonomic units is well documented. The differences and similarities are regarded as parameters of biodiversity, as well as distance or closeness of affinities. The mechanisms and pathways of alterations in chromosome complement are also reflected in the karyotype, which provides with an index of advanced or primitive status. The evidences in the plant kingdom show that, just as evolutionary progress from symmetry to asymmetry and long metacentrics to short acrocentrics have occurred in majority of the groups, the reverse process involving Robertsonian fusion is true for certain families of flowering plants. Such diverse evolutionary pathways in the progress of biodiversity could be measured principally through chromosome analysis.

Further refinements in methods bringing out cryptic details of chromosomes, have resolved molecular diversity and genetic polymorphism at the lowest strata of taxonomic hierarchy, that is, at the population level. In several species of grasses including crop relatives, such genotypic difference at the molecular level has been manifested. The resolution of all the details in chromosome structure has proved to be essential for the study of biodiversity, delineation of phylogeny and affinities, and finally for a strategic breeding programme.

The ultrastructural and fluorescence patterns, differential banding, refined staining and autoradiographic analysis have helped in understanding the fine structure of chromosome, and its subcellular details, which cover nucleosomes, synaptonemal complex and scaffold skeleton. The knowledge of these basics in structure has provided an insight into chromosome dynamics, both in terms of structure and function—the prerequisites for manipulation.

The study of the amount of chromosomal DNA following *in situ* estimation has yielded data of evolutionary importance. Both increase and decrease

4

in amount of DNA have taken place in evolution. The enormous data so far gathered show that total nuclear DNA can be utilized in different groups, as parameters of phylogeny, status of species and determination of genetic diversity *vis-à-vis* biodiversity.

Simultaneously with clarification of chromosome structure and its chemical components, molecular hybridization *in situ* has emerged as a powerful tool in localization of genes at the chromosome level. Extensive use of fluorescent and radioactive probes in this strategy have led to precise delineation of functional loci in the chromosome. The value of this technique can be judged by the very fact that it enables the identification of gene sequence and its function at the microscopic level. Its impact, both in the study of biodiversity on the one hand and gene manipulation on the other, can hardly be overrated.

CHAPTER II.1

PREPARATION OF MATERIAL FOR
ANALYSIS OF CHROMOSOME AT
STRUCTURAL LEVEL

PRE-TREATMENT

Pre-treatment of the dividing tissue with specific chemicals for the study of chromosomes is generally performed for: (a) clearing the cytoplasm, (b) separation of the middle lamella, thus softening the tissue, or (c) bringing about scattering of chromosomes with clarification of constriction regions. It may also be needed to achieve rapid penetration of the fixative by removing undesirable deposits on the tissue. The first two applications involve removal of extranuclear contaminants, while the third and most important one affects the chromosomes directly.

In order to clear the cytoplasm from its heavy contents, acid treatment has often been found to be very effective. Short treatment in normal hydrochloric acid or other acids brings about transparency of the cytoplasmic background. The limitations of acid treatment include the need for—(a) thorough washing to remove excess acid; (b) mordanting to counteract reduction of basophilia of chromosomes.

Of several enzymes applied for clearing the cytoplasm and cell separation through digestion, most common ones are pectinase, cytase from snail stomach, ribonuclease and a complex enzyme preparation 'Clarase'. In the authors' laboratory, treatment with cellulase has given very satisfactory preparations in difficult dicotyledonous materials. A mixture of 2.5% cellulase and pectinase (1:1) is very effective.

Alkali solutions as pre-treatment agents are effectively employed for materials with heavy oil content in the cytoplasm since alkalis such as sodium hydroxide or sodium carbonate remove the oil by saponification. Thorough washing in water is necessary after the tissue has been kept in alkali solution.

Pre-treatment may also be needed to remove deposits of secretory and excretory substances from the surface of the tissue which may hinder the access of the fixing fluid. The best example is the application of hydrofluoric acid to remove siliceous deposits prior to fixation in bamboos; or a very short treatment in Carnoy's fluid, containing chloroform, to remove deposits on the cell walls before fixation, in a number of plant materials.

6

For separation of chromosome and clarification of constrictions

The underlying principle is the viscosity change in the cytoplasm. Since spindle formation is dependent on the viscosity balance between cytoplasmic and spindle constituents, a change in cytoplasmic viscosity brings about a destruction of the spindle mechanism with the chromosomes remaining not attached to any binding force within the cell. Pressure applied during squash or smear scatters chromosomes throughout the cell surface. Changes in cytoplasmic viscosity simultaneously affect the chromosome, which undergoes differential hydration in its segments. Due to this differential effect, constriction regions in chromosomes appear well clarified.

Pre-treatment also aids in securing a high frequency of metaphase stages through spindle inhibition. Colchicine is the most active substance known, and also compounds having similar properties, such as chloral hydrate, gammexane, acenaphthene, vinblastine sulphate (Velban) and vincaleuco-blastine, amongst others (see Table 1 for list).

The concentration used and the period of treatment have to be strictly controlled, as also the temperature in most cases. Prolonged treatment leads to narcotic effects, including chromosome breaks. For further details, see Sharma and Sharma (1980).

Colchicine ($C_{22}H_{25}O_6N$) was first isolated from the roots of *Colchicum autumnale* by Zeisel in 1883. It is the methyl ether of an enolone containing three additional methoxy groups and acetylated primary amino group and three non-benzenoid double bonds. The threshold regions of colchicine-mitotic activity are identical for both crystalline and amorphous forms. Colchicine is soluble in water (500.00 in 10^{-6} mol/l) and is very active at an extremely low concentration.

Its reaction is chemical. It brings about a change in the colloidal state of the cytoplasm, causing spindle disturbance. With regard to the exact reactive groups in the colchicine molecule (a) at least one methoxy group in ring A is necessary for its action; (b) ring C must be 7-membered and the hydroxyl group should preferably be replaced by an amino group; (c) esterification of amino group in ring B increases the activity; and (d) isocolchicine and its derivatives are less active. A proper distance should be maintained between esterified side chains of rings B and C. A number of sulphydryl poisons, such as iodoacetamide, dimercaptopropanol (BAL), mercaptoethanol, sodium diethyldithiocarbamate act in the same way as colchicine implying a chemical combination with an intracellular group.

In the study of chromosomes, without inducing polyploidy, colchicine has to be applied in a low concentration, such as 0.5% for 1 h, thus straightening the chromosome arms to allow a thorough study of the constriction regions. It is especially effective for long chromosomes.

Colchicine is active within a very wide range of temperatures. The period necessary for the manifestation of effect varies in different groups. The range of concentrations is also wide, between 0.001 and 1%. It is applied through soaking, plugging and injection. It can also be used in lanolin paste and in agar. In artificial culture colchicine is added to the medium. In general, the drug is exceptionally suitable for the study of chromosome structure and metaphase arrest, under strict control over the concentration and period of treatment in almost all plant tissues.

Acenaphthene, a naphthalene derivative, has the same property as colchicine of arresting metaphase though its use is limited. As its structure is quite different from that of colchicine, several other aromatic compounds were tried and various derivatives of benzene and naphthalene were found to be effective. Since it is sparingly soluble in water, a saturated solution is used, particularly in pollen.

Chloral hydrate [$CCl_3CH(OH)_2$] has been used as a pre-treatment chemical in cytology since 1935. It has effects similar to colchicine, and is less expensive. However, it is not suitable for all plants.

Gammexane, γ-hexachlorocyclo-hexane, has been effectively used in plants. In addition to metaphase arrest, polyploidy and fragmentation have also been recorded in some cases. Its *O*-isomer was found to be similarly effective. It is most active at low temperatures and the reaction is chemical in nature.

Coumarin and its Derivatives   The importance of coumarin in chromosome analysis was pointed out by Sharma and Bal (1953). It is the *o*-coumaric acid lactone found mostly in the glycoside form in plants. Several natural and synthetic coumarins have been screened for their pretreatment effects in this laboratory. It is effective even at a temperature of 30°C, though cold treatment gives better action. Being sparingly soluble in water coumarin is used as a saturated solution and has been used to study the chromosomes of a large number of monocotyledonous species.

Aesculine is a derivative of coumarin and is extracted from *Aesculus hippocastanum* Tourn. It is directly derived from aesculetin, in which the H from the -OH group in the 6-position is replaced by $C_6H_{11}O_5$. It has been found to be suitable for different groups of plants having chromosome numbers ranging from very high to very low. Its approximate solubility is 0.04% in water, and for chromosome analysis both saturated and half-saturated solutions are applied. It is effective only at very low temperatures, ranging between 4 and 16°C.

Isopsoralene, another natural derivative of coumarin, gave excellent results in different groups of plants. Chaudhuri, Chakravarty and Sharma (1962) inferred that the position of the furan ring is an important controlling factor in the manifestation of karyotype clarification.

Umbelliferone, another derivative of coumarin able to clarify chromosome structure, is extracted from several species belonging to the family Umbelliferae. Its application is very limited.

A majority of the coumarin derivatives can be employed for chromosome analysis, but their applicability is restricted mostly to specific groups of plants, with the exception of aesculine which has a very wide application.

Since viscosity change constitutes the principal basis of action of all of these compounds, attempts were made to find out the degree of change in viscosity induced by these derivatives through their action on protoplasmic movement in *Vallisneria spiralis* leaf or *Tradescantia paludosa* staminal hair *in vivo*. Sharma and Chaudhuri (1962) observed that daphnetin increased viscosity of the plasma within a very short period while coumarin, aesculine and aesculetine differed with respect to this property, the effect of coumarin being least. These results indicate the differential efficacy of different chemicals as pre-treatment agents.

Oxyquinoline (OQ) is a member of the most important group of compounds used in studying chromosome structure—the quinoline complex. Tjio and Levan (1950) utilized 8-hydroxyquinoline in chromosome analysis. It causes mitotic arrest, thus demonstrating its c-mitotic property, and is also endowed with certain characteristics not shared by colchicine. The spindle is inactivated and permits the chromosomes being spread out during squashing; the chromosome arms contract equally. Unlike colchicine, OQ allows the metaphase chromosomes to maintain their relative arrangements at the equatorial plane. OQ is specially suited for plants with long chromosomes and best effects are obtained at low temperatures.

Phenols can be effectively employed for the study of chromosome morphology when applied at concentrations below the one causing chromosomal abnormalities. All the three types of phenols, mono-, di- and trihydric, induce clarification of chromosome morphology. Dihydric forms are the most effective, even at a very low concentration. Amongst these types, such as resorcinol and hydroquinone, the location of the -OH groups is possibly responsible for their differential activity.

p-Dichlorobenzene ($C_6H_4Cl_2$) is the most suitable compound for chromosome work amongst the benzene derivatives. It is sparingly soluble in water. It causes spindle inhibition and leads to clarification of chromosome constrictions due to the contraction and differential hydration of chromosome segments. It has a comparatively wide application, and complements with both long and short chromosomes appear to respond equally, even though

9

the period of treatment may require modification. The limitations of its use are the prolonged period of treatment (3 h) and the specificity of the temperature needed (10–16°C) for optimal results.

Monobromonaphthalene has been effectively used for chromosome studies after 10–15 min treatment at low temperature in saturated aqueous solution in certain groups of plants. For aquatic angiosperms especially, pre-treatment with this chemical is effective possibly due to rapid rate of penetration into the tissues.

Veratrine ($C_{32}H_{40}NO_9$) is an alkaloid, sparingly soluble in water, which has been shown to possess the property of clarifying chromosome structure. Cold treatment for 30–40 min is found to give the best results.

Several other chemicals, including various hormones, heterocyclic bases and vitamins, as also aqueous solutions of inorganic metallic salts, in specific concentrations, have been recorded to be able to disturb the spindle and produce clear chromosome preparations at metaphase. The effects are, however, selective for specific plants. In general, the concentration of the pretreatment chemical and the duration of exposure required for optimal clarification of mitotic chromosomes are different for different species but related species with similar genotypes are susceptible to a common pretreatment protocol.

FIXATION

Fixation may be defined as the process by which tissues or their components are fixed selectively at a particular stage to a desired extent. The purpose is to kill the tissue without causing any distortion of the components to be studied.

A large number of fixing reagents have been devised by various workers since plant species differ widely in their response to a fixative. All of them possess certain common characteristics. Each fixing chemical is lethal to the tissue. The structural integrity of the chromosome must be maintained intact. Precipitation of the chromatin matter is essential to render the chromosome visible and to increase its basophilia in staining. Under living conditions, the phase difference between different components of the cell is not adequate for observation as distinct entities. Coagulation of protein and consequent precipitation markedly alter the refractive index of the chromosomes, so that they appear as differentiated bodies within the cell. All fixatives, so far used, have the property of crosslinking proteins. The primary requisite of a fixative for the study of chromosomes is therefore the *property of precipitating chromatin*.

Another important property is that of rapid penetration so that the tissue is killed instantaneously, the divisional figures being arrested at their respective phases. *Immediate killing* is essential as otherwise the nuclear division may proceed further and attain the so-called 'resting' or metabolic phase.

Death of the cells may lead to certain changes which are detrimental to the preservation of chromosome structure, the most important being the autolysis of protein. A fixative should therefore be able to *check autolysis of proteins.* With the onset of lethality, bacterial action causes the tissue to decompose. Another prerequisite is to *prevent this decomposition* by maintaining an aseptic condition in which bacterial decay cannot take place. Certain chemicals, though possessing these properties, yet hamper staining by reducing basophilia of chromosomes. A proper fixative should, therefore *enhance the basophilia* of the chromosome. A mixture or a chemical which fulfils all these conditions can be considered to be a truly effective fixative for chromosome study. But since all these properties are rarely found within a single chemical, a fixative is generally a combination of several compatible fluids which jointly satisfy all requirements.

An alternative to chemical fixatives is fixation through freezing at low temperature, followed by drying the tissue. The principle involves rapid cooling of the tissue to a low temperature, followed by extraction of water in a vacuum. This method allows a life-like preservation of the tissue. The cooling process must be so rapid that the water cannot crystallise, which may lead to distortion of cellular components and consequent misinterpretation of chromosome structure. Water can be frozen into amorphous ice in a cooling bath of $-175°C$, secured by condensation and liquid nitrogen. Following fixation by freezing, water is removed in vacuum at a low temperature or by passing it through a stream of dry cool gas. The material can be sectioned directly or may be infiltrated with paraffin or some other medium.

The advantages of the freezing method of fixation are, (a) minimum distortion of the tissue after its death, (b) least possibility of diffusion, (c) no significant effect on the enzyme system and (d) the tissue can be directly embedded in paraffin without dehydration or clearing. Its inherent drawbacks include the distortion of cellular components during sectioning and the relatively high cost in setting up the apparatus. Some of these limitations have been eliminated in the process of freezing-substitution. The specimen is rapidly frozen, followed by dehydration at a very low temperature $(-20$ to $-78°C)$ through any one of the following reagents: n-propanol, n-butanol, methanol, ethanol, methyl cellosolve, or the chemical fixatives. After complete dehydration, the material is brought back to room temperature slowly.

The freeze-drying method of fixation is particularly accurate for the study of the effect of chemical and physical agents on the chromosome, where the immediate effect has to be analysed, but it is not very useful for the study of the structure and behaviour of chromosomes. Other advantages of chemical fixation, such as increase in the basophilia of chromosomes and differential precipitation of chromatin matter in its different segments are also obtained by the freezing-substitution technique. These factors, taken together, have contributed largely to the wide use of chemical fixatives in routine work on chromosome studies, and only under special circumstances, is freezing-substitution applied.

The fixing chemicals, in general, may be classified into two categories, based on their property of precipitating proteins within the cell. The best examples of precipitant fixatives are chromic acid, mercuric chloride, ethanol, etc. Among the non-precipitant fixatives can be included osmium tetroxide, potassium dichromate, etc. Certain fixatives undergo chemical combination with proteins, some of which precipitate out proteins while others do not. A fixative with a strong precipitating action, is usually counteracted by the addition of other reagents.

Principal chemicals used as ingredients of a fixing mixture may be (a) non-metallic: ethanol, methanol, acetic acid, formaldehyde, propionic acid, picric acid, chloroform or (b) metallic, such as chromic acid, osmic acid, platinic chloride, mercuric chloride, uranium nitrate, lanthanum acetate. Several of these compounds are also used as vapour fixatives, converting the soluble substances into insoluble ones, before coming into contact with water or other solvents, so that *in situ* preservation is maintained. Most

Table 1 Some common fixatives

Chemical	Molecular weight	Used alone or in mixture	Concentration commonly used
Acetic acid	60.05	Both	100 and 45%
Acetone	58	Mixture	100%
Chloroform	119.59	Mixture	100%
Chromic acid	100.01	Mixture	1 and 2%
Dioxane	88.11	Mixture	100%
Ethanol	46	Both	50–100%
Ether	74	Mixture	Absolute
Formaldehyde	30	Mixture	40%, 5%
Formic acid	46	Mixture	1%
Hydrochloric acid	36.5	Both	Normal
Isopropanol	60	Mixture	100%
Methanol	32	Mixture	100 and 95%
Mercuric chloride	272	Mixture	—
n-butyl alcohol	74.12	Mixture	100%
Nitric acid	63	Mixture	10%
n-Propyl alcohol	60	Mixture	100%
Osmic acid	255	Both	0.5–2%
Platinum chloride	518.08	Mixture	0.5–2%
Potassium dichromate	294	Mixture	2–7.5%
Propionic acid	74.08	Mixture	45 and 100%
Trichloracetic acid	163.40	Both	—

non-metallic fixatives, have one advantage over the metallic ones; that no washing in water is required after fixation. In materials where the cells are loosely scattered in a suspension, fixation is improved by decreasing the amount of fluid surrounding the cells. The cells, therefore, are centrifuged into a small pellet and the supernatant liquid removed. The fixative is added and allowed to remain undisturbed for up to 30 min. The cells may be centrifuged into a pellet again, followed by the addition of fresh fixative. Cells on a coverglass or a slide are fixed usually for not less than 5 min or more than 24 h.

The more common fixing fluids are mentioned below:

Non-metallic fixatives

Ethanol is used extensively as a constituent of chromosome fixatives, in percentages 70 to 100%. An important advantage is its capacity for immediate penetration and dehydration. It precipitates nucleic acid and causes an irreversible denaturation of proteins. The only limitation is a hardening effect on the tissue.

Being a reducing agent, it undergoes immediate oxidation to acetaldehyde and then to acetic acid in the presence of an oxidiser, and so cannot be used in combination with many metallic fixatives. It is used principally in combination with acetic acid, formaldehyde or chloroform. For enzyme studies on chromosomes, chilled 80% ethanol fixation for 1 h or more is recommended and for monolayer cultures, fixation in absolute or 96% ethanol for 1–15 min is often applied.

Acetic acid can be mixed in all proportions with water and ethanol or methanol from very low concentration to even glacial (100%). It has a remarkable penetrating property, even higher than alcohols. It is, in general, an ideal fixative for chromosomes, and has been observed to maintain the chromosome structure intact. Since chromosome arms become smaller on exposure to acetic acid alone, it should be used in combination with alcohols, or similar chemicals, which shrink and harden the tissue.

It is also a good solvent for aniline dyes, and is a necessary component of staining-cum-fixing mixtures, like acetic–carmine, acetic–orcein, acetic–lacmoid, etc.

Formaldehyde in its commercial form, known as 'formalin', contains a 40% solution in water. It is a bifunctional compound capable of forming cross-links between protein end groups. For the fixation of chromosomes, 10–40% solution of commercial formalin in water is used.

Tissues, placed in fixatives containing formaldehyde, often show well scattered chromosomes, especially after sectioning from paraffin blocks. The cell volume increases considerably, resulting in spreading of the chromosomes over a larger area. The constriction regions appear slightly exaggerated due to contraction of the segments, possibly related to its action on the chromosome proteins. In cytochemical work, washing after formalin fixation is essential so that the reactive groups of proteins remain unmasked to combine with the reagents.

A serious disadvantage of formaldehyde as a fixing agent is the fact that the tissue treated with this reagent is difficult to smear due to exceptional hardening.

Methanol is occasionally used in chromosome studies of plants. While ethanol causes heavy shrinkage of chromosomes, methanol causes swelling and this property has been utilized in fixatives where a swelling agent is needed to compensate for the shrinking effect of other chemicals. Its effective concentrations are the same as ethanol.

Propionic acid has been used extensively in the fixation of chromosomes. It is miscible with water, ethanol and ether in all proportions, and is a good solvent for aniline dyes. It is generally used as a substitute for acetic acid. Its penetration is not as rapid as that of acetic acid but it causes much less swelling of the chromosomes.

Chloroform is miscible in all proportions with alcohols, ether and acetone. In the study of plant chromosomes, it is generally used in the fixative to dissolve the fatty and waxy secretions from the upper surface, facilitating the penetration of the fixative. A judicious use in fixing mixtures is recommended, as an excessive dose or long period of treatment may be toxic.

Dioxane dissolves resins, fatty oils, etc. ensuring rapid penetration of the fluid, and also because of its ready miscibility with most solvents, it forms a good ingredient of fixing mixtures.

Commercial diethyl ether mixes well with alcohols or liquid hydrocarbons in all proportions, being a solvent for fats and oils and has been used in a few fixing mixtures, like Newcomer's fluid.

Isopropanol has the same effect as ethanol on chromatin. It has been preferred in some materials due to its comparatively less drastic action.

Acetone is miscible with water, ethanol and ether in all proportions. It is a good solvent for cellulose acetate and for many organic compounds, and when acting in a fixing mixture serves the same purpose as chloroform or ether in clearing the cytoplasm by dissolving the organic matter.

Metallic fixatives

Osmium tetroxide is a strong oxidising agent and should never be mixed with formaldehyde or alcohol. In solution, initially the entire molecule combines with the amino groups of proteins. In the secondary phase, the compound formed undergoes oxidation, during which the residual part of osmium tetroxide is reduced to a lower oxide or hydroxide. Due to this, the tissue fixed in osmium tetroxide turns black. In general, it fixes homogeneously, maintaining a life-like preservation of the tissue. It does not cause much shrinkage of the tissue, but there is a slight swelling.

The effect of osmium fixation depends to a significant extent on the pH, toxicity and temperature of the fixing mixture. It preserves chromosomes during the divisional cycle, but cannot be recommended for the study of the interkinetic nuclei. Moreover it often results in protein loss.

The best result is obtained if the fixative is applied in the form of vapour. This can, however, be applied only on small materials, such as prothalli of ferns, unicellular objects or materials having no cellulose wall. In electron microscopy, osmium fixation is widely utilized.

Materials, after osmium fixation, require bleaching to remove the black precipitate produced by fats, usually with hydrogen peroxide when the slide is brought down to water prior to staining. The slide is transferred from 80% ethanol to a jar containing H_2O_2 and 80% ethanol in equal proportions and kept from 1 to 12 h. For bleached tissues, pre-mordanting in 1% chromic acid solution is necessary.

Platinum chloride solution in water is often applied in place of osmium tetroxide, especially in the somatic tissue of plants. It is compatible with formalin and can be used as a substitute for chromic acid in chromic–formalin mixed fixing fluids, but its application is rather limited. Fixation requires bleaching of the fixed tissue as well.

Chromic acid is a strong precipitant of protein. Protein undergoes denaturation and precipitation by the primary action of chromic acid, and the secondary action results in hardening. Chromic acid penetrates the tissues slowly and the hardening induced by this acid makes the tissue resistant to hardening by ethanol in subsequent processing. Materials fixed in this acid require thorough washing in water, at least over-night. This fluid may be used as a fixative for squash preparations, only when softened by some strong acid. Basic dyes adhere closely to tissue fixed in chromic acid.

In general, chromic acid is considered an essential ingredient of several fixing mixtures. It imparts a better consistency to the tissue and aids staining better than osmium tetroxide.

Potassium dichromate is not as soluble in water as chromic acid. Over pH 4.6, $K_2Cr_2O_7$ can maintain the structure of chromosomes, whereas, with less acidity, only the cytoplasmic structures are preserved. On the whole it is good fixative for lipids. It has a rapid rate of penetration and shrinkage is not very marked.

It does not harden the tissue significantly. After dichromate fixation cellular constituents respond well to acid dyes, and the response of chromatin to basic dyes can be maintained if the fixation is performed at an acidic level. Because of its rapid rate of penetration, it is often preferred to chromic acid. It can be washed away with dilute solution of chloral hydrate in water.

Mercuric chloride is compatible with the majority of the fixing fluids. Its rapid penetration and strong protein reaction can be advantageously employed in mixtures with compatible fluids. Chromosomes respond well to most of the dyes after this fixation. A serious limitation is the formation of a needle-like precipitate of mercurous chloride in the tissue following this fixation.

Water soluble inorganic salts of lanthanum, uranium and iridium have also been used in fixing fluids, usually replacing chromium salts but the application is very limited.

Fixing mixtures

The different types of mixtures employed for chromosome studies can be divided into two major categories. Both metallic and non-metallic fluids have been included in the first category, whereas the other is constituted purely of non-metallic fixing fluids. See Table 2 for list.

STAINING

The structure and behaviour of chromosomes can be studied only after they are made visible under the microscope. In order to maintain normal activity at the time of observation it is best to mount the dividing cells in the body fluid of the organism and to observe chromosome movements under a phase contrast microscope. The evaporation of the fluid can be prevented by paraffin oil, which has the added advantage of being oxygen-solvent and non-toxic to the tissue. Staining of chromosomes can be classified as *vital* and *non-vital*. For vital staining, non-toxic dyes are applied to the living tissue so that the latter can be studied without being killed. Methylene blue has been found to be the only vital stain effective in demonstrating cell division in tissue culture. It is a basic dye of the thiazine group, $C_{16}H_{18}N_3SCl$, and is soluble in water.

In non-vital staining, the coloration of the chromosomes in the fixed tissue is caused by certain chemical agents which are insoluble in the chromosome substance. The principal dyes used to stain chromosomes are synthetic organic dyes. The colour of a dye is due to a chemical configuration, known as *chromophore*. The retention of the colour in the tissue is due to a certain chemical configuration in the dye molecule itself, known as the *auxochrome*.

Table 2 Some common fixing mixtures

Name	Components	Proportion	Modification	Treatment period	Application
With ethanol					
Carnoy's fluid and its modifications	glacial acetic acid absolute ethanol	1 3	1:2, 1:1, with chloroform 1:6:3 and 1:1:1; add corrosive sublimate to saturation; wash out with 70 to 95% ethanol; mordant with ferric ammonium sulphate (1–3%) for 3–12 h after fixation	15 min to 24 h	effective for all materials; with variations in period and concentrations.
Chromo-nitric acid	aq. nitric acid absolute ethanol aq. chromic acid	10%, 4 parts or 20%, 3 parts 3 or 4 parts 0.5% or 1%, 3 parts		4–5 h, followed by washing in 70 and 100% ethanol for 2–3 days after fixation in dilute mixture or 20–30 min followed by washing for 1 h after fixation in strong mixture	embryonic nuclei
Propionic acid modification instead of acetic acid	(a) propionic acid: 95% ethanol (1:3)			24–36 h, washed out with 70% ethanol	perithecia of Ascomycetes
	(b) propionic acid: 95% ethanol (1:1) + ferric-hydroxide (0.4 g)			if needed add a few drops of carmine	plants with small chromosomes
	(c) propionic acid: chloroform : abs. ethanol = 1:1:2 other proportions are 1:4:2			12–24 h	*Plantago* chromosomes
Iron acetate modification	(a) acetic acid : abs. ethanol (1:3) + small amount of iron acetate			12 h followed by keeping for 15 min in sat iron acetate solution in 45% acetic acid : 45% acetic acid : 1% formaldehyde soln (3:5:2); rinse in 45% acetic acid	anthers with small chromosomes

Table 2 (*Continued*)

Name	Components	Treatment period	Application
	(b) ethanol (95%) : acetic–carmine soln with added iron acetate (3 : 1) or a flake of rusted iron	12–24 h followed by washing and storage in ethanol (95%) with iron flake for 5–10 days	flower-buds, mainly of Cucurbitaceae
Newcomer's fluid	isopropyl alcohol : propionic acid : petroleum ether : dioxane 6 : 3 : 1 : 1 : 1	4–24 h	very stable fixative; can be used in combination with pretreatment chemicals perithecia of *Venturia*
Lactic acid modification	glacial acetic acid : abs. ethanol : lactic acid (1 : 6 : 1)		
With formalin			
Navashin's fluid	solution A: chromic anhydridge : glacial acetic acid : dist. water (1.5 g : 10 ml : 90 ml) solution B: 40% aq. formaldehyde : dist. water (40 : 60) mix in equal proportions just before use.	24 h, followed by washing in running water for 3 h modifications include altering the relative amounts; replacing acetic acid by propionic acid	both block and squash preparations of plant material
Bouin's fixatives	sat. aq. picric acid : 40%; aq. formaldehyde : glacial acetic acid : urea : chromium trioxide (75 ml : 15–25 ml : 5–10 ml : 1–2 g : 1–1.5 g)	1 h 30 min–12 h followed by repeated washing in 70% ethanol or ethanol grades and aniline oil	fungi, otherwise limited application; substitution of urea with maltose used in ongraceae
Levitsky's fluid	aq. chromic acid (1%) : aq. formaldehyde (10%) in different proportions (3 : 2, 4 : 1, 1 : 1, 1 : 2, 1 : 3)	12–24 h in cold or room temperature followed by washing for 3 h for higher concentrations	root-tips
Mixtures with acetic/propionic acid and ethanol/methanol	acetic acid : aq. formaldehyde (40%) : ethanol (95% or 100%) in different proportions (1 : 12 : 30ml + water) (1 : 6 : 14); (1 : 10 : 85) acetic acid can be replaced by propionic acid and ethanol by methanol	1–2 h	plant tissues

Battaglia's 5111 mixture	ethanol (95%) : chloroform : acetic acid : aq. formaldehyde (40%) = 5 : 1 : 1 : 1	5–10 min	plant tissues
With osmic acid		
Flemming's fixatives	aq. chromic acid soln. (1–2%) : acetic acid (5–100%) : aq. osmic acid soln. (2%), mixed in different proportions and diluted with dist. water; mixed just before use.	1 h to overnight in cold, followed by washing for 1 h to overnight in running water	plant tissues

Several modifications are devised with different proportions of the basic components and additions of NaCl/urea/maltose

La Cour's fixatives	aq. chromic acid soln. (2%) : aq. potassium dichromate soln. (2%) : aq. osmic acid soln. (2%) : acetic acid (10%) : aq. saponin soln (1%) : dist. water in proportions of (100 : 100 : 60 : 30 : 20 : 210); (100 : 100 : 32 : 12 : 10 : 90); (100 : 100 : 120 : 60 : 10 : 50); mixed just before use	24 h in cold followed by washing in running water for 3–12 h	mainly plants
Smith's fluid	aq. chromic acid (1%) : aq. osmic acid (2%) : acetic acid (5%) : potassium dichromatic : saponin : distilled water (100 : 35 : 25 : 0.5 : 0.05 g : 50) or (75 : 25 : 125 : 1 g : 0.05 g : 46)		first set suitable for early prophase and second set for diakinesis and metaphase stages in p.m.c.s.

The best example of a chromophoric group is the quinonoid ring. Auxo-chromes are mostly -NH$_2$ and -OH groups which convert the non-dyeing coloured substance into a form which undergoes electrolytic dissociation in water and is capable of forming salts with acids or bases.

The dyes are generally termed basic or acidic on the basis of their chemical nature and behaviour. In an *acidic* or *anionic* dye, the balance of the charge on the dye ion is negative while in a *basic* or *cationic* dye, it is positive. Most acid dyes are prepared as metallic salts and are generally neutral or slightly alkaline in reaction, but they react with, and stain substances with, a basic reaction. A basic dye is a salt of mineral or aliphatic organic acids, and stains substances which are acidic. Several dyes used in the study of the chromosomes are amphoteric, such as orcein. The majority of the chromosome dyes are triphenyl methane or aniline derivatives, though other dyes have also been used.

The general procedure for staining is to over-stain the chromosome, followed by the removal of the excess stain—a process called *differentiation*. Differentiation allows the stain to adhere to specific sites of the chromosomes. Similarly, acidic dyes may stain chromatin by following suitable differentiation, where quite likely the dye reacts with the protein moiety of the chromosome. For chromosome staining, basic dyes are applied since chromatin is strongly acidic. The terms basophilic and acidophilic are based on the affinity for basic or acidic dyes, chromosomes being basophilic and cytoplasm acidophilic.

The adherence of a dye to the tissue may also be accelerated through the process of mordanting. Certain metallic hydroxides referred to as *mordants* form compounds with the dye which attach the dye to the tissue and are called the 'lake' for the particular dye.

In cases where the 'lake' is insoluble in water, the general practice is to immerse the tissue in the mordant first, followed by staining. In carmine staining, iron salts are often used as mordants and are added to the stain itself. The process of combination, between the dye and the mordant, is known as *chelation*.

Whenever the dye or the mordant is used separately, the latter is employed either to modify the isoelectric point of the tissue or to form a chemical link between the stain and the chromosome. In principle, it changes the surface conditions of the fixed chromosomes.

Post-mordanting helps to retain the stain for a prolonged period and also clears the cytoplasm. This effect is due to the acidity of the mordant, such as iodine in ethanol, which, being higher than the cytoplasm, removes the stain from its surface. Chromatin, on the other hand, having a stronger acidity, retains the stain. An oxidising post-mordant oxidises the dye, present at certain sites, to a colourless substance.

The term 'mordant' should preferably be restricted to those agents which are applied before staining and which form a complex with the dye or the tissue. Agents, when applied after staining, act more as differentiating chemicals. Some common stains and their preparation are given in Table 3.

Fuchsin Of all staining methods employed for the study of chromosomes, the Feulgen reaction is considered to be the most effective. In 1924, Feulgen and Rossenbeck devised a method based on the Schiff's reaction for aldehydes which stains the nucleic acid of the chromosomes specifically and, as such, has been effectively employed for the visualization of chromosomes.

Feulgen solution or fuchsin sulphurous acid is prepared from the dye, basic fuchsin, which belongs to the triphenyl methane series. The commercially obtained 'basic fuchsin' is a mixture of three compounds, namely p-rosaniline chloride, basic magenta (rosaniline chloride) and new magenta (new fuchsin). All three compounds are characterised by quinonoid arrangements within the molecule.

The principle of preparing the Feulgen or Schiff's reagent is to treat the basic fuchsin solution with sulphurous acid, the product obtained being colourless fuchsin sulphurous acid or Schiff's reagent, utilized by Feulgen and Rossenbeck (1924) for the demonstration of the DNA component of chromosomes. The procedure involves the preparation of a basic fuchsin solution in warm water, followed by cooling at a particular temperature and the subsequent addition of hydrochloric acid and potassium metabisulphite, needed for the liberation of SO_2, prior to storage in a sealed container in a cool, dark place.

Preparation: Dissolve 0.5 g of basic fuchsin gradually in 100 ml boiling distilled water. Cool to 58°C. Filter, cool the filtrate down to 26°C. Add it to 10 ml N HCl and 0.5 g potassium metabisulphite. Close the mouth of the container with a stopper, seal with paraffin, wrap the container in black paper and store in a cool dark chamber. After 24 h, take out the container. If the solution is transparent and straw-coloured, it is ready for use. If otherwise coloured, add to it 0.5 g of charcoal powder, shake thoroughly and keep overnight in cold temperature (4°C). Filter and use.

Alternatively: After dissolving the dye, bubble a stream of SO_2 through the solution. Filter and store. The addition of a pinch of activated charcoal removes the yellowish impurities and a transparent colourless reagent can be obtained.

The colour of Schiff's reagent varies, depending upon the type of dye used, the hydrogen ion concentration and SO_2 content. It loses SO_2 on continued exposure to air and becomes coloured again. The basic fuchsin or p-rosaniline chloride solution, undergoes conversion to leucosulphinic acid, which is colourless, by the addition of sulphurous acid across the quinonoid

Table 3 Some common stains for plant chromosomes

Name and active principle	Michrome no.	Empirical formula	Mol. wt.	Applied to
Basic fuchsin is a mixture of:				
p-rosaniline chloride	722	$C_{19}H_{18}N_3Cl$	328.815	specific for DNA; widely used;
basic magenta	623	$C_{20}H_{20}N_3Cl$	227.841	period of hydrolysis varies
new magenta	624	$C_{22}H_{24}N_3Cl$	365.893	according to the plant species and tissues
Carmine				
carminic acid	214	$C_{22}H_{20}O_{13}$	492.38	widely applicable, mainly to lower groups and meiotic preparations as acetic or propionic carmine
Orcein	375	$C_{28}H_{24}N_2O_7$	500.488	widely applicable to somatic chromosomes as acetic or propionic orcein
Chlorazol black E	92	$C_{34}H_{25}N_9O_7S_2Na_2$	781.738	auxiliary stain before aceto-carmine staining
Gentian violet is a mixture of:	417			
Crystal violet	103	$C_{25}H_{30}N_3Cl$	407.971	pmcs and sectioned tissue
Pentamethyl p-rosaniline chloride		$C_{24}H_{28}N_3Cl$	393.945	
Tetramethyl p-rosaniline chloride		$C_{23}H_{26}N_3Cl$	379.919	
Haematoxylin Haematein	360	$C_{16}H_{12}O_6$	300.256	mordanting essential, limited application
Lacmoid		$C_{12}H_6NO_3Na$	235.173	root tip, embryosac, pollen grain
Brazilin brazilein		$C_{16}H_{12}O_5$	284.256	similar to haematoxylin but weaker

Name	CI No.	Formula	M.W.	Notes
Azure A	718	$C_{14}H_{14}N_3SCl$	291.799	similar to basic fuchsin but less effective, staining heterochromatin at resting stage and telophase and chromosomes at other stages
Orange G used in combination with		$C_{16}H_{10}N_2O_7S_2Na_2$	452.382	
Aniline blue			737.736	not widely used for chromatin
Giemsa contains Methylene blue Azure A Azure B Eosin Y	144 in different proportions	$C_{32}H_{25}N_3O_9S_3Na_2$		
Toluidine blue	641	$C_{10}H_{16}N_3SCl$	305.825	wide application
Fluorochromes for fluorescent staining Quinacrine dihydrochloride Ethidium bromide		$C_{23}H_{30}ClN_3O_2 \cdot 2HCl \cdot 2H_2O$ 2,7-diamino-10-ethyl-9-phenyl anthridinium bromide		consistent results with plant chromosomes

nucleus of the dye. Sulphurous acid is obtained through the action of HCl on potassium metabisulphite. The excess of SO_2 undergoes reaction with leucosulphinic acid to produce bi-N-aminosulphinic acid, popularly known as Schiff's reagent.

Staining: The procedure involves hydrolysis of the fixed tissue in normal HCl at 56–60°C, for a period varying from 4 to 20 min before immersing the material in Schiff's reagent. The colour develops within a short time and the chromosomes take up a magenta colour and can be observed after mounting in 45% acetic acid. If the recommended procedure is followed carefully, the chromosomes appear as specifically coloured bodies against a clear cytoplasmic background. Prolonged keeping in Schiff's reagent may result in further hydrolysis. A rinse in sulphite solution or SO_2 water is often helpful to remove excess of colour, if any, in the cytoplasm.

The chemical reaction takes place in two principal steps:

(1) By hydrolysis with normal HCl, the purine-containing fraction of deoxyribonucleic acid (DNA) is separated from the sugar, unmasking the aldehyde groups of the latter.
(2) The reactive aldehyde groups then enter into combination with fuchsin sulphurous acid to yield the typical magenta colour. Feulgen reaction is, therefore, based essentially on the Schiff's reaction for aldehydes. After removal of the base, carbon atom 1 of the furanose sugar is so arranged as to form a potential aldehyde, capable of reacting with fuchsin sulphurous acid. The ribose sugar, with -OH in place of -H at carbon 2, is not hydrolysed by normal HCl and so does not react with fuchsin sulphurous acid. In the pyrimidine-sugar linkage, on dissociation, the aldehyde groups are not free to react, unlike the open and reactive aldehydes obtained after breakdown of the purine-sugar linkage. Since different factors are involved in the development of the colour it is always necessary to keep a strict control over temperature and duration of hydrolysis and the type of fixation. With careful control, the specificity of the test is unquestionable.

Certain other basic dyes of the triphenyl methane series are able to replace basic fuchsin to a certain extent as observed in our laboratory. The methods of preparation and staining are similar to that followed for leucobasic fuchsin (see following table).

Carmine, one of the most widely used dyes for chromosome staining, is prepared from the dried bodies of *cochineal*, females of *Coccus cacti*, a tropical American Homoptera. It is a mixture of substances, the composition of which often varies on the basis of the method of manufacture. The active

Dye	MI no.	No. of batches	Active group	Colour of nuclei
Basic fuchsin	421	3	$-NH_2$	Mauve
Dahlia violet	105	2	$-C_2H_5$	Mauvish violet
Magenata roth	624	1	$-NH_2$	Magenta
Methyl violet 6B	180	2	$-CH_3$	Violet
Crystal violet	103	2	$-CH_3$	Deep violet
Brilliant green	315	1	$-C_2H_5$	Blue-green
Malachite green	315	1	$-CH_3$	Green
Light green		1	$-CH_3$	Yellow-green

principle of carmine is carminic acid, which belongs to the anthraquinone group. The chromophoric property is attributed to its quinonoid linkage and auxochromes are also present. It is soluble in water in all proportions, and is a dibasic acid and claimed to be nearly insoluble at its isoelectric point, pH 4–4.5. If it is dissolved on the acid side of its isoelectric point it acquires a positive charge, behaves like a basic dye and stains chromatin, but if dissolved in alkaline solution it can behave as an acid dye.

In chromosome studies, carmine is dissolved in 45% acetic acid, and the stain is known as acetic–carmine. It serves the double purpose of fixation and staining, since acetic acid is a good fixative for chromatin and is a rapidly penetrating fluid. 1% solution of the dye is prepared in hot 45% acetic acid and in some cases ferric hydroxide may be added to acetic–carmine during its preparation to allow the formation of a lake needed for the intensification of colour.

The common procedure of using acetic–carmine as a stain is to squash the tissue in a drop of the dye solution. In the case of bulk compact tissues, such as root tips, leaf tips, etc., materials can be treated in hot acetic–carmine and hydrochloric acid mixture which serves the double purpose of softening and staining. The use of a 2% iron alum solution for a few min prior to staining may act as mordant and help in the intensification of colour. While squashing, the best way of adding iron is to tease the tissue in a drop of carmine with the help of a scalpel, or penetration can be aided by slight warming. Being present in the form of an acetic solution, carmine is not a suitable stain for sectioned materials. In certain cases acetic acid is substituted by propionic acid.

Tissues which present difficulties in Feulgen staining may be mounted in acetic–carmine after Schiff's reaction. In such cases, hydrolysis in normal HCl and treatment with fuchsin sulphurous acid clear the cytoplasm allowing specific coloration of the chromosomes. The application of carmine as a chromosome stain is widespread from the lower plants like algae, to all other advanced groups.

Preparation of acetic–carmine, acetic–orcein, acetic–lacmoid
Materials required   For 2% solution: Carmine, orcein or lacmoid—2 g; glacial acetic acid—45 ml; distilled water—55 ml.

For 1% solution: the same amounts are used except for 1 g of the dye.

Add distilled water to glacial acetic acid to form 45% acetic acid solution. Heat the solution in a conical flask to boiling. Add the dye slowly to the boiling solution, stirring with a glass rod. Boil gently till the dye dissolves. Cool down to room temperature. Filter and store in a bottle with a glass stopper. Keep the mouth of the flask covered with cotton wool while the solution is being heated. Store the stain as 2.2 g, dissolved in 100 ml acetic acid. Dilute as needed.

Procedure   Acetic–carmine: use 1% solution directly for staining and squashing. Acetic-orcein or acetic–lacmoid : use 1% solution directly for staining. *Alternatively*, heat the tissue for a few seconds in a mixture of 2% solution and normal hydrochloric acid mixed in the proportion of (9 : 1) and then squash in 1% solution. The preparation and application of propionic–carmine and propionic–orcein are similar to those of acetic–carmine and acetic–orcein except that propionic acid is used instead of acetic acid. Lactic–propionic–orcein, prepared by dissolving 2 g natural orcein in 100 ml lactic and propionic acids mixture (1 : 1) and diluted to 45% with water, is very effective for p.m.c.s. For mitotic preparations, maceration in 1 NHCl at 60°C for 5 min between fixation and staining is necessary.

Orcein   Orcein was first employed as a chromosome stain by La Cour in 1941. It is a deep purple-coloured dye, obtained from the action of hydrogen peroxide and ammonia on the colourless parent substance *orcinol*, or, 3,5-dihydroxytoluene. It is available both in natural and synthetic forms. In nature, it is obtained from two species of lichens, *Rocella tinctoria* and *Lecanora parella*.

Orcein is soluble in water as well as in ethanol. Under certain conditions, it can behave as an amphoteric dye. In the study of chromosomes it is used in the form of acetic–orcein, that is, 1% solution in 45% acetic acid or, propionic–orcein (1% dye in 45% propionic acid). It can be used in the same way as acetic–carmine with the added advantage that no iron mordanting is necessary. In our experience the intensity of the stain, specially for meiotic materials, is not as good as carmine, though it is effective where carmine staining fails. It has been found to be a very effective stain for the chromosomes of mosses. For the study of root tip and leaf tip chromosomes, the use of a stronger hot solution of acetic–orcein and normal HCl, mixed in a specific proportion, is necessary for softening the tissue before mounting in a dilute solution of acetic–orcein. In various species of fungi, especially Ascomycetes, hydrolysis in normal HCl for a few min at 60°C, after fixation and prior to staining, has been found to be very effective. It is found to be specially useful in the study of somatic chromosomes though applied frequently for meiotic chromosomes as well. It has, however, to be applied with extreme caution, since overheating in orcein–HCl mixture has been found to induce chromosome breakage.

Lacmoid or resorcin blue is a blue acidic dye of the oxazine series. Similar to carmine, it can be used as an acid–base indicator and, when dissolved in acetic acid, it behaves as a basic dye. Unlike carmine, it is fairly soluble both in water and alcohols. Darlington and La Cour used it in place of carmine, and it was found to be very effective for the chromosomes of root tips, embryo sacs and pollen grains. For comparatively compact tissues of plants, like root tips, similar to orcein, heating in acetic–lacmoid–HCl mixture is needed prior to squashing for dissolution of the pectic salts of the middle lamella. However, it has a comparatively limited application and may be tried on those materials where other stains have failed.

Chlorazol black E is an acid dye of the tris-azo group. In ethanol, it has been applied as an auxiliary stain for chromosomes, along with acetic-carmine. This dye was applied after fixation prior to acetic–carmine staining and proved effective for species of Rosaceae. It is highly soluble in water and sparingly in ethanol. Being an acidic dye, the basis of its stainability with chromosomes is not clear, but it is probable that the protein component is stained. It is effective in materials where protein components of the chromosomes are high.

Crystal violet Newton (1926) used gentian violet to stain chromosomes. It is a mixture of crystal violet and tetra- and penta-methyl *p*-rosaniline chlorides. Crystal violet itself is hexamethyl *p*-rosaniline chloride. Gentian violet is a basic dye belonging to the triphenyl methane series. Crystal violet, one of the most adequate stains for chromosomes, is a bluish violet dye. The dye is closely allied to basic fuchsin from which it can be derived by the replacement of six hydrogen atoms of three amino groups by six methyl groups. It is soluble in both water and alcohols. In chromosome studies, aqueous 1% solution is used. In Newton's crystal violet-iodine technique, after the application of the stain to the sections or smears, the excess dye is first washed off in water. Then the slides are processed through iodine and potassium iodide in ethanol mordant, followed by dehydration in ethanol; differentiation in clove oil and cleaning in xylol before final mounting in balsam. To obtain a proper colouration of the chromosomes in difficult materials, iodine mordanting, which is normally carried out for a few seconds, should be further reduced, but in no case should this step be omitted, as acidic components of the cytoplasm also take up the colour which is removed by iodine in ethanol. Slow differentiation in clove oil is an essential step. Clove oil obtained from the flower buds of *Eugenia caryophyllata*, consists principally of eugenol, a guaiacol derivative. Differentiation in clove oil completes the dehydration, clears the cytoplasmic background and imparts a bright colour to the chromosomes. Final clearing in xylol is essential to remove clove oil completely before final mounting. For materials that are difficult to stain, slides can be mordanted in 1% chromic acid and washed prior to staining. For materials having a heavy cytoplasmic content, the slides can be further mordanted in chromic acid between the different ethanol grades, after mordanting in iodine. Crystal violet is widely used as chromosome stain. It is most effective on pollen mother cell smears or for mitotic and meiotic studies from sectioned tissues. It cannot be applied effectively on tissues to be squashed after staining.

For preparation of stain, dissolve 1 g in 100 ml water with constant stirring and boiling. Filter. Allow it to mature for a week before use.

Haematoxylin is obtained from the heartwood of *Haematoxylin campechianum*. The dyeing property of haematoxylin is attributed to its oxidation product, haematein. The presence of quinonoid arrangement in haematein and its absence in haematoxylin is clear. The process of oxidation, which is otherwise known as ripening, may take several weeks, spontaneously, but may be hastened by the use of oxidising agents such as sodium iodate, hydrogen peroxide, chloral hydrate, potassium permanganate, etc.

In view of the necessity of oxidation in the preparation of haematein from haematoxylin, the aqueous solution of haematoxylin is prepared and allowed to ripen for several weeks. Without the use of a mordant, haematoxylin solution is entirely ineffective in staining chromosomes. For chromosome studies, potassium aluminium sulphate and iron alum are widely used, the latter being more effective. The potash alum lake of haematoxylin is used for progressive staining, whereas iron alum is utilized in regressive staining. Progressive staining implies gradual addition of the stain till the maximum colour is obtained, whereas regressive staining involves overstaining the material and subsequently washing off the excess stain.

Haematein, after ferric mordanting, has a strong tendency to accumulate around densely stained materials and has most often been used in chromosome studies. The sections or smears in water are first mordanted in a strong solution of iron alum (4%) followed by washing in water and staining in haematoxylin. Differentiation is carried out in dilute solution of iron alum or picric acid. In properly controlled differentiated preparations, chromosomes appear intensely black. After washing once more in water, the tissue is dehydrated through ethanol, cleared in xylol and mounted in balsam. For plant chromosomes, haematoxylin staining is not very effective, due to the heavy cytoplasmic content.

Preparation Dissolve ammonium alum in distilled water to prepare a saturated solution. Dissolve haematoxylin in absolute ethanol and add slowly to the former. Expose to air and light for one week. Filter, add 25 ml each of glycerol and methanol. Allow to stand, exposed to air, until the colour darkens. Filter. Store in a tightly closed container. Allow the solution to ripen for a month before use.

Wittmann's acetic–iron–haematoxylin schedule

Treat plant material fixed in acetic-ethanol (1:3) in 6 ml of a mixture of HCl and ethanol (1:1) for 10 min, then in Carnoy's fluid (6:3:1) for 10–20 min, squash in a drop of stain containing 4% haematoxylin and 1% iron alum in 45% acetic acid and heat gently.

Toluidine blue a basic dye of the thiazine series, is bluish violet in colour. It can be used selectively for DNA if normal hydrochloric acid hydrolysis is performed for a very short period prior to staining. Pelc (1956) utilized toluidine blue for staining through film in autoradiographic procedure, applied in an aqueous solution which is soluble both in water and alcohols. It is not recommended as a general stain for chromatin.

Giemsa is a mixture of several dyes, namely methylene blue and its oxidation products, the azures and eosin Y. The combination stain, Giemsa, is generally prepared by dissolving the powdered mixture in glycerin and methanol, and in staining, chromatin is stained red and cytoplasm blue.

Preparation Add 3.8 g of powdered Giemsa (R 66-Gurr) to 250 ml glycerine or alternatively, 1.0 g to 66 ml glycerine. Keep at 55 to 60°C for 1.5 to 2 h and add an equal quantity of methanol. For phosphate buffer, pH 6.4, prepare—Solution A: 11.336 g/100 ml distilled water or 56.68 g/100 ml distilled water. Solution B: 8.662 g/100 ml distilled water or 43.31 g/100 ml distilled water. Mix 5 ml of each of the solutions, make up to 1.0 l with distilled water, adjust pH to 6.4 with 0.1 N HCl.

Before staining, add 2 ml of the stock to 2 ml of the phosphate buffer and make up to 50 ml with distilled water.

Fluorochromes

Staining with fluorochromes binds to a particular component of the nucleus, followed by observation under ultraviolet light.

Acridine Orange (AO, Gurr) has been used in staining chromosomes as a 1:1000 solution in ethanol; the optimum period is 10 min. It fluoresces green in combination with double stranded and red with single stranded nucleic acids. The methods of fixation affect the secondary nature of deoxyribonucleoprotein complex and interfere with AO-binding capacity of chromosomes.

Quinacrine dihydrochloride (Winthrop or Sigma) is employed widely. The period of staining ranges from 4 to 6 min, followed by washing for 2 min in distilled water and differentiation in McIlvaine's phosphate buffer solution at pH 5.5. Commercial antimalarial atabrine powder (atabrine dihydrochloride or atabrine hydrochloride) can be used in routine studies instead of chemically pure quinacrine dihydrochloride as 100 mg to 20 ml glass distilled water.

Quinacrine Mustard (QM) was initially synthesized by E. J. Modest at Boston, giving highly specific banding patterns. It is regarded to bind DNA both through the alkylating group reacting primarily with the guanine content of DNA (Casperson *et al*., 1968) and by intercalation of the quinacrine group in the double helix of DNA. The amount of fluorescence exhibited by different segments of a chromosome stained with QM is controlled by the quantity of DNA and by the qualitative differences in QM-binding capacity of the DNA in different segments. The pattern is determined by irregularities in DNA distribution, reflecting the chromosome pattern. Weisblum and de Hasseth (1972), suggest that strong fluorescence with QM reflects the presence of DNA with high A-T content against Caspersson's contention that it indicates local differences in G-C content.

Ethidium Bromide (EB, 2,7-diamino-10-ethyl-9-phenyl anthridinium bromide) reacts specifically with both DNA and RNA by intercalation, to form relatively stable complexes with markedly increased fluorescence. Some plant chromosomes fixed in acetic-ethanol (1 : 3) and stained with EB gave a reversed pattern to QM while other plants stained uniformly. 0.005% EB at pH 6.8, colours bright orange with native DNA and dull orange with denatured DNA.

Fluorescein-tagged reagents have been employed in antinucleoside antibody binding methods. Base-specific antinucleoside antibodies react with specific nucleoside bases in single stranded DNA *in vitro*. They attach to fixed chromosomes on being treated with denaturing agents like NaOH. A banding pattern may be obtained with antibodies which react specifically with only one nucleoside base in DNA.

Hoechst 33258 is used as a stock solution of 50 μg/ml in water.

Quinacrine conjugates: Quinacrine derivatives of polylysine stain chromosomes in a banded fluorescence pattern similar to QM.

DNA-Binding nucleoside specific antibiotics: DNA-binding guanine-specific antibiotics, chromomycin A_3 (CMA) and mithramycin (MM) have been used as chromosome fluorescent dyes, as also the A-T specific fluorochrome 4',6-diamidino-e-phenylindole (DAPI). Non-fluorescent dyes may be used as counterstain—methyl green with CMA and actinomycin D (AMD) with DAPI.

Some miscellaneous double stains

Gallocyanin and other stains: Tissues warmed in gallocyanin solution for 2–4 min can be counterstained in Biebrich scarlet, phloxine or eosin Y.

Safranin O and Aniline blue: Root tips are stained 15 min in 1% aqueous safranin O and rinsed in distilled water. They are then stained in 1.0% aniline blue in 95% ethanol for 2 min.

Orange G and Aniline blue: have been used for both mitotic and meiotic chromosomes. Sections fixed in chromic–formalin (1 : 1) are rinsed in potassium citrate buffer, stained in a mixture containing 2 g orange G and 0.5 g aniline blue dissolved in 100 ml potassium citrate buffer for upto 3 min, washed in the buffer, dehydrated and mounted.

Ruthenium red and Orange G after Fuchsin staining: Stem tips, after 30 min hydrolysis, are stained in fuchsin solution for 24 h, rinsed, stained in aqueous ruthenium red solution for 30 min, dehydrated, stained for 1.5 min in orange G in absolute ethanol and clove oil, passed through clove oil and xylol and mounted in balsam. Chromosomes take up deep purple stain.

PROCESSING AND MOUNTING

After fixation, the tissue is processed for further study. Different schedules are followed for block and smear preparations.

Block preparation

Materials which cannot be squashed or smeared have to be dehydrated and embedded in a suitable medium. By embedding, small or delicate objects can be surrounded with some plastic substance which supports it on all sides and allows sections to be cut without distortion. It is also useful for showing the arrangement of cells in a tissue and the sequence of the stages of division.

The procedure entails: washing, dehydration, clearing, infiltration and embedding, microtome-section cutting and removal of embedding material.

Different periods of **washing** in running water, from 1 h to overnight, are employed, depending on the nature and the thickness of the tissue and the fixing fluid used. Comparatively hard tissues, like flower-buds or root tips, are kept in perforated corked porcelain thimbles under running water. Very small or delicate tissues are kept in the original tube, the fixing fluid is drained off and the tissue is washed in successive changes of warm (44°C) water at half-hourly intervals. Alcoholic fixatives should be washed out with alcohols of approximately the same percentage as that of the original solution and reagents containing picric acid with ethanol.

Since the embedding material usually does not mix with water and aqueous solutions, it is necessary to **dehydrate the tissues** before they can be embedded. The tissue is passed through a series of solutions, each containing a mixture of the dehydrating agent and water, with the concentration of the former increasing gradually, on to the pure agent. The most suitable and economical dehydrating agent is ethanol. It also has a hardening effect on the tissue and the schedules are arranged as to utilize both dehydration and hardening effects. The tissue is generally passed through successive grades containing 30, 50, 70, 80, 90, 95% and absolute ethanol, the period of treatment being variable, depending upon the nature and thickness of the specimen. For tissue of the size generally used for chromosome study, 1 h in each is quite long enough, while for thicker plant tissue, overnight treatment in each of 70% and absolute ethanol is found to be most effective. The tissue can be stored in 70% ethanol, if necessary.

Alternatively, the concentration of the medium containing the tissue can be gradually increased by adding drops of strong ethanol at fixed intervals. The disadvantages of using ethanol as a dehydrating agent are: the length of the schedule; excessive hardening and shrinkage of the tissue and since ethanol does not mix with paraffin or celloidin, an intermediate *clearing* agent is needed, which lengthens the process.

Several alternative **dehydrating agents** are known, some of which are miscible with the embedding material. Some of these are given below:

Tertiary butanol was used by Johansen (1940). The tissue, which had been dehydrated up to 30, 50 or 70% ethanol is transferred to a mixture of distilled water, ethanol and tertiary butanol, then passed through grades containing decreasing proportions of distilled water and ethanol and increasing proportions of butanol, till a mixture of butanol and ethanol in proportion 3 : 1 is obtained. Finally the tissue is given three changes in pure tertiary butanol (one overnight).

Dioxane (*diethyl dioxide*) mixes with water, ethanol and xylol, and dissolves balsam and paraffin wax, and can therefore be used as a substitute at any stage of the usual ethanol-xylol dehydrating and clearing schedule.

n-Butanol: The material can be transferred, after partial dehydration in a low concentration of ethanol, to a mixture of ethanol and butanol, with successive changes in mixtures containing increasing proportions of *n*-butanol, followed by changes in *n*-butanol alone, and kept overnight.

Iso- and normal propanols can also be used effectively as substitutes for ethanol. The schedule followed is similar. Isopropanol has been used for dehydration, removal of paraffin and clearing before mounting in balsam through treatment in 60% followed by three changes in 99% isopropanol and two changes in molten paraffin. Two methods for dehydration are known, (1) through primary dehydration by glycerol; and (2) through isopropanol alone (60, 85 and 99%). Tissues are placed over solid paraffin in a phial and heated to 56–58°C.

Other chemicals used as combined dehydrating and clearing agents include amyl alcohol and methylal-methylene dimethyl ether followed by paraffin oil.

Glycerol is found to cause less distortion than ethanol; 95% ethanol removes some glycerol, sets the protoplasm and improves the staining.

Clearing is required when paraffin is the embedding material. For celloidin-embedding, no separate clearing is required.

Since paraffin does not mix with many dehydrating agents, an intermediate medium which is miscible with both the dehydrating agent and paraffin is necessary, this medium performing the function of ridding the tissue of the dehydrating agent. Most of these reagents render the tissues translucent, as their refractive indices are close to that of the proteins of the tissue and rays of light can pass through without refraction; therefore they are also called 'clearing' agents or antemedium and the process 'clearing the tissue'.

The ideal antemedium should have rapid penetrating power and should mix equally well with both the dehydrating and embedding agents. An immediate transfer from the pure dehydrating agent to the pure antemedium may cause shrinkage or distortion of the tissue and so mixture of the two fluids should be used in varying proportions before transfer to pure antemedium. The most satisfactory and most widely used antemedium is chloroform. The tissue is usually passed, after dehydration in ethanol, through a series of ethanol–chloroform grades, 3:1, 1:1 and 1:3, being kept for 1 h, or more if necessary, in each, the period of treatment depending on the thickness of the tissue. Finally it is transferred to pure chloroform.

Another effective clearing agent, benzene, can be used in a series of grades in combination with the dehydrating agent. Xylol is also used extensively, but has a tendency to shrink and harden the tissue. Various organic oils, like Bergamot, cedarwood, clove and aniline oils have also been used as antemedia.

Two **major methods for embedding** used in the study of chromosomes are the paraffin method and the celloidin or collodion method. A modification is the 'double' embedding method. If the material is to be cut in a fresh condition, the frozen section technique can be employed.

Paraffin method is most extensively used for the study of chromosomes. As the embedding mass is removed before staining, a wider selection of stains can be used. Its chief drawback is the lengthy procedure.

The most commonly used schedule is to dehydrate the tissue by ethanol and then clear it by passing it through ethanol–chloroform grades, until finally pure chloroform is reached. The tissue is kept in pure chloroform for 10–30 min, and then small chips of paraffin wax, of a melting point lower than the one desired for embedding, are added to the chloroform containing the tissue. The tube is kept at 35°C for periods ranging from 2h to overnight, depending on the nature and thickness of the tissue. Later, the tube is transferred to 45°C and kept overnight. It can be stored in this temperature for an indefinite period. The tissue is finally transferred to a hot bath maintained between 55 and 60°C, at the same or higher temperature than the melting point of the embedding paraffin. Two more successive changes are given with molten embedding paraffin at intervals of 15–30 min, the final change being given only when no trace of the smell of chloroform is left.

An alternative method is gradually to warm, on a hot bath, the chloroform containing the tissue up to the melting point of the paraffin employed and, during warming, to add by degrees small pieces of paraffin to the chloroform. As soon as the bubbles of the tissue cease, the addition of paraffin may be stopped. This process, being a gradual one, minimises the danger of shrinkage, but cannot be recommended for tissues which are to be treated for chemical study. Numerous alternative infiltration techniques exist to suit the different dehydrating and clearing schedules. The time of infiltration varies with the thickness of the tissue. Each technique can be followed with other dehydrating and clearing agents, such as xylol, normal or secondary butanol or an essential oil. With dioxane as dehydrating agent, little chips of Parowax are added gradually to the dehydrated material in pure dioxane. The mixture is kept in a warm bath till dioxane, which is not a good paraffin solvent, is saturated with Parowax. The later steps are similar to chloroform–paraffin infiltration. Acetone can be used as a substitute for ethanol in dehydration, and clearing can be done through acetone–chloroform grades.

During embedding, after the complete infiltration of the tissue with paraffin, the molten paraffin with the tissue is poured into a suitable receptacle. The tissue is arranged in parallel rows and rapidly cooled.

In the cellodion method, the tissue is transferred to a mixture of absolute ethanol and ether (1:1) and kept overnight. A hole of a suitable size is made in a piece of junket to hold a part of the tissue and both the tissue and junket are transferred to 2% celloidin solution and kept overnight.

Frozen section technique is based on the principle of freezing the tissue directly to harden it, and cutting sections while the tissue is frozen. It takes up much less time than the paraffin and celloidin methods. Since the tissue is not dehydrated, the cells retain a life-like appearance and the tissues can be sectioned, if necessary, without any fixation at all. The material can be cut fresh or after fixation and any one of the usual fixatives can be used, the tissue being washed thoroughly. It may be kept overnight in thick gum arabic, if necessary.

The piece of tissue is trimmed into a size within 3 × 3 cm and is placed in a little water on the microtome table freezer to freeze the tissue to the table.

Tissues with compact cells should be frozen at about -10 to $-15°C$ while others can be cut at -20 to $-30°C$. The knife is oriented with its edge close to the tissue at right angles, and the tap controlling the CO_2 cylinder is opened and closed several times until the tissue is congealed to the table and frozen right through. In order to regulate the freezing, only short jets of CO_2 should be allowed to escape at intervals on the material.

The tissue is left for a few minutes and then the knife is passed over it, till sections begin to cut. If the sections have fine cracks and are brittle or roll up, it means that the tissue has been overfrozen and it should be allowed to warm before being cut again; but if the sections are too soft and disintegrate the tissue is underfrozen and must be exposed to further jets of CO_2. Sections should be cut very rapidly and allowed to accumulate in a mass on the knife, and then be transferred to a petri dish containing distilled water. The loose sections can be lifted on clean slides and observed unstained under a coverglass.

A freeze-drying apparatus devised with liquid nitrogen, permits section cutting within 5 h after the fresh tissue is obtained. Different variants are available. The sections may be attached to slides by gelatin or by their own coagulated juice.

Gelatin-coated slides are prepared by smearing them with a thin film of specially prepared gelatin (2 g of gelatin soaked in 100 ml of distilled water and heated to 50–60°C) and dried. The section is lifted out of the water on a gelatin-coated slide, flattened out by pressing on it with pad through a piece of wet cigarette paper and the paper is peeled off.

In a modification of this schedule, the frozen sections are soaked for 5 min or longer in a mixture of 1.5% aqueous gelatin and 80% ethanol (1:1), teased on to a slide and blotted with filter paper dampened in rectified spirit. The gelatin congeals, anchoring the section to the slide.

After cutting the sections and mounting them on slides, the next step is to remove the embedding material and gradually **bring down the tissue to the medium** in which the stain is dissolved, usually water. The steps followed are usually the reverse of the process leading to embedding through dehydration and infiltration and the schedules differ depending on the material used for embedding.

For removing paraffin wax from sections, the most effective chemical is xylol; treatment in pure xylol is followed by a mixture of xylol and absolute ethanol in equal proportions, then pure absolute ethanol and ethanol grades with decreasing percentage of ethanol to water. If necessary, the slides can be preserved for an indefinite period in 70% ethanol. If bleaching is necessary for tissues fixed in heavy metallic fixatives containing osmium or platinum, the slides are transferred from 80% ethanol to a jar containing hydrogen peroxide and 80% ethanol (1:1) and kept overnight or as long as needed on a hot plate at 35°C. Afterwards the slides are passed through 70, 50 and 30% ethanol and brought down to water.

For **softening paraffin-embedded tissues**, several procedures are available. Dilute hydrofluoric acid, used alone and with glycerol and ethanol, softens plant materials embedded in paraffin. Tannins and phlobaphene compounds can then be removed by treating the sections for 12–48 h in a mixture of aqueous chromic acid, potassium dichromate and glacial acetic acid. Exposure to a mixture of glacial acetic acid and 60% ethanol (1 : 9 or 2 : 8) for 2–5 days also gives very good results. Paraffin-embedded plant specimens can be soaked, after exposing one side of the tissue, in a mixture of glycerol, 10 ml; Dreft 1 g and water, 90 ml for 2–3 days at 37°C.

Sections embedded in celloidin are usually stained as such, without removal of the celloidin matrix, which does not interfere with staining. The slides are brought down to water from 70% ethanol and then stained.

In smears, the cells are directly spread over a slide prior to fixation, and no treatment is necessary to secure cell separation. Pollen mother cells from anthers are the most convenient objects.

In squashes, special treatments are needed for dissolution of the pectic salts of the middle lamella so that separated individual cells can be obtained from a compact mass of cells, this treatment being carried out after fixation or even after staining. After passing through the required steps, the softened bulk material or small tissue can be neatly squashed on a slide by generally applying pressure or tapping with a needle over the coverglass. It is the best way to study mitotic behaviour of chromosomes of somatic tissue.

The term *smear* is commonly applied to cases where cells have been spread on the slides before fixation, while *squashing*, on the other hand, is used for the process performed after fixation or staining.

The general procedure for preparing smears of pollen mother cells is to squeeze out the fluid from the anther on to a dry slide, spread it with the aid of a scalpel and invert it in a tray containing the fixative. The entire process should be rapidly executed and must not take more than 4–5 s. Quick handling is essential, as otherwise the fluid tends to dry up, resulting in chromosome clumping. The use of a scalpel aids in the addition of iron which acts as a mordant.

Very small anthers should be smeared complete. The anther debris is removed after fixation with a needle, keeping only the pollen mother cells. When the buds are too small to dissect out the anthers, the entire bud is fixed in Carnoy's fluid and chromosome preparations are made following section-cutting.

Pollen mother cells can also be smeared or squashed in a fluid which serves the double purpose of fixing and staining e.g. Belling's (1926) *Iron acetic carmine schedule*, when the anther is directly smeared in a drop of a solution containing carmine dissolved in acetic acid. This method can be applied both in smears and squashes in plant tissues after modifications, principally involving intensification of colour.

For certain materials it may be necessary to fix the flower buds in Farmer's fluid (acetic–ethanol 1 : 3) prior to smearing in carmine in order to secure cytoplasmic clearing and for fixation. Before staining, the material is kept in 45% acetic acid for 15–30 min to cause swelling, to counteract the effect of ethanol and to soften the tissue.

In squashes, the most important step is the softening of the tissue. The different schedules can be divided into two categories, namely, softening performed prior to staining, and softening, clearing and staining accomplished in the same fluid. Within the first category are included the various types of chemical agents employed by different authors for this purpose, including enzymes.

The most important agent for softening the tissue is dilute hydrochloric acid. In Feulgen staining, this step is essential to secure Schiff's reaction for aldehydes. In addition to liberating aldehydes of sugar, normal hydrochloric acid at 58°C dissolves pectic salts of the middle lamella, thus helping in cell separation (Sharma and Bhattacharjee, 1952), and clears the cytoplasm. With dilute hydrochloric acid (10%), the treatment should be carried out in a slightly warm temperature (58–60°C) for 4–5 min until softening and the acid washed off either in 45% acetic acid solution or water before staining.

Softening and maceration of the tissue can also be achieved during fixation by the use of a mixture of equal parts of 95% ethanol and concentrated hydrochloric acid as the fixative. No warming is needed and even after 5 min treatment the tissue becomes fixed and softened at the same time. If necessary, hardening the tissue for 10 min in Carnoy's fluid after this treatment can also be carried out. For hard materials, ethanol and hydrochloric acid mixed in the proportion of 3 : 1 is more effective. This acidified ethanol treatment is specially effective for materials with thick walls, such as pollen grains, leaves, etc., where slowly penetrating fluids are ineffective.

The most reliable method of softening and clearing without injury to cellular parts is *enzyme treatment*. *Cytase*, together with other enzymes from the stomach extract of snails, *Helix pomatia* has been found to be very useful in a wide variety of plants, including fungi and pectinase for dissolution of pectic salts of the middle lamella. A 5% solution in 1% aqueous peptone was most effective. This procedure requires at least 2–5 h of treatment. Harris and Blackman (1954) treated Feulgen-stained roots in 2% pectinase solution (pH 6.6) for 12 h, followed by a commercial pectin product 'Certo', and then suspended the cells by suction and expulsion through a pipette. The present authors have, however, noted that a 2% aqueous solution of pectinase, if applied for half an hour at 37°C, results in considerable softening of plant materials.

Of all the methods so far devised for cell separation and softening, treatment with dilute HCl, in spite of its limitations, is most commonly employed because temperature and period of treatment can be varied as necessary and principally because of the low cost, easy availability and rapidity of the schedule.

In a number of schedules, the softening is carried out together with staining. The tissue is heated, after fixation, over a flame for a few seconds in a mixture of one of the acetic dyes and hydrochloric acid. The commonly used acetic solutions of dyes are acetic–orcein, acetic–lacmoid and acetic–carmine. Normal hydrochloric acid is mixed with a 2 or 1% solution of the acetic-dye, usually in the proportion of nine parts of dye to one part of acid. The proportion of acid may be increased if the tissue proves difficult in squashing. The material is directly squashed under a coverglass in the mounting medium which is either 45% acetic acid or 1% solution of the dye dissolved in 45% acetic acid solution.

Summarising, smearing is a comparatively easier schedule than squashing and should be performed prior to fixing only on cells lying in a fluid medium, such as pollen mother cells. Squashing involves different steps, such as pre-treatment, fixation and softening. Softening can be carried out by acid, alkali or enzyme treatment after fixation but prior to staining in most cases. In a number of schedules, staining and softening are carried out in the same fluid and following staining, the materials are either first teased with the needle in the mounting medium prior to mounting or mounted directly in the medium under the coverglass. Final squashing is performed by applying pressure over the coverglass on a blotting paper before sealing for observation.

Modifications have been developed for both smear and squash techniques to obtain better preparation, better preservation and easier schedules. Preservation of plant tissues by storage at or below $-10°C$ after fixation in acetic–ethanol (1:3) and chloroform–ethanol–acetic acid (4:3:1) mixture gives very good staining with carmine squash even up to 6 months. Squashing the stained root tip in a drop of stain between two pieces of plastic, and lamination in an electric press eliminates dehydration. Glass coverglasses can be replaced in squashes by a wet square of cellophane. The slide is exposed to formalin vapour for 45 min before stripping off cellophane and staining. In another method, the coverglass is directly glued on a slide by a rubber solution or 10% polyvinylacetate in acetone–ethanol (1:1) mixture. After smearing, fixation and staining on the coverglass, it is detached from the slide by a suitable solvent. Smears can be made on 35 mm photographic film, instead of a glass slide, and fixed and stained as usual by handling on a photographic film developing reel. The slides are sprayed with a plastic cement and examined, using a special holder to keep the film base flat.

A further development is of **airdrying** or **flamedrying** cells for processing. The method can be applied to tissues which can be converted easily into cell suspensions. Longterm cultures can be handled as monolayers or converted into cell suspensions. The cell suspension is pretreated and fixed prior to airdrying.

Clean slides are stored in absolute methyl alcohol in cold before use. The fixed material in suspension is centrifuged and most of the supernatant is discarded so that the fluid at the bottom of the tube consists of a small quantity of fixative with a high proportion of suspended cells. A slide is wiped carefully and one surface rinsed with the fixative. A small amount of cell suspension is drawn out by a Pasteur pipette. The slide is tilted at an angle and a drop of suspension allowed to fall at the upper end of the slide on the wet surface. As it runs down the slide, the slide is shaken vigorously to dry the suspension quickly. Alternatively, the slide is kept slanted and the suspension allowed to spread and dry by blowing hot air across it or even passing it through flame.

A recent modification has been the enzyme treatment maceration technique, followed by airdrying, developed principally for studying banding patterns in root-tips (for details, see chapters on Representative schedules and Banding patterns).

Mounting media

Temporary mounts are mainly squashes and smears. The medium for mounting is either the stain itself or its solvent. In the former case, the processes of staining and mounting are carried out simultaneously. The tissue is then lifted on to the slide containing a drop of the stain and squashed. In the latter, the tissue is stained and squashed on a slide in a solvent of the stain. In both cases, the coverglass is placed on the material and medium and suitable pressure applied to flatten the tissue and then ringed with paraffin wax.

Permanent mounts involve mounting the tissue, after suitable processing, so that the preparations can be kept for a long period without appreciable distortion of the structure or intensity of stain.

Permanent mounts of sections from paraffin blocks
The different steps in the process usually depend on the medium in which the stain is dissolved. In general, the entire process is based on first dehydrating the tissue, then impregnating it in successive grades with the solvent of the mounting medium and finally mounting with the medium chosen.

The most commonly used dehydrating agent for paraffin sections is ethanol, though acetone and various other alcohols are also used. After sectioning, the tissues are brought down to water for staining through successive xylol : ethanol grades. If the stain is dissolved in water, the section can be transferred directly to absolute ethanol after staining and mordanting. If, however, a counter-stain in a lower grade of ethanol is applied, the tissue has to be passed through the required grade before transfer to absolute ethanol.

The slides can be transferred directly from absolute ethanol to the mounting medium or to a mixture of the solvent of the mounting medium and absolute ethanol in equal proportions, then to the pure solvent. Alternatively, they may be transferred to differentiating medium for removing excess of the stain before transfer to the solvent of the mounting medium.

The choice of differentiating or clearing medium usually depends upon the stain used. The most suitable differentiating agent for crystal violet is clove oil. It removes superfluous stain from the cytoplasm, thus rendering the stain in the chromosomes brighter, and also completing the dehydration. As soon as the surplus stain is washed off, the slides are transferred to the pure solvent for the mounting medium. The clearing agent may be a solvent for the stain or a mordant.

CHAPTER II.2

MICROSCOPY FOR STRUCTURAL AND ULTRASTRUCTURAL ANALYSIS

ORDINARY LIGHT MICROSCOPY

The underlying principle is to obtain a real, inverted, and enlarged image of the material by means of the objective lens, followed by the formation of virtual image by means of the eyepiece lens. In the study of chromosomes, the compound microscope should have at least the following attachments:

(1) Apochromatic objective and oil immersion lenses (×100) (1.3–1.4 n.a.);
(2) Sub-stage aplanatic and achromatic condensers—1.4 n.a.;
(3) Compensating eyepieces (×10 ×15 ×20);
(4) Fitted mechanical stage.

Special lenses are used principally to eliminate aberrations. Spherical aberration is inherent in lenses with spherical surfaces, where the rays passing through the periphery of the lens focus at a different point to those passing through the entire lens or close to the axis. In *aplanatic* lenses, which are compound and constituted of different kinds of glass, all the rays are brought to a common focus by suitable corrections.

Chromatic aberration implies that ordinary sources of illumination, being composed of light of different wavelengths, of different colours, focus at different points. In suitably constructed lenses, these aberrations are eliminated so that preferred rays are made to focus at a common point. In *achromatic* objectives the chromatic aberration is corrected for two colours, and the spherical aberration for one colour. In *semi-apochromat* or *fluorite* lenses, a higher degree of correction is achieved and in *apochromatic* lenses, the chromatic aberration is corrected for three colours and spherical aberration for two.

Different colours undergo different degrees of magnification. This inequality of colour magnification is corrected by means of compensating eyepieces, while apochromatic lenses are used as the objectives. The separation of finer details in an image is dependent on the resolving power of lenses. This power, or resolution, depends on the wavelength of the illuminating source as well as on the *numerical aperture* of the lens. Resolution is always proportional to the numerical aperture, which is $n\sin\mu$, n being the refractive index of the coverglass, and μ the maximum angle to the optical axis formed by any ray passing through the specimen, before the formation of total internal reflection.

With an objective of n.a. 1.4, good resolution can be achieved if the intervening minimum distance from the specimen is 0.24×0.001 mm. Brightness increases with increase in numerical aperture, but decreases with increase in magnification. Oil immersion objectives are used for critical work where the scattering of light, due to its passage through media of different refractive indices, is to be avoided. If a fluid of refractive index similar to that of glass and Canada balsam is used to bridge the gap between the coverglass and objective lens, then a homogeneous medium can be achieved for the path of light, avoiding loss of light as far as possible. Cedarwood oil is generally used for the purpose, its refractive index (1.510) being close to that of glass (1.518) and balsam in xylol (1.524) but several synthetic media are now available. For the source of illumination, Pointolite, ribbon filament, mercury arc, or even a 100 W ordinary lamp may be used. Wratten yellow-green filters for violet-stained preparations and blue filters for red-stained preparations are suitable.

For photography, 35 mm camera may be used. The camera is fitted over the eyepiece. Cameras with microphotographic attachments are available. Both slow- and high-speed films are satisfactory, depending on the requirements, the former requiring a longer exposure period.

PHASE AND INTERFERENCE MICROSCOPY

The two types are identical in the sense that the purpose is to bring about visible change in intensity from an undetectable phase change. Interference systems, being more plastic, allow variable phase changes. The phase change is represented as $\phi = (n_p - n_m)t$, where n_p and n_m are the refractive indices of the object and the immersion medium respectively and t is the thickness of the object. The advantages of phase microscopy are:

(1) Simple and easily adjustable arrangement.
(2) Low cost of the apparatus.
(3) Insensitivity to slight variations in slide and coverglass.
(4) Internal details are often better resolved through zone of action effect.

The *principal limitation* of the phase system is that it is not possible to carry out quantitative measurements conveniently and the presence of a halo prevents proper resolution to some extent. The direct light is allowed to fall on a conjugate area (annulus) here, while the diffracted light is separated and falls on the entire phase plate. Two types of phase contrast microscopes are generally available, (i) based on the principle of negative phase contrast, and (ii) depending on positive phase contrast. In the positive phase contrast system, the slightly retarding object details appear brighter against a lighter background, thus resembling visually stained preparations; while in the negative system, the object is lighter than the background. For routine use in cytology, the former is usually preferred due to the excellent contrast of living cells in aqueous media.

The convenience of securing quantitative results is the principal reason for the development of interference microscopy. The light is split into two beams by a beamsplitting mirror, one beam being transmitted through the object and the other passing some distance to the side of it. The interference is produced by the two beams combining at the semireflecting mirror.

The advantages of interference microscopy are (i) Phase changes of the material can be measured. (ii) Bright colour effect can be secured. (iii) The contrast can be varied, allowing a proper contrast to be selected with respect to the object, to secure the intracellular details. The system is elastic and variable phase contrast can be obtained. (iv) 'Halo' and 'Zone' effects are absent. (v) Mass per unit area of the cell can be conveniently measured. The flatness of the image obtained and glare are serious disadvantages of the interference system.

FLUORESCENCE MICROSCOPY

The utilization of the fluorescence shown by some of the cell constituents, as well as of some special dyes, against ultraviolet light, forms the basic principle of fluorescence microscopy. In this system, intracellular constituents are detected either through their own property of autofluorescence, or by secondary fluorescence due to the adherence of labelled or fluorescent dye.

The general practice in the majority of laboratories is to utilize the property of secondary fluorescence obtained by fluorochrome preparations, but such dyes become effective only in long blue or near ultraviolet light. These agents are either acidic, basic or amphoteric in nature and can be used in accordance with their application in cytological practice; but special technique is necessary for the detection of secondary fluorescence of specific compounds within the cell. The method involves the observation of frozen sections or squashes against ultraviolet light and through a standard microscope fitted with adequate condensing lenses and filters. A special advantage of this technique is that unfixed materials can be studied, thus eliminating the artefacts which often arise due to fixation.

Conjugated planar dye molecules, in solution, form complexes having properties different from those in monodispersed form. The metachromatin dye polymer complexes allow observation of fluorescence and absorption spectra. One of the planar dye molecules is acridine orange, dissociating near pH 7. This dye, as a monomer at pH 6.0, has fluorescence and absorption peaks at 535 and 490 nm, and as a polymer the peaks are at 660 and 455 nm respectively. It is difficult to distinguish spectroscopically between acridine orange-DNA and acridine orange-RNA complexes. However, in fixed preparations, it may be possible to distinguish DNA and RNA on the basis of metachromasia of acridine orange, a red colour specifying RNA, and the yellowish-green colour indicating DNA.

Several azo dyes become fluorescent after combining with tissues. Aromatic compounds in the dye control its fluorescent property. In other classes of dyes, ring closure, coplanarity of chromophores, accumulation of ring system and low dye concentrations exert an inhibiting influence. Other factors also control the fluorescence of triphenyl methane, anthraquinone and quinone-imine dyes.

Quantitative measurements can be carried out by fluorescence microspectroscopy. Long exposure time for photographs should be avoided as the intensity may fade with continuous exciting irradiation of the fluorochrome-stained cell.

In preparing tissue sections meant for observation under the fluorescent microscope, chemical fixatives are avoided as far as practicable, freeze-dried or chilled preparations of unfixed materials being preferable. Frozen sections of unfixed tissue can be cut in a microtome and maintained in a refrigerated cabinet at $-20°C$. Freeze-drying, during which the temperature of the tissue should be maintained at about $-40°C$ by means of a slush of diethyl oxalate, can also be employed.

The most adequate light source, emitting ultraviolet rays, is the carbon arc, preferably with a direct current and fitted with an electromagnetic field. In all fluorescence microscopes, barrier filters for cutting off undesirable wavelengths and exciter filters for excitation of fluorescence are used. The latter may be of wide and narrow band types. Barrier filters are chosen in appropriate combination with exciter filters.

Fluorescence microscopy has also been applied to detect specific antigens. In principle, the method involves the forced production of antibodies by the injection of antigens labelled with fluorescein isocyanate into the tissue. The conjugation of the gamma globulin of the antibody with fluorescein isocyanate in the antigen, makes visible the site of the antibodies within the tissue. For the study of the effect of chemicals on the nucleus and cytoplasm, fluorescence microscopy is often recommended. In the study of cellular nucleoproteins and nucleic acids, diaminoacridine dyes, such as acriflavine, acridine orange and acridine yellow have been employed because of their affinities. With acridine orange, RNA and DNA stain differentially, the former appearing as red and the latter as green under fluorescence.

CHROMOSOME IMAGE ANALYSIS

Chromosomes, both long and short, show variations in condensation within the arms, and the pattern is termed as the *Condensation Pattern* (Fukui and Mukai, 1988). Image analysis permits quantitation of the condensation pattern and density profile and brings out the cryptic differences existing at the intra and interchromosomal levels (Fukui and Kakeda, 1994; Kamisugi *et al.*, 1993; Nakayama *et al.*, 1995). To secure effective analysis of image, well spread, well stained prometaphase and metaphase chromosomes are essential. The photography involves three exposures—(i) normal exposure, (ii) over-exposure and (iii) under-exposure. The normally exposed photograph is used for visual inspection and the other two are subjected to image analysis. The overexposed photograph is utilized for preparation of contour line of chromosomes, including the terminal positions. The under-exposed photograph is meant for determination of condensation pattern or density profile along the midrib of the chromatid. The contour lines, extracted from over-exposed photographs, are super-imposed on under-exposed ones. The density profile and condensation pattern are studied under three different exposure conditions. Density distributions of the same area under three different exposure conditions are compared. The under-exposed photograph yields better resolution of the density difference at compact segments of chromosomes. As such, the underexposed photograph is essential for analysis of condensation pattern and the normal photograph for measurement of chromosome size. For detailed analysis of CHIAS, please consult Iijima and Fukui (1991) and Nakashima *et al.* (1995).

System architecture of the CHIAS (Chromosome Image Analysis System)

The CHIAS was designed for the analysis of plant chromosomes in metaphase spreads, ranging in number from four (e.g. *Haplopappus gracilis*) to 120. A man–machine interactive system was adopted as operation system for the chromosome studies. Researchers can interact in any analytical procedure and modify the image manipulation based on their knowledge and experience on the chromosomes. Routine processes in the analysis were automatized as much as possible. For the excision of chromosome images from a total image frame or a photograph scissors are used in conventional methods. CHIAS can carry out this procedure automatically for every chromosome under identical conditions within a second by detecting the chromosomal contour lines of the same gray level. Microscope automation was designed as a basic function of the CHIAS. An automatic scanning stage and an automatic focussing unit were equipped, and operated by the host computer. All the processes of the analysis were carried out as an on-line

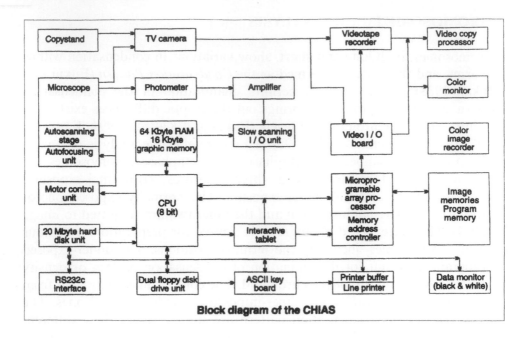

Block diagram of the CHIAS

system. i.e. no photographic procedure is interrupted in any analytic process. The image analysis system involves a fluorescence microscope with autoscanning state and autofocussing unit. A photometer and a filter system are also attached. A high resolution TV camera is directly mounted on the top of the microscope. Images on the photographic paper or film can be taken through a zoom lens and a close up lens. Image analysis units are selected as the image processing component of the system. For output of data and images, a printer and a colour image recorder are chosen. A videotape recorder can also be selected as additional input (Zeiss Oberkochen-Photomicroscope III).

Rice has been shown to be a suitable material (Iijima and Fukui, 1991):

Pretreat roots with aqueous colchicine solution (0.1%) or colchicine (0.1%) in a 1/15 buffer phosphate for 3.5 h at 6°C. *Alternatively*, pretreatment can be carried out in cold water or cold buffer for 7 h at 0°C; *or* in 8-hydroxyquinoline (8 mM in water or buffer) for 3.5 h at 6°C; or in KCl (7 mM in water or buffer) for 3.5 h at 6°C; or in a mixture of ethidium bromide (5 μg/ml) and colchicine (0.05%) for 3.5 h at 6°C.

Fix in methanol–acetic acid (3 : 1) for 1 h to overnight. Macerate on a slide in a moist chamber using a mixture (0.3% pectolyase Y-23, 2% cellulase Onazuka RS; 1% macerozyme-R-200, 1 mM EDTA, pH 4.2) for 15 min at 37°C. Keep the tips in 0.2% Triton X-100 for 15 min. Break the samples into five pieces with forceps in fixative. Air dry. Stain in Wright's stain (5% phosphate buffer, pH 6.8).

Alternatively, wash in distilled water for 15 min and macerate at 37°C for 50 min in enzyme mixture (4% cellulase Onazuka RS, Yakut Honshu Co. Tokyo, 1% pectolyase Y-23 Seishin Pharmaceutical Co. Tokyo, pH 4.2). After thorough washing, place a few drops of fixative (acetic–methanol 1:3) on a slide, place a root-tip in fixative, chop up into pieces, air dry, stain in 3% Giemsa (pH 6.8) for 40 min and dehydrate in 100% methanol for more than 10 min. Photograph images of good prometaphase chromosomes and analyse by the CHIAS. The density or condensation profile along the midrib of each chromatid (CP) is measured. Compare the CP of pretreated and not pretreated chromosomes.

Representative schedules using CHIAS

(1) *Barley* (Fukui and Kakeda, 1990, 1994)

(i) Germinate seeds of barley at 27°C. Excise roots about 1 cm long and pretreat in aerated distilled water at 0°C for 24 h. Fix in methanol–acetic acid (3:1) for 1 h. Wash through a methanol series of 50 and 25% followed by distilled water. Macerate on a glass slide with the enzymatic mixture (4% cellulose Onozuka RS, 1% pectolyase Y-23, 75 mM KCl, 7.5 mM EDTA, pH 4.0), at 37°C in a moist chamber for 40 min. Rinse thoroughly in water. Tap the root-tip with forceps in a few drops of the fixative and briefly flame-dry. Stain some slides with Giemsa 1% (in 1/15 M phosphate buffer, pH 6.8) for 20 min and mount in Eukitt. Observe remaining unstained slides under phase contrast microscope after mounting with a coverglass without any mounting fluid.

(ii) Use an automatic stage (pitch size 0.25 μm) and an auto-focussing unit integrated in the CHIAS for automatic detection. They are controlled by the host computer in the CHIAS system. Divide the central area of an 18 × 18 mm coverglass into four scanning regions, each consisting of a 22 × 22 matrix of subfields. They appear one by one on a high resolution TV monitor with a 512 × 512 pixel matrix image. Each pixel has 256 steps of the gray values. Each subfield is checked under an optical magnification of 25.6×. When the CHIAS detects regional disorder in the neighbouring scanning lines of the TV monitor, which characterizes the existence of the putative metaphase spreads, the system proceeds in the next automatic discrimination step of the metaphase spreads for the objects, using the following parameters: (a) area range of the objects, (b) minimum number of objects to be included in a circle to be defined and (c) radius of the circle. The coordinates of the location of putative metaphase spread are stored in the CHIAS. The spreads are finally selected by the researcher.

(iii) For karyotyping of chromosomes, store good phase contrast images from the unstained preparations described earlier at a total magnification of 4000× (objective lens 100×, opto-bar 2×, TV factor 20×) digitally in floppy disks. Store also reference or blank images. The reference images can be obtained by N-banding of the samples to identify individual chromosomes. Subject selected plates, where all chromosomes can be completely identified, from the stored digital images, to semiautomatic image processing to obtain numerical data for 250 metaphase plates. The parameters for the one, two and virtual 3-dimensional volumes of the chromosomes in the images are defined as the length, area and density volume respectively. Density volume is obtained by multiplying area by the average density of the chromosome. In case of banded chromosomes, quantify position and length of each band on each chromosome in 20 plates.

(iv) Recover the image from the floppy disk. Generate the original digital image automatically after shading correction using a stored reference image. Improve the contrast with a normalization digital filter. Enlarge the image two-fold. The different levels of gray indicate the levels of density and can be coloured automatically. Idiograms can be prepared both of unbanded and banded preparations from the detailed data automatically.

(2) *Mapping of C-banded Crepis chromosome by CHIAS*
(Fukui and Kamisugi, 1995).

(i) Pretreat root tips of *Crepis capillaris*, about 1 cm long, in 0.05% aq. colchicine soln for 2 h at 18°C. Fix in acetic acid–methanol (1 : 3) for 0.5 to 2 days. Wash. Macerate in the enzymatic mixture (1% macerozyme R 200, 0.3% pectolyase Y-23, 2% cellulase Onozuka RS, 1 mM EDTA, pH 4.2) for 15 min at 37°C on glass slides in a humid chamber. Wash. Chop up root tips in fine pieces in fresh fixative and air dry. Bake the samples at 80°C for 1 h in a vacuum oven. Treat with 0.2 N HCl–45% acetic acid (2 : 1) for 3 min at 55°C. Then treat with 5% barium hydroxide for 5 min at room temperature. Dip the slides in 2× SSC for 30 min at 55°C. Then stain with 8% Wright's solution for 30 min to 1 h. Take photomicrographs.

(ii) Directly record a photographic image of the C-banded chromosome plate into the Chromosome Image Analysing System, CHIAS. Freeze the image in image memory (512 × 512 pixel matrix with 256 grey levels; 0 = black, 255 = white), using a high resolution TV camera. Average the source image over 20 frames to eliminate effects of voltage fluctuations. Eliminate illumination distortion through shading correction using a blank image as a reference. Equalize the grey histogram to enhance the contrast of the image. Apply a median filter to smooth the texture of the image. Use the resulting image as the original grey image for the analysis. Extract the chromosome contour line as an overlay line from a chromosome binary image that had been produced by setting the upper and lower thresholds of the grey levels for the original image. Introduce the overlay line into the original grey image with a grey value of 0, so as to delimit the borders of the chromosomal area. Thereafter draw an overlay line interactively on a midrib of a chromatid in the original image by the aid of pseudocoloration. Measure the density distribution under the overlay line.

Determine the contour lines of the chromosomes and cut the fused bands at their borders. Gather the image information of chromosome contour lines and the adjusted bands into a single image. Draw chromatid midrib lines interactively on pseudocoloured image and determine positions of centromeres by the aid of pseudocoloration.

Limit the midrib lines to the edge of the bands, the centromeric positions and the borders of chromosomes. Colour the lines differently to distinguish between adjacent fragments. Measure the fragment lengths and then edit the individual chromosome figures on the display of the CHIAS and rearrange in order of length to prepare an idiogram.

Imaging methods have been applied in several plant species. The method aids in standardizing and optimizing the information obtained by semi-automatic means though there is still ample scope for improvement. A major limitation is the dynamic nature of the chromosome, since its morphology changes according to condensation rate even during mitotic metaphase.

CONFOCAL MICROSCOPY

Confocal microscopy with proper scanning device permits an analysis of three-dimensional image of chromosomes and cellular constituents with practically no scope for out of focus limitation. The video image can be digitally displayed, through computer arrangements. If it is combined with fluorochromes specific for GC sequences and observed through laser scanning device, where the resolution is high, the position of chromosomal segments at interphase *vis-à-vis* their association and relation with nuclear membrane can be precisely understood. The use of this method has been able to demonstrate clearly the non-random arrangement of chromosomes and chromosome segments at interphase, polarity of chromosome ends and specially association of functional regions of chromosomes with nuclear membrane.

Confocal images are optical sections through X–Y plane of the object. The light from above and below the plane-in-focus is prevented from forming images through a pinhole diaphragm. As such, the contrast and resolution are very high. At present, Zeiss, Leica and other manufacturers are supplying laser scanning confocals. In laser scanning, a focussed argon laser source of 1.0 to 1.5 micrometer is passed through the object. For details, please consult Noguchi, J. and Fukui, F. (1995) *J. Plant Res.* **108**: 209–216.

Representative schedule

(i) Pretreat fresh root-tips of *Hordeum vulgare* or *Secale cereale* with a mixture of 2 mM 8-hydroxyquinoline and 0.05% colchicine (1 : 1) for 3 h. Fix the tips in 45% acetic acid at 0 to 4°C for 10–20 min. Preincubate in McIlvaine buffer containing 10 mM $MgCl_2$ (pH 7.0) at room temperature for 10 min. Macerate the tips in 45% acetic acid at 60°C for 5 to 30 min after incubation with a solution of 4% cellulase Onozuka RS and 1% pectolyase Y23 in 15 mM EDTA at 37°C for 10 to 15 min. Stain with 0.04% chromomycin A for 5 to 15 min. Squash in McIlvaine buffer (pH 7.0).

(ii) For study of interphase nuclei, fix root-tips in a mixture of ethanol, chloroform and acetic acid (2 : 1 : 1) at 4°C for 2 h without pretreatment. Macerate and stain as described for (i). Disperse the stained cells in a mixture of glycerine and McIlvaine buffer (9 : 1) and remove excess moisture with filter paper.

(iii) For laser scanning microscopy, a confocal laser scanning microscope system (MRC-500; BIORAD, USA and LSM; Carl Zeiss, Germany) with an argon laser at 488 nm is used.

Record optical images obtained by consecutive focus steps using the pre-programmed operation mode with 0.5–1.5 µm steps along the Z-axis. In *Secale cereale*, stereo pair images can be obtained, based on a complete stock of optical sections, by shifting views gradually to the right and left with 70 depth and 80 transparent levels. The method is used successfully in studying chromatin arrangements in intact interphase nuclei.

ELECTRON MICROSCOPY

Ultrastructural analysis of chromosomes is carried out through electron microscope which brings out finer details at a level beyond the resolution of the light microscope. In electron microscopy, electron or magnetic fields are shaped to refract electrons producing the image. The image of electrons is transformed to light image through a fluorescent screen.

Basically, the electron microscope has a vertical column, illuminating source, specimen chamber, objective, intermediate and projecting lenses, and a viewing chamber with inbuilt arrangement for photographing electron images. The vacuum in the column is maintained at a specific pressure through oil and mercury diffusion pumps. The source of the electron is a hot tungsten filament serving as cathode and an anode with a surrounding Wehnelt cylinder, with a hole for the passage of the beam of electrons. The image can be viewed on a fluorescent screen or in a photograph. In order to observe with the help of the electron microscope, ultrathin sectioning is necessary, which is cut in an ultramicrotome. Several makes of good electron microscopes are available in the market, including those of Phillips, Jeol, Siemens, Hitachi and others. Ultramicrotomes of Cambridge, Huxley, LKB and others are widely used. In order to obtain ultrathin sections for ultrastructural analysis, several steps, specially designed for EM analysis, are involved including fixation, dehydration, embedding, sectioning and mounting.

Fixation

Low temperature is normally not required for chromosome research excepting for studies on enzyme distribution at the chromosome level, such as that of alkaline phosphatase. *A chemical fixative* should simulate the natural milieu of a living system especially with regard to pH, osmolarity and ionic concentration, in order to prevent swelling, shrinkage or extraction. The osmolarity is normally adjusted by changing the buffer concentration or by addition of sodium chloride, polyvinyl pyrrilidone glucose, sucrose and such other non-ionic compounds. The chemical fixative consists of a fixing agent and a vehicle which is normally a buffer solution, with salts. *Of all the metallic fixatives* so far tried, sodium tetroxide (OsO_4) is the most widely used.

Of the non-metallic fixatives, formaldehyde is widely used. Another aldehyde with extensive use in electron microscopy is glutaraldehyde, or more precisely, glutaric di-aldehyde, $(CH_2)_3CHO$. The monomeric glutaraldehyde is the principal reactive compound but the commercial forms contain polymers as impurities. It is often necessary to purify the chemical by charcoal or distillation. Charcoal purification is performed by shaking a 25% solution of glutaraldehyde with 10% (w/v) activated charcoal at 4°C for 1 h before filtration. In fixatives containing glutaraldehyde, the pH should be maintained at 7.5 to check polymerization. Phosphate buffer is always preferable. For 2% solution, take 2.26% $NaH_2PO_4 \cdot H_2O$ in water, 64 ml; 25% glutaraldehyde in water, 8 ml; 2.25% NaOH in water for adjusting pH. Make upto 100 ml with distilled water after the adding the required amount of NaOH.

For immediate fixation, take the tissue in a drop of fixative on a sheet of dental wax, cut into small pieces of required size with a blade and transfer with the aid of pipette or tweezers to small glass vials containing the fixative.

Acrolein, otherwise known as 2-propenal or acrylic aldehyde, is another bifunctional aldehyde having the capacity of forming crosslinks between end groups of proteins. By distillation the polymerized material can be removed. The fixative is usually prepared by 10% acrolein in 0.025 or 0.05 M phosphate buffer.

Dehydration

Water is usually removed through increasing grades of solutions of ethanol or acetone. Dehydration is carried out in acetone for embedding in polyester resins. For embedding in epoxy resins, either ethanol or acetone can be used. As these resins react well with propylene oxide, the latter is often used at the last stage of dehydration. Propylene oxide in increasing concentrations of aqueous solution can also be used as the dehydrating agent itself, specially for Epon embedding.

Embedding

The choice of a suitable embedding medium depends on its stability in the electron beam, uniform polymerizing capacity, convenience in sectioning, low viscosity in the monomer form and solubility in the dehydrating agent. Of the three types of embedding media normally employed, i.e. epoxy resins, polyester resins and methacrylates, the former is the most widely used.

The embedding medium also contains a freshly prepared accelerator or activator for infiltration. The final embedding is carried out in polyethylene or gelatin capsules. The intermediate steps for the complete removal of the dehydrating agent are carried out through a sequence of solutions in the glass vials containing the fixed material. These intermediate solutions are removed and replaced with the aid of a pipette and finally with pure embedding medium. The vials are kept overnight with the stoppers open to accelerate evaporation of the dehydrating fluid.

For final embedding, gelatin or polyethylene (BEEM or TAAB) capsules are available of different sizes. The completely dried capsules are kept for a minimum period of overnight at 60°C before use, are then half filled with final embedding medium and the specimens are gently transferred to the bottom of the capsules.

The *epoxy resins* are polyaryl ethers of glycerol with terminal epoxy groups. Their viscosity may be different from one another. They are polymerized by bifunctional agents which criss-cross the epoxy groups, forming ultimately a three-dimensional structure. The common epoxy resins are Araldites (CY212, 502, 6005, CIBA 506) and Epons (812, 815 Shell) which may be used separately or mixed together. All epoxy resin embedding media have, in addition to the resin, a hardener and an accelerator. The hardeners of the block are often softened by the addition of additives, plasticizers or flexibilizers. The period required for infiltration is directly proportional to the viscosity of the medium.

The common hardeners in the market are: (a) dodecyl succinic anhydride (DDSA—290 cps at 25°C), (b) hexahydrophthalic anhydride (HHPA, m.p. 35°C, (c) methyl nadic anhydride or nadic methyl anhydride (MMA or NMA—175–275 cps at 25°C), and (d) nonenyl succinic anhydride (NSA—117 cps at 25°C). They differ from each other in viscosity. The common accelerators are: (a) 2,4,6-tridimethyl amino methyl phenol (DMP 30), (b) benzyldimethyl amine (BDMA) and (c) dimethylamino ethanol (DMAE). The additives generally used are: (a) dibutyl phthalate—plasticizer (DBP), (b) carbowax 200 (polyethylene glycol 200), (c) triallylcyanurate (TAC), (d) polyglycol diepoxide—flexibilizer (DER 736, Dow), (e) diglycidyl ether of propylene glycol (DER 736, Dow), (f) long chain mono-epoxide, flexibilizer (Cardolite NC 513) and (g) polythiodithiol—liquid polymer of low viscosity, flexibilizer (Thiokot LP 8).

Some of the common epoxy embedding mixtures are: (i) Epon (812)—10 m, DDSA—8 ml, BDMA 1%. (ii) Epon (812)—10 ml, HHPA—1 ml, MNA—8.5 ml, DMP 30—0.15/0.3 ml. (iii) Epon (812)—10 ml, NSA—13 ml, DMP 30—1.5/2%. (iv) Epon (812) and (815) (ratio determining hardness)—10 ml, DDSA—14–16 ml, DMP 30 or BDMA—2 ml.

Protocol for embedding

Drain off the solvent with a pipette. For flushing off, it is preferable to pour it down with a large volume of water in a fume chamber. If propylene oxide is used, keep the material slightly moist. Add a mixture of solvent and embedding medium (1:1) in the vial, shake for thorough mixing. Keep the vial for 30 min to 1 h at 24–28°C. Remove the fluid with the help of a pipette and add the final embedding medium. Remove the cap from the vial and keep for 16–24 h at room temperature. Take several dry polyethylene (BEEM) capsules (dried overnight at 60°C), fit them in a cardboard box with punched holes of proper size and half fill with embedding medium. Transfer the material with the aid of tweezers into the capsules. Fill the capsules with the embedding medium and allow polymerization overnight or more at 60°C. Before sectioning, keep the capsules for a few days, if possible at the same temperature.

A number of water soluble embedding media have been devised. Such aqueous fixatives meant for cytochemical work can be used for the selective extraction of chromosome components, including Aquon which is the water soluble constituent of epon.

Some common water soluble epoxy media are: (i) Durcupan, 5 ml; DDSA, 11.7 ml; DMP 30, 1 ml; (dibutylphthalate DP, 0.2 ml may be added). (ii) Durcupan, 100 ml; MNA, 120 ml; DMP 30, 1.5% (Thiokol LP8, 20–35 ml may be added in requisite proportion to secure adequate hardness). (iii) Aquon, 10 ml; DDSA, 25 ml; BDMA (benzyldimethyl amine), 0.35 ml; (iv) Epon 812, 20 ml; hexahydrophthalic anhydride (HHPA), 16 ml; BDMA, 1.5%.

For aquon, the material is fixed in 10% formaldehyde-veronal acetate buffer (pH 7.3) at 3–4°C and washed. It is dehydrated through increasing concentrations of aquon in water to pure aquon at 4°C and finally kept immersed in the embedding mixture for 4 h. Curing is performed by transferring the material to a gelatin capsule with fresh embedding mixture and keeping it at 54–60°C for 4 days.

Sectioning

For ultra thin sectioning (0.1–0.01 μm), *several models of ultramicrotomes are at present available*. In the ultramicrotome of LKB-Producter AB-Stockholm, fluctuation in thickness of the sections is eliminated to a significant extent, the principle being based on a thermal advance system. One end of a cantilever arm holds the specimen block, the other end being attached to a leaf spring joined to the base of the microtome. This spring causes the up and down motion of the bar. Thermal control of the cutting arm guides the advance of the block against the knife. The gravitational force controls the cutting stroke and a motor regulates the motion and the upward movement. An electromagnetic force, which acts during the return stroke, causes the flexing of the base below the knife holder, necessary to ensure the *bypass of the cutting surface and knife edge during the return stroke*.

A standard method for preparing the glass knife most commonly used to obtain ultrathin sections is outlined below:

Make a 1.27 cm score mark with a sharp cutter on a clean 20 × 20 cm sheet of plate glass, at right angles to the base of the glass plate. Position the scored edge of the plate to overlap the edge of the working surface by about 0.63 cm. Keep the edge of the glass parallel to the edge of the table. Take a pair of glass breaking pliers with wide parallel jaws. Place a narrow strip of adhesive tape on the inner surface from the cutting edge to halfway to the middle of the bottom jaw. Place two lateral strips at the edges of the inner surface of the top jaw. Cover inner surfaces of both jaws with wide pieces of adhesive tape, smoothly. Keep the jaws open. With the central piece of tape of the bottom jaw centred beneath the score mark on the glass, push the face of the bottom jaw flush against the table. Gently squeeze the pliers to produce a slow, even and straight break, with two smooth new surfaces, which will be free of artefacts except for the short line where the initial score was made. Turn the two pieces of glass through a right angle so that the smooth edges are away from the table edge. Score one piece of glass in the centre of the old long edge and repeat the procedure to have two 10×10 cm plates. Repeat the process till 2.54×2.54 cm squares are obtained, each with at least two smooth edges meeting at a 90 degree angle. Choose the best adjacent edges for the final break. Start a diagonal score 1 mm or so from the apex of the angle where the faces meet and extend to bisect the opposite corner. Carefully centre the pliers halfway along this line and gently increase pressure till the glass breaks to give a triangular knife. A good knife should have an even and straight cutting edge, an absolutely flat front surface and a back face with either a right or left-handed configuration when viewed from above. The part of the knife edge closest to the top of the arc formed by the back surface is best for thin sectioning. A good 45 degree angle knife is usually suitable for cutting tissue embedded in media of average hardness. An angle of 55 degrees has been recommended.

The trough is needed for section cutting with diamond knives. With a glass knife, prepare a trough with adhesive-backed cloth or paper tape, which is disposable. Coat the exposed adhesive surface of the trough with paraffin to prevent contamination with the trough liquid. Seal to the glass with melted paraffin. The liquid in the trough should be able to detach the section from the knife, eliminate all electrostatic charges, spread the sections through solvent action, and should have an adequate surface tension to penetrate the layer between section and knife facet. *For trimming and sectioning*: the principal controlling factors are the cutting edge of the knife, the embedding material, the cutting face of the block and operating speed. In case of hard epoxy-embedded blocks, files and jeweller's saws are required for trimming, followed by a final finish with an acetone or chloroform washed razor blade. To prepare the block for sectioning, trim the side walls of the portion delimited from the tissue. A surface area about 0.3×0.08 mm is usually desirable. Orient the block in the microtome with the long side parallel to the knife edge. After trimming, the final shape of the block should be that of a truncated pyramid or, in the case of larger materials, like a roof-top. In general one side of the pyramid, or the long face of the roof-top shaped top, is adjusted parallel to the knife edge. Dip the blocks in a filtered mixture of Carnauba wax and paraffin (1:2) and keep at 80°C. Orient the tissue specimen by mounting the block on holder made of a wooden dowel rod 5–16 mm in diameter, which fits well in a Porter-Blum microtome.

When the block is fitted in the chuck, the front edges of the jaws should clamp it strongly and the projecting portion alone should not be more than 3–4 mm. The knife should be tilted so as to have a 1–3 degree clearance angle and a rake (knife) angle of about 30 degrees. The entire process of sectioning should be performed very gently at uniform operational speed.

When the sections have been cut, detach the ribbon from knife edge and transfer to a trough in which the level of the liquid is controlled with a hypodermic syringe fitted at the base with a plastic tube. The level of the fluid is generally maintained over the knife edge, forming a well-rounded meniscus, and the ribbon can be detached with a fine hair brush. In order to estimate the thickness of the section correctly, it is always preferable to use reflection of interference colours while the sections are floating in the trough. The light should be adjusted to allow total reflection on the liquid surface. Peachey (1958) published a detailed account of the thickness of the sections and the corresponding interference colours.

Mounting

The removal of the ribbon from the fluid needs special care. Mount on a specimen grid with a backing film (Parlodion and Formvar are the common films used because they provide good supporting media). They dissolve quickly and become tough when the solvent evaporates. Parlodion is the trade name of nitrocellulose plastic (prepared by Mallinck Rodt Chemical Works, St. Louis). Polyvinyl formal plastic of Shawinigan Products Co., New York is called Formvar. Gay and Anderson's method (1954) is suitable for serial sectioning. The principal implement is a thin film of Formvar supported by a small wire. These loops can be inserted in the liquid of the trough in a tilted position. By suitable adjustment, centre the ribbon across the diameter. When the loop is raised, the sections adhere to the Formvar, after which directly transfer them to the supporting grids for examination. Place the grids on a combination of transparent plastic discs, fitted on the top of an adjustable condenser in a standard microscope. By lowering the condensers, keep the grid below the stage and the ribbon suitably arranged. Contact is achieved by lifting the condenser. Epon-embedded materials can be mounted on 300 mesh copper grids. For araldite-embedded materials, the sections do not require any support for mounting, even a carbon film may be adequate.

Staining

Positive staining technique involves treatment with components which increase the weight density, whereas in negative staining, the material is surrounded with a structureless material of high weight density. Good negative staining can be obtained with sodium tungstate, uranium nitrate or disodium hydrogen phosphate. The number of staining methods available is increasing gradually. Uranyl nitrate staining requires a filtered aqueous saturated solution at pH 4.0. For combined staining, first moisten the section on the grid with a drop of distilled water, followed by staining (in an inverted position) in 7.5% uranyl acetate solution for 20 min at 45°C. Dry the sections on filter paper, again moisten and finally stain with 0.2% aqueous lead citrate solution for 10–60 s.

In situ fixation and embedding

Methods have also been developed for fixation and embedding *in situ*, thus avoiding any distortion or displacement of the structure. A comparatively simple technique involves culturing of cell on coverglasses sprayed with teflon. Monolayers attached to coverglasses are fixed and dehydrated in staining dishes before final embedding in epon or other media using a silicon rubber mould. Coverglasses are then separated from the block by immersion in liquid nitrogen. The technique allows mass harvesting of cells, since several coverglasses with monolayers can be obtained from a single culture flask. Synthetic substrates may be used for culture, which can be easily separated from embedding medium. For ultrastructural analysis of chromosomes at different stages of division, falcon dishes have been used. The procedure involves fixation in glutaraldehyde, dehydration in hydroxypropyl methacrylate and *in situ* embedding in epon. Polymerized blocks can be observed and photographed under light microscope and the selected area embedded in BEEM capsules for sectioning. This technique allows rapid analysis of a large number of materials.

Representative schedules

EM chromosome analysis of slime mold—Echinostomum minutum De Bary. Fix plasmodia in a mixture of 3% glutaraldehyde + 1 µM $CaCl_2$ buffer (pH 6.8) with 0.05 M Sörensen's phosphate buffer for 1 h. Post-fix in 1% OsO_4 in 0.05 M phosphate buffer (pH 6.8) for 1 h. Dehydrate in ethanol series (30, 50, 70, 95% and absolute). Transfer to propylene oxide and follow the usual procedure of embedding in Epon 812. Select metaphase cells under phase contrast microscope, cut out the selected portion and remount in epon stubs. Cut 0.5 µm thick sections in ultramicrotome, and mount on Formvar-coated grids. Stain in 2% uranyl acetate in methanol and later in lead citrate for 45 and 20 min respectively.

Desirable: Apply a thin layer of carbon to the grids by a screened carbon source.

EM chromosome analysis of algae (*Acetabularia* sp.)

Fix the algae in 5% glutaraldehyde solution in 0.1 M sodium cacodylate buffer (pH 7.2) at 5°C for 2 h. Rinse in cold buffer. Dehydrate through ethanol grades and follow the usual procedure for epon embedding. Cut 1.3 μm thick sections in ultramicrotome. De-eponise one set and observe following haematoxylin staining.

EM analysis of dinoflagellate chromosomes

Centrifuge exponentially growing cells of dinoflagellates for 10 min at 1200 rpm. They can then be sectioned or mounted whole. For sectioning, fix the cells for 1 h at 4°C in 2% glutaraldehyde in sodium cacodylate–saccharose buffer. Wash in the same buffer for 30 min at 4°C. Post-fix in 1% osmium tetroxide in veronal acetate for 45 min at 4°C. Embed in Epon, cut sections 80 nm thick and stain with uranyl acetate and lead citrate.

For whole mounting, further centrifuge the cells at 1200 rpm for 5 min. Cover the non-disrupted pellet with 2 ml of isolation buffer (1% citric acid, 1% Triton X-100, 6 mM $MgCl_2$) and keep at room temperature for 30 min. To isolate chromosomes, suspend the cell pellet in the isolation buffer (2×10^4 cells/ml) in a Dounce homogenizer and give 10 strokes of a loose pestle. Centrifuge the homogenate at 1200 rpm for 5 min through a 2 ml cushion of isolation buffer onto Formvar-coated grids. Wash in 50, 75 and 100% ethanol, keeping 10 min in each. Immerse the grids in baths of amyl acetate twice for 5 min each and airdry. Examine and photograph with Zeiss 109 Turbo EM operated at 50 kV. For negative staining and positive contrast of whole mounted chromosomes, place a drop of 2% phosphotungstic acid (PTA) at pH 7.0 on the grid with chromosomes after chromosome isolation and centrifugation through isolation buffer as described above. Keep for 40 s. Remove stain with filter paper. Air dry the grids at 37°C for 30 min. Positive contrast can be obtained by 2% PTA staining at pH 5.0 for 40 s (Costas and Goyanes, 1987).

SYNAPTONEMAL COMPLEX

This complex is the ultrastructural pattern of meiotic synapsis. It is found in meiotic prophase between two paired homologues represented in 10 nm fibres arranged in superstructure. They are considered to be necessary for synapsis, to align the two homologues, pairing at the molecular level. It is very prominent in the pachytene stage. There are lateral elements, and the central element connected by the transverse elements, which connect the central core with the lateral elements. It has the appearance of a ladder and each synaptomer is supposed to be spaced at minimum of 20–30 nm. In the lateral elements, both the nucleic acids and proteins have been found but the central core is rich in RNA, in addition to the much lower amounts of DNA and protein.

The sectioning permits a study from one end of the chromosome to other. Development of the technique of spreading the complex has facilitated its analysis to a great extent (Gillies, 1983). The surface spreading method enables a detailed study of the configuration through wide spreading of the skeleton revealing details of architecture. It involves spreading the material in dilute solution of sucrose followed by transfer to plastic-coated slide and fixation and hardening in formaldehyde added fixative. The fixation within a short period checks overspreading of the material. After air drying, the slides can be retained and plastic discs can be mounted on copper grids for observation under electron microscope.

Dissect out fresh anthers. Place in a drop of spreading medium (0.1% bovine serum albumin and 2 mM disodium EDTA salt in Eagle's minimum medium, pH 7.7). Cut each anther in half. Squeeze out meiotic cells and remove anther debris. Take up the suspension of cells in micropipette and place one drop on the convex surface of 0.5% NaCl solution in a black dissection dish, of about 35 mm diameter. Pick up the meiotic cells spread out on the surface of the NaCl solution by touching the surface with a plastic coated slide (prepared previously by dipping in a solution of 0.6% Falcon optilux petri dish dissolved in chloroform). Fix slide with attached spread cells for 5 min in 4% paraformaldehyde (pH 8.2), containing 0.03% sodium dodecyl sulphate (SDS); then in 4% paraformaldehyde without SDS for 5 min and then rinse for 20 s in 0.4% Photoflo (Kodak pH 8.0) and air dry in a vertical position. Cover dried slide with coverglass and observe under phase contrast to detect SC and nucleolus at 40X. Stain suitable slides for electron microscopy with either phosphotungstic acid (PTA) in 75% ethanol for 30 min or ammoniacal silver nitrate solution.

A. From *Allium cepa* and *A. fistulosum*

Tap out pollen mother cells from fresh anthers in prophase I in a digestion medium (0.1 g snail gut enzyme LKB cytohelicase; 0.375 g polyvinyl pyrrolidone; 0.25 g sucrose, in 25 ml sterile distilled water) and keep for 5 min. Transfer a single drop onto a single drop of detergent solution (0.5% Lipsol) on plastic coated slide. Add after 5 min, 5 to 6 drops of paraformaldehyde and dry at 20–25°C for 6 h. Rinse and air dry. For PTA staining, immerse slide in 1% ethanolic PTA for 10 min, rinse in 95% ethanol and air-dry. For silver staining, place a few drops of $AgNO_3$ solution on slides covered with patches of nylon cloth (instead of coverglass) and incubate in a moist chamber at 60°C for 40–45 min. Transfer suitable surface spreads to EM grids by scoring and floating the plastic film on a water surface. Mark the positions of nuclei on underside of slide, and place a grid over each before floating the film off. Examine and photograph spread and stained nuclei using Phillips EM301.

B. Ultrastructure of synaptonemal complex (SC)

Remove and check by aceto-carmine staining, an anther from fresh bud of *Zea mays* to determine suitable stage (early-late pachytene). When the anther shows the correct stage, transfer the remaining anthers into a depression slide, containing 0.1 ml solution of 0.9 M sorbitol, 0.6 mM KH_2PO_4, 1.0 mM $CaCl_2$, 1.6 mM $MgCl_2$, 0.1 mM potassium citrate buffer (pH 5.2) and 0.3% potassium dextran sulphate. Adjust pH of final solution to 5.1 with 0.1 KOH or HCl. Cut anthers in half transversely with a razor blade and incubate for 5 min. Squeeze out the contents of the anther with a steel dissecting needle. Remove anther walls. Add 1 mg of desalted β-glucuronidase. Incubate for 10–15 min to digest cell walls. Draw out protoplast suspension into a micropipette and transfer onto a hydrophilic Falcon plastic-coated glass slide. Add a siliconized coverglass. Draw 5 μl of distilled water under coverglass toward a paper on the far side. The protoplasts swell and burst in the resultant hypotonic medium. Remove coverglass by dry ice method. Air dry the spread protoplasts. Fix the slides for 10 min in fresh ice cold 4% formaldehyde solution (pH adjusted to 8.5 with borate buffer). Then dip the slides in 0.4% Photoflo 200 (Kodak) at pH 8.5. Airdry briefly. Stain by silver nitrate technique of Goodpasture and Bloom (1976).

For electron microscopy, place 50-mesh grids on spreads of SCs. Float plastic with grids on top onto water and pick up with glassine weighing paper. Dry. Stabilize the plastic with a light coat of evaporated carbon. Examine the grids and photograph in a AE-1 EM 6 B.

NUCLEOSOME

Nucleosome represents the first stage of DNA packaging. The linear array of nucleosome cores is involved in the fibre of 100 Å thickness. Each core nucleosome lies in contact with one another. The supercoiling of 100 Å fibre leads to higher orders of thickness reaching 200–300 Å in diameter. In the 300 Å fibril, the nucleosomes are arranged in a solenoid with a 100 Å pitch (Carpenter *et al.*, 1976). The interaction between nucleosomes is stabilized or influenced by H_1 linker in the spacer region between the nucleosomes. The spacer regions may vary in length.

The eukaryotic chromosome structure is beaded in appearance, in which DNA is packed into repeating chains of nucleosome subunits made up of octamers of histones complexed with almost 200 base pairs of DNA. Of these base pairs, nearly 140–160 bp remain associated with H2A, H2B, H3, H4 around a central histone octamer and further 20 base pairs remain associated with H1 in the linker region adjoining the core. To some extent, the heterogeneity of the nucleosome is achieved by the histone variants but the relative amount of nonhistone protein is more variable then the histones.

In addition to the function of non-histones in gene action, their role in chromosome structure is established (Kornberg, 1981). Two proteins in the nucleus are deeply involved in the nucleosome assembly, namely DNA Topoisomerase I, which is a nicking closing enzyme interacting with DNA followed by addition of histone, and the other nucleoplasmin which promotes also histone–histone interaction (Laskey and Earnshaw, 1980). The nucleosomes can be visualized through spreading technique and electron microscopy. The spreading technique of Miller and associates permits the study of chromatin from a variety of chromosomal types. In principle, it involves swelling and dispersion of nucleoplasmic chromosomal material with low ionic solution followed by centrifugation. Ultimately, the three dimensional structure becomes two-dimensional on fitting over the surface of electron microscope grid. In order to unwind and disperse the chromatin, low ionic solution such as distilled water or water at alkaline pH is employed, which removes some chromosomal proteins including H1. This permits the higher order fibres to unwind into the basic strand of beads representing nucleosomes.

Schedule: Isolate interphase nuclei from tissues. Wash and centrifuge twice in CKM buffer and once in 0.2 M KCl, suspend in 0.2 M KCl at a concentration of approximately 10^8 nuclei per ml and dilute 200-fold into distilled water. Allow nuclei to swell for 10–15 min. Make 1% in formalin (pH 6.8 to 7). Fix for 30 min. These steps are carried out in cold (0–4°C). Centrifuge aliquots of the fixed nuclei through 10% formalin (pH 6.8 to 7) onto carbon-covered grids, rinse in dilute Kodak Photoflo and airdry. When examined after positive staining, chromatin fibres are observed to stream out of ruptured nuclei.

CKM buffer (= 0.05 M sodium cacodylate, pH 7.5; 0.025 M KCl; 0.005 M $MgCl_2$ and 0.25 M sucrose).

CHROMOSOME SCAFFOLD

The chromosome structure contains both histone and nonhistone proteins in addition to nucleic acids. The concept of nucleosomes has undoubtedly unravelled the structural relationship between DNA and histone in chromatin. But data on the relationship of nonhistone protein with other nucleic acid components is relatively meagre. The chromosome, if depleted of histone, reveals a scaffold or core and a halo of DNA loops attached to it (Stubblefield and Wray, 1971). The nucleosomic complex is connected to a nuclear filament about 12 µm, in diameter. This is further coiled to yield a 30 µm fibre of chromatin, the organization of which has been presented in different models (Van Holde, 1988). The presence of nonhistone protein in scaffold is undisputed (Lewis and Laemmli, 1982). The scaffold structure has also been manifested as intergenic structures of chromosomal filament, which serve as structural and functional entities through the spatial ordering of axis and have a linear structure (Haapala, 1985). The shape of the chromosome is maintained in core structure. The study of scaffold can be carried out by (i) isolating DNA and histone-depleted chromosome and (ii) spreading after special treatment, thus permitting the analysis of nonhistone protein. The other method is the staining of the core structure in chromosome through silver staining. It is always preferable to have good resolution by silver staining and correlated light and electron microscopic analysis of chromosomes in mitosis. The core has been demonstrated as a compact network of fibres. The axial element appearing as specific segments is interspersed along the length of the chromosome.

The existence of chromosome scaffold or skeleton originally recorded in the animal system has been much debated. Clear evidence of the presence of non-histone protein and RNA in a structure termed as Residual chromosome was recorded in 1947 by Mirsky, mainly through chemical treatment and enzyme digestion pattern.

In the plant system, such as in *Allium cepa*, and *A. sativum*, scaffold of chromosome has been observed to be a silver-stained structure and is made chiefly of nonhistone protein resistant to DNAse and trichloracetic acid (TCA), but digestible by trypsin and urea. It was noted in experiments with *Allium cepa* before silver staining, that if the slides containing metaphases are first treated with $0.2\,N\ H_2SO_4$ and digested with DNAse (100 µg/ml) to remove histone and DNA, the silver-stained structure could be seen which is unaffected by RNAse but disappears following trypsin treatment. It has also been noted that this component is a complex structure, composed of compact fibres and granules, distributed throughout the chromosome. The chromosomal deoxynucleoprotein (DNP) combines with non-histone protein. The latter forms a large part of the combined complex and can retain the intact morphology of the chromosome after histone-DNA depletion. Kinetochore is connected with the scaffold by kinetochore proteins which have a higher affinity to silver nitrate than the scaffold proteins.

SCANNING ELECTRON MICROSCOPY

Scanning electron microscopy permits manifestation of three dimensional structure of chromosomes. In transmission system, with the conventional method of fixation and uranyl acetate staining of ultrathin sections, it is often difficult to distinguish the nucleic acid and protein components of chromosomes. The visualization becomes all the more difficult because of its highly condensed solenoidal composition and its superimposition. In order to eliminate these limitations, scanning ultrastructural analysis is carried out. It not only gives a three-dimensional topography but with specific stain for DNA, the differentiation between DNA and protein and their distribution can also be resolved.

Of the stains for scanning electron microscopy, uranyl acetate yields chromosome contrast, but differentiation can be secured with Platinum Blue stain. It is water-soluble and the concentration used is 20 mM at 9 pH. It is ideal for scanning electron microscopy of chromosomes under sterilized and denatured conditions in view of the wide pH range of the specific reaction determined by the precipitation and stability in room temperature. The affinity of chromosome with platinum compounds has lately been emphasized in view of GC specificity of some Pt compounds which react selectively with DNA. It is likely that such studies may yield confirmatory three dimensional image of chromosome bands. Moreover, at high magnification, platinum bluestained DNA strands can be located selectively against unstained proteinaceous components.

Representative schedules

A. *Preparation of chromosomes*
Pretreat roottips in a suitable pretreating chemical. Fix in acetic acid–ethanol (1:3). Squash in 45% acetic acid. Remove coverglass by freezing in liquid nitrogen or solid CO_2. Immerse the slides in a 2.5% glutaraldehyde fixative buffer (either 50 mM cacodylate, 2 mM $MgCl_2$, pH 7.2 or 100 mM sodium phosphate, pH 7). Wash three times in buffer and then once in distilled water. Treat with 1% osmium tetroxide and a saturated solution of thiocarbohydride (or 1 mM dithioerythritol), followed by osmium tetroxide (twice). Wash with distilled water between steps. Then dehydrate the specimens through a graded acetone series (20–100%). Critical point dry with liquid CO_2 and sputter with 3–5 nm of gold or palladium. View with a Hitachi S-800 FESEM (after Wanner *et al.*, 1991).

Reagents used
Sodium hydroxide–borate buffer: 3.1 g of boric acid in 250 ml of glass-distilled water to give approximately 0.2 M solution. Adjust pH to about 9.5

by adding 30 ml of 1 M NaOH. *Formaldehyde–sucrose*: Dissolve 4 g of paraformaldehyde powder in pH 8.5–9 water to a final volume of 100 ml. Stir and warm to dissolve. Do not boil. Cool and add 3.4 g of sucrose (Grade I, Sigma cat. S9378) to make an approximately 0.1 M solution. Filter through nitrocellulose bacterial filter. *pH 9 water*: Bring glass distilled water to pH 9.0 by adding NaOH–borate buffer. Prepare just before use. *Phosphotungstic acid* (PTA): 0.2 g (Polaron Equipment Ltd, Cat. no. NC 3009) dissolved in 5ml of glass distilled water. Filter through nitrocellulose filter. Dilute one part stock with 3 part 95% ethanol immediately before use.

DNA imaging by high resolution scanning electron microscopy

The procedure is DNA-specific staining, using a blue platinum organic dye, which allows DNA imaging of chromosomes by detection of back scattered electrons in the scanning EM (after Wanner and Formanek, 1995).

For preparation of stain, mix (2 g) potassium tetrachloroplatinate with (3 ml) acetonitrile in 40 ml distilled water at room temperature. After keeping for 10 days, precipitation of pale yellow crystals of platinum dichloro-diacetamide is nearly complete. Decant the liquid. Air dry the crystalline complex and shake with the same weight of silver sulphate and a 5-fold volume of water. Keep for several hours. When the blue colour reaches maximum density, indicating complete conversion to platinum blue, add a 10-fold volume of methanol. Filter the solution. Precipitate the platinum blue by adding diethyl ether. Collect the dye by filtration and airdry.

Prepare chromosomes for EM study as described elsewhere (Martin *et al.*, 1994).

Fix chromosomes with 2.5% glutaraldehyde in 75 mM cocodylate buffer and then incubate at room temperature for 30 min with 10 mM platinum blue in 50 mM Tris buffer containing 2 mM $MgCl_2$ at pH 7.5. Then wash the preparation three times with water and dehydrate with a graded series of acetone concentrations and critical point dry from liquid CO_2. Coat the specimens with about 3 nm of carbon by evaporation. Examine with Hitachi S-4100 field emission SEM. Monitor back scattered electrons (BSE) at 15 kV with an Autrata detector of the YAG type.

Platinum blue forms oligomers; the molecular weight of the monomer is 311. The dye is readily soluble in water to give a concentration of 20 mM and a pH of 9. As a solid, or in solution, it is stable at room temperature for at least 1 month. Platinum blue and related platinum organic compounds selectively react with nucleic acids, especially with DNA. Large chromosomes, such as those from the genus *Lilium*, incorporate so much platinum blue that they appear pale blue in the light microscope; however, there is no significant change in image characteristics as shown by observation with phase-contrast, interference-contrast, epifluorescence or epipolarization microscopy.

When viewing platinum blue-stained chromosomal preparations by BSE microscopy at low magnification, the nuclei and chromosomes are clearly visible and appear very bright at all stages of condensation. Interphase nuclei exhibit a fibrous network mixed with nodular elements. The nucleolus and heterochromatic regions appear much brighter than euchromatin. During all stages of condensation from early prophase to telophase, the chromosomes are easily monitored by the electron microscope in the BSE mode; in secondary electron (SE) mode, however, the chromosomes are only detected when liberated from karyoplasmic material. Even when chromosomes are embedded in a dense proteinaceous network, good contrast makes all their characteristic details clearly visible, including chromatids and primary and secondary constrictions.

When superimposing chromosome images, a good fit is generally obtained between the BSE and SE images, although the chromosomes appear smaller in diameter in the BSE image. The most striking difference between the images occurs in the centromeric and satellite region; both are frequently much thinner and darker than the chromatids and often appear as a black gap. Higher magnification reveals thin filaments (after Wanner and Formanek, 1995).

CHAPTER II.3

CYTOCHEMICAL ANALYSIS THROUGH ISOTOPE UPTAKE AND CYTOPHOTOMETRY

Chemical analysis of chromosomes *in situ* can be carried out through the use of radio-isotopes (autoradiography) and estimation of nuclear DNA (cytophotometry).

LIGHT MICROSCOPE AUTORADIOGRAPHY

The objective of autoradiography is to locate radio-active material in a specimen with the help of photography. In autoradiography, the radioactive material is detected by a photographic process of development. The method is based on the principle that when a photographic emulsion is brought into contact with radioactive material, the ionizing radiation so converts the emulsion as to show spots at certain points after being developed.

The radioactive specimen is brought into contact with the film for a certain period of exposure, with consequent decay of radioactive atoms. The radiation thus emitted affects the emulsion, activating silver halide crystals. The final result is the formation of a latent image which can be developed to denote the location, intensity and distribution of the radioactive material.

In the stripping film method, the preparation is covered with a sensitive emulsion having a bottom layer of inert gelatin against a glass plate.

The emulsion technique involves the application of the emulsion in the liquid form and has been found to be most suitable for autoradiography. It is now widely used for plant, animal and human tissues and also for their cultures and allows the formation of a monolayer of emulsion on the tissue. In the study of high resolution autoradiographs (*see* section on high resolution autoradiography), the liquid emulsion technique has been found to be the most convenient one.

When a radioactive molecule is used as a tracer, the distribution of specific precursors of the molecule is studied within the cell. For this purpose, in the study of chromosome metabolism, labelled uridine is used for the detection of RNA, thymidine for the detection of DNA, and specific amino acids for the detection of proteins. With ^{32}P alone, the localization of DNA is difficult, since phosphorus is incorporated in different metabolic products of the cell unless such products are extracted or digested. To obtain an accurate autoradiograph, the tissue should be treated with at least a certain minimum concentration of the labelled substance. Approximately 10 grains/100 μm² of emulsion is just enough, requiring an exposure of $10/\delta/\beta$-particles (where δ denotes the yield of developed grains per particle hitting the field), which were obtained from the decay of double the number of radioactive atoms at the time of exposure. The half-life should also be considered. *Minimum concentration needed for short-lived isotopes*: half-life C in days, in 5 μm sections and emulsions, and f denoting the proportion of labelled to unlabelled tissue; $C = 11.5f/d^{\delta}$ in μCi/ml and *for long-lived isotopes*: $C = 12.5f/d^{\delta}$ in μCi/ml. The isotope being long-lived in this case, the decay during the time of exposure (d) is negligible.

The different steps followed in autoradiography are: (a) administration of the tracer into the tissue; (b) fixation; (c) paraffin embedding or smearing; (d) staining; (e) application of the photographic emulsion; (f) drying; (g) exposure; and (h) photographic process. Staining can be done before or after the application of the emulsion. For details, please see section on high resolution autoradiography.

The radioactive tracer can be obtained as specific salt solutions or tagged with metabolic precursors. For instance, ³²S is available as H_2SO_4 in dilute HCl solution or as sulphate in isotonic saline solution. Tagged isotopes are available in the form of tritium (³H) labelled thymidine, ¹⁴C-labelled adenine, thymine, uracil, etc., representing nucleic acid bases. For proteins ³²S-labelled methionine, phenylalanine, etc. are used.

Fixation can be performed either through freeze substitution or through a number of non-metallic fixatives. As far as practicable, metallic fixatives should be avoided. Ethanol, a mixture of ethanol and acetic acid, neutral formalin, and formol-saline can all be employed for fixation, but the most reliable method so far found is freezing.

In order to secure a high resolution, *paraffin sections* should not exceed 5 μm in thickness, should preferably be cut in a freezing microtome, and should be mounted on slides coated with a film of alum gelatin (0.5% aqueous solution of gelatin and 0.1% chrome alum) or with egg albumen. Absolute drying for at least 48 h is necessary before mounting the sections. Similar slides should preferably be used for *smears* as well. *The preparations* can be stained prior to, or after, the application of the photographic emulsion. In the former, the protection of the material by a very thin layer of impermeable substance, like celloidin, may be necessary.

The photographic emulsion can be applied over the section either in the form of a liquid smear, or as a film pressed on the material, being composed principally of a fine layer of gelatin, containing numerous silver halide crystals. The formation of radioactive spots is based on the principle of the production of ion pairs due to an electronic event, ultimately manifested in the single grain of black silver.

The tissue, after being coated by emulsion, either in liquid form or as stripped film, must *be dried quickly*, preferably in a strong current of air in a cold chamber. Immediate drying is necessary to prevent the formation of air bubbles within the tissue and also the production of artefacts in the presence of excess moisture.

The period of exposure in a cool dark chamber depends on the type of isotope used and its concentration, being the period required for a specified number of ionized particles to heat a unit area.

The photographic process to secure autoradiographs includes the development of the latent image, fixation, washing and drying. In autoradiography the use of an acid hardener fixing bath becomes necessary for the protection of film or plate. Continuous washing in running water is necessary to remove traces of any excess sodium thiosulphate, and a temperature of $15.5-21.0°C$ speeds the washing and prevents softening of the emulsion.

Representative schedules for light microscope autoradiography

Method of administration of isotope
For the study of somatic chromosomes, grow young seedlings of *Vicia faba* in medium containing $2\,\mu Ci$ of ^{32}P (as orthophosphate)/ml in tap water. Fix young healthy roots at different intervals, ranging from two days to one month, in acetic–ethanol or chilled 80% ethanol. Follow the standard protocol for squash preparations described under general methods. Enzyme treatment is preferred for softening the tissue.

For the study of meiosis in plants, $20–75\,\mu Ci$ of ^{32}P in 1 ml of water can be administered on flower buds of *Tradescantia paludosa* or *Rhoeo discolor* or $4\,\mu Ci/ml$ labelled ^{32}P in White's medium, on inflorescence of *Lilium henryii*. Anthers from flower buds of suitable size are smeared on slides.

Liquid emulsion autoradiography
In order to study materials administered labelled amino acids or nucleosides, acetic–ethanol (70–95%) $1:3$ fixative is recommended. Stock solution of the emulsion should be kept at 22–24°C. Kodak NTB series is used with different grain sizes. With decrease in grain size, higher resolution and sensitivity are expected. The experiment should be carried out in dark.

Melt the emulsion at 42–45°C using a constant temperature water bath. To avoid background effect, also develop a clean dry plate, without any material, as control. Subbed slide is not necessary if Kodak emulsion is used. Fit two slides with materials back to back and keep in emulsion, in a long trough, for 4–5 s. Drain off the emulsion. After separating the slides, dry in racks against a stream of air and keep in slide boxes, sealed with tape, in a cool dark chamber, for the required period of exposure (2–3 weeks).

After exposure, develop the slides in Kodak Dektol, D-11 or D-19 developer for 2 min. Fix in Kodak Acid-Fixer for 2–5 min. Rinse in running water for 20 min. Give a final rinse in distilled water and dry. Any of these stains can be used: 0.25% aqueous solution of toluidine blue (pH 6); methyl green-pyronin; Giemsa stain. Wash in 95% ethanol, make air-dry preparations and mount in euparal.

Stripping film autoradiography

All operations should be performed in the dark room in safelight at 20–22°C. Cut 40×40 mm squares of AR 10 film mounted on glass slides and keep the plate for 3 min in 75% ethanol. Transfer to another tray containing absolute ethanol. With forceps, lift a single square of film and float, with emulsion side down, in a tray of distilled water to spread out the film. Bring the glass slides containing the material below the film under water and lift the slides so that the film adheres neatly on the surface of the slide with the material on it. Dry in a stream of air and store in a slide box, sealed with tape, containing silica gel desiccant for the required period in the cold. Prior to developing, paint the reverse side of the slide, with a paint and dry. Develop with Kodak developer D-196 for 5 min. Treat with the acid fixer as usual, wash and dry. Detach paint and film with a blade and follow the usual procedure for mounting. If the material had not been stained earlier, stain in either of the dyes, rinse in distilled water and mount.

FIBRE AUTORADIOGRAPHY

Autoradiography of the spread fibres of DNA is otherwise termed fibre autoradiography. It permits a study of replicons and their spacing as well as the duration of the "S" phase. It also enables the analysis of the degree of synchrony of the onset of replication. The duration of the "S" phase in different tissues is a function of development and differentiation. However, the space between the replicons has been shown to be an important factor in controlling this duration.

At 2 h before termination of suspension cultures, give two pulse treatments of thymidine: 2 h before harvesting, hot pulse (Sp. act. 45 Ci/mmol, 100 Ci/ml Radiochem, Amersham) for 1 h and warm pulse (Sp. act. 12.7 Ci/mmol : 20 μCi/ml) 1 h.

Stop ^{3}H TdR uptake by repeated washing in calcium or magnesium-free phosphate buffered saline (PBS). Suspend the cells in PBS highly diluted with water for swelling and spreading. Take a subbed slide and add 1% SDS on a drop of suspension containing sufficient number of cells. Spread the suspension and air dry. Treat the slides with cold TCA for 5 min. Wash in running water and rinse in ethanol. Coat the slides with equally diluted E_4 emulsion. Keep in dark for 6–10 months. For developing, use Kodak D19 developer for 10 min at 10°C and photograph autoradiograms. Measure the track length in photographs and measure replicon spacing.

HIGH RESOLUTION AUTORADIOGRAPHY

The combined use of electron microscopy and autoradiography, as involved in high resolution autoradiography, also facilitated the study of ultrastructure as correlated with function.

High resolution technique and its combination with immunofluorescence have allowed the delimitation of functionally specialized segments of chromosomes. In general, this aspect of the subject has been much refined through the use of tritiated compounds of high specificity available in forms with specific activity even more than 15 Ci/mM. Moreover, the use of fine-grained nuclear emulsions, with grains as small as 0.1 μm (Ilford Nuclear Research Emulsion K5-L4) has facilitated the preparation of emulsion of uniform thickness and rendered the technique more convenient. It has been found necessary to observe thicker sections (0.3–0.5 μm) under light microscopy, prior to the study of high resolution ultrathin autoradiographs, principally to get an idea about the approximate exposure time needed for ultrathin autoradiographs and for their comparative assessment. Exposure time needed for ultrathin sections is about 10 times more than that required for comparatively thicker sections. Light microscopy is also required to obtain a demarcated picture of the area to be studied in ultrathin section. For both these purposes, the desirable thickness of the sections is between 150 and 500 nm.

Fixation and embedding

For analysing chromosome structure, fixation in osmium tetroxide solution buffered to pH 7.2 to 7.4 is often recommended. The addition of divalent cations like calcium (10^{-2} M) in the fixative, checks swelling and helps in maintaining uniformly the packed macromolecular configuration. Very satisfactory results were obtained with chromosomes of *Vicia faba*, by fixing in 1% unbuffered (pH 6.0–6.4) solution of OsO_4 in double distilled water, to which varying amounts of calcium chloride were added. Freeze-drying methods can also be adopted for dehydration after quick freeze fixation. For embedding, any of the usual media, such as methacrylate, epoxy resins, araldite, Epon or polyester, Vestopal W, can be employed.

Before section-cutting
Clean slides with frosted ends should be dipped in the subbing solution (1 g Kodak purified calfskin gelatin is dissolved in 1 liter hot distilled water, cooled, 1 g chromium potassium sulphate is added and the solution stored in cold). The subbed slides are dried in a dustfree chamber and stored in boxes.

To secure thick sections meant for predicting the exposure time needed for ultra thin section, as well as for comparative assessment, the block is trimmed so as to obtain a much larger face than that needed for ultrathin section. After mounting on the microtome, sections of desired thickness (150–500 nm) can be cut by setting the section indicator and observing the interference colour by adjustments of the water level and illumination (see Reid, 1974). When a ribbon with 2–4 sections is cut, the sections are picked up with a damp, fine-haired, clean nylon brush and transferred to a drop of water placed near the edge of a subbed slide. The slides are then dried at 40°C or at 60–80°C.

In order to locate specific regions in ultrathin sections, the block is trimmed in such a way that ultrathin sections can be obtained. First a thick section (120 nm) is cut and transferred to the slide by the method given previously, followed by several ultrathin sections (60–100 nm) which are shifted on the boat, prior to cutting another thick section of the original thickness. The latter is also mounted on the slide and both are observed. If the desired region is present in both the thick sections, the intervening thin sections are mounted on the grid, since the presence of the desired zone or material in the ultrathin section has been ensured.

Coating of ultrathin section
Section embedded in methacrylate, epon, araldite or Vestopal may be used. Selection of proper emulsion is one of the most important factors in high resolution autoradiography. Its sensitivity depends on the extent to which it can register and develop the latent images formed by electrons in their path on silver halide. It is measured by the number of grains developed per unit distance in the track of particles with minimum ionization. The particle energy and the distance the electron has to traverse through the silver halide, control the formation of the latent image, Normally, ^{32}P, ^{131}I, ^{14}C and ^{35}S have long range ionizing particles. With tritium, nearly all electrons emitted into the upper hemisphere can be developed. During exposure, oxidation of the latent image affects sensitivity. Protection against oxidation becomes essential with smaller crystals and fine-grain development. With tritiated compounds, the problems of sensitivity and resolution are not so severe since the β-particles emitted are heavily scattered within one silver halide crystal of Ilford L4 emulsion. With finer grained emulsions, such as

Gevaert 307 or Kodak NTE, multilayered crystals add to the sensitivity. More suitable are Ilford L4 (crystal size 120 nm) for electron microscopic and Ilford K5 (180 nm) for light microscopic autoradiography. A closepacked monolayer of silver halide adds to resolution by preventing the spread of electrons from the source.

For mounting the sections, both the collodion film and the sections must be perfectly smooth. Electroplated Athene-type copper grids are used for coating the collodion with a thin carbon layer. A thin film is spread so that resolution is not hampered and at the same time, breakage is avoided during the procedure. Sections are mounted on the grid which is attached at one edge with a piece of scotch tape (double coated) to a slide. Several grids can be placed on one slide.

Several methods have been devised for applying the emulsion so as to form a uniform layer. It may be applied, either by immersing the slide in the emulsion or by dropping the latter on the slide (5 ml of distilled water per 1 g of emulsion). A thin layer may be allowed to form on a specially constructed loop before applying on specimen grids. The emulsion may be centrifuged directly on the specimen grids or may be finely layered on agar before application on sections.

Thickness of emulsion layer
The most appropriate method for ascertaining the thickness needed for quantitative work is to use a developer (Devtol) which does not affect silver halide crystals. After developing, and prior to fixation, the slides can be air-dried and viewed in white light for interference colours.

Developing
After adequate exposure, developing of the photographs should be carried out in clean and dustfree conditions. It is always essential to find out the optimum period of development which would permit the maximum number of grains to be developed with least background effect. The grids must always be kept in absolute ethanol for 3–4 min before developing. This hardening schedule is an essential step since it prevents the sudden swelling caused by aqueous developer and the resultant loss of grains. Several developers are in vogue and their adequacy depends on the type of emulsion used for coating. *Chemical developers*, which reduce silver halide crystals, result in a coil of silver filament of 0.3–0.4 μ diameter in certain developers like D19, or a long filament as in Microdol X. Athene-type grids are placed on filter paper for drying after being lifted from ethanol with forceps. The dried grids are then floated in an inverted position on the convex surface of the developer and kept for 6 min at 22–24°C. After immediate transference to a watch glass containing distilled water, for a few seconds, the grids are

placed in the fixer with the sections facing upwards. The fixer effectively removes all unexposed silver halide crystals and helps in the later removal of the gelatin. With *physical developers*, the highest possible resolution can be obtained. The principle is to *dissolve completely the silver bromide crystals*. Only the latent image with silver ions is kept, with the use of 1.0 M sodium sulphite and 0.1 M 4-phenylene diamine, on silver nitrate in varying proportions. Development for even 1–2 min at 20°C is sufficient. With this method, the latent image can be localized with minimal error.

Staining

Staining can be performed even before applying the emulsion. A serious drawback of keeping the gelatin intact is the possibility of disruption of the grains through shattering of the gelatin layer by the electron beam.

In a modified technique where gelatin is dissolved before staining, the grids are first floated for 3 min on the convex surface of distilled water, kept at 37°C and then transferred to 0.5 N acetic acid at 37°C and kept for 15 min. They are rinsed in a stream of distilled water and then floated in a second change of distilled water at 22–25°C for 10 min. Staining is performed by first wetting the sections with drop of distilled water; staining in an inverted position in 7.5% aqueous uranyl acetate for 20 min at 45°C; subsequent drying on filter paper; wetting again with distilled water and further staining with 0.2% lead citrate for 10–60 s to a few minutes. Uranyl acetate changes the properties of the macromolecules in such a way that they bind better with lead citrate. The excess stain is immediately flushed off with a stream of distilled water.

Protocol for spreading and autoradiography

For short labelling, treat the tissue for a very short period with ^{3}H (80–100 µCi/ml, sp. act. 25–30 Ci/mmol CEA or Amersham) and *for longer incubation* 1 h and more, with tritiated precursor 50–30 µCi/ml. After labelling, transfer to ice-cold phosphate sucrose solution, centrifuge and resuspend in same solution. Add 1 ml Nonidet P 40 in 0.2 M EDTA, pH 7.4 with stirring. Dilute cell lysate with 0.2 M EDTA.

Layer lysate on 4% formaldehyde, 0.1 M sucrose soln., pH 8.5, in a translucid plastic chamber, the bottom of which contains freshly glow-discharged copper or gold EM grid, coated with Formvar-coated membrane. Centrifuge the material at 2400 g at 4°C for 10 min. Remove the grid from the chamber and treat for 30 s in 0.4% Photoflo 600 (Kodak), pH 7.5–7.9. Airdry.

Stain with 1% PTA in 70% ethanol for 1 min. Dehydrate in 95% ethanol for 2 s and air dry. Rotary shadow the grids (angle 7°C–10°C) to obtain thin layer of platinum. Apply Ilford L4 emulsion using loop technique, prepared the previous day by dissolving gel shreds in water at 40°C and kept in cold. Develop in preparation following Elon-ascorbic acid procedure followed by 2–8 months exposure. Bring the material to room temperature and use safe light. Dip the holder into the intensification bath for 5 min. Transfer into distilled water bath for 10–20 s. Transfer into Elon ascorbic acid developer for 7.5 min. Transfer to fixing bath for 2 min. Wash in three successive baths of distilled water. Observe in EM using suitable objective aperture (30–40 μm).

MICROSPECTROPHOTOMETRY

Under ultraviolet light

Ultraviolet light, when used instead of visible light, has the unique advantage of clarifying unstained living cells, due to the strong ultraviolet absorption by nucleoprotein. It also aids the quantitative estimation of the cell nucleoprotein owing to the characteristic absorption of purine and pyrimidine components of nucleic acid at 265 nm. A linear relationship has been shown between absorption and section thickness and the concentration of DNA in nuclei. Ultraviolet absorption spectra of cytological objects generally exhibit absorption between 20–40 nm; nucleic acids at 260 nm, and proteins free from nucleic acids at about 280 nm.

The principal difference between an ordinary light microscope and an ultraviolet microscope lies in the fact that in the latter, transparent fused quartz lenses are used in place of optical glasses, which are opaque to shorter ultraviolet wavelengths. The source of ultraviolet rays is generally the mercury vapour lamp. The slides and coverglasses are made of quartz. In order to obtain a monochromatic beam, a quartz monochromator is fixed between the source and the microscope. The photographic image is obtained by using a photographic plate and a photoelectric cell. Focussing is generally carried out in visible light or on a fluorescent screen. Computation of the images can also be obtained. The commonly accepted method is to have the source as illuminator through the exit slit of the monochromater. The aperture of the condenser is generally kept at about 0.3 or below. Discharge lamps, such as hydrogen and deuterium lamps and low pressure mercury lamps, with good achromatic objectives, give a wide band at 254 nm and can be used with suitable monochromators or interference filters.

Most equipments utilize Xenon compact arcs, which combine intensity with output through the ultraviolet range at 260 nm. In the monochromators meant for selecting the particular wavelengths, a band of energy is emitted. Its wavelength distribution is controlled by the dispersion of the elements showing diffraction and refraction and the size of the slits. Several microspectrophotometers are equipped with grating monochromators where the change in wavelength is done by grating rotation. A number of interference filters have been developed, to serve as protective filters and as wavelength selecters. Photography may be adopted for the integration of absorption of objects and the negatives may further be scanned through densitometry. Photoelectric recording is utilized for a study of the series of absorption spectra needed for each wavelength. The most convenient method is to allow the light to pass through a selected area for final recording in a photomultiplier tube. The intensities may be recorded at different wavelengths and the background intensity measured by removing the object and allowing the light to pass through the empty space. In principle, to measure changes in the quantity of nucleic acid and protein, the microscope is generally used as spectrophotometer. The monochromatic beam of ultraviolet light may be split into two beams, in the split beam device, one falling directly on the photoelectric cell (the blank) and the other passing through the ultraviolet microscope to another photoelectric cell. The sample to be measured is placed in the path of the beam passing through the microscope. The light passing through the material is reduced in intensity; this reduction is calculated by counting the difference in the photoelectric current yielded by the two beams, as indicated by a galvanometer. The data can also be electronically computed and televised. The measurement of absorption is based principally on Lambert-Beer's law which may be stated as follows:

$$I_x = I_o \times 10^{-Kcd},$$

where I_x is the changed intensity of the beam of I_o; I_o the incident intensity, after the ray passes d; d the thickness (in cm) of c; c the concentration (in g/100 ml) of the absorbing molecules; and K the extinction coefficient. As in cytomicrospectrophotometry, relative amounts are obtained, the constant K is ignored, since absolute values are not necessary. The value of K, when necessary, can be worked out from biochemical data.

The values of I_x and I_o can be obtained without changing the position of the material. With the aid of these values, the percentage of transmission (T) can be worked out (I_x/I_o) and the presence of the components per arbitrary units, showing ultraviolet absorption, can be computed. A limitation of ultraviolet microscopy is that the ray may have deleterious effects on the absorbing material, but in the above method the period of exposure to ultraviolet is very much reduced.

For a correct assessment of the absorbance data, it is desirable to follow the extraction of specific cellular components simultaneously. Such procedures, in addition to aiding identification of the absorption of particular chemical constituents, may also serve as controls. Further, the most significant use of extraction is to secure a *blank*, so that non-specific light loss and light scatter can be corrected. These purposes are served through digestion with proteases or nucleases. In mounting the specimens in ultraviolet microspectrophotometric work, media like pure glycerine, glycerine-water mixture, 45% zinc chloride and paraffin oil possess the essential prerequisites for checking the non-specific light loss.

During the applicaton of the ultraviolet rays, caution is recommended, since continuous exposure for even 10 min at 257 nm in a Köhler microscope may result in the loss of absorption capacity by the chromosomes treated with acidic fixatives. Techniques for the ultraviolet microscopy of cells in culture have been developed to study conditions and changes *in vivo*, for which perfusion chambers with quartz coverslip windows are generally used. In the scanning procedure, improvements have been devised in which large suspended cell populations are quickly analysed with the rapid cell spectrophotometer.

Under visible light

Microdensitometry or cytophotometry in visible light is based on the principle that between 400–700 nm light is partially absorbed by the matter due to interaction with outer constellation of electrons. This interaction mainly depends on the physical and chemical nature of the matter and the wavelength of the light. The process follows Bougner-Beer law which holds that the absorbance is dependent on the concentration of the matter and the pathlength. It is expressed as logarithms of the reciprocal of transmitted light after absorption. It is in fact the optical density (OD) or extinction (E) i.e. $\log I_o/I_s = \mathrm{OD} = E = \mathrm{A}$. As the absorbance is dependent on the concentration of the absorbing matter, the technique has been adopted for quantitative estimation of cell *vis-à-vis* chromosome constituents. Absorption measurements are carried out on materials stained with Feulgen solution, methyl green pyronin, azure B, Millon dye and such other compounds capable of staining specific cell constituents.

For the fixation of materials, Carnoy's fluid or 10% neutral formalin is widely used to measure chromosomal DNA. Staining may be carried out in Feulgen solution or in pyronin–methyl green mixture or in any other specific dye for binding with nucleic acids or proteins.

Microscope and accessories
The appliances needed for visible light photometric or densitometric work are quite simple. For example, in Leitz MPV, they include, in principle, a strong tungsten light source; monochromatic unit, fitted with condensing lens; a microscope (Aristophot) having condenser with low numerical

aperture; phototube, measuring variable field diaphragm, a photomultiplier, a power supply unit and galvanometer. Alternate arrangements are available for observing specimens in ordinary light (without monochromator) as well as for photography. Similarly, through suitable accessories, the data can be recorded by a recorder or oscilloscope, instead of galvanometer. In Reichert Zetopan photometer, interference filters are used between tube and photocell over the microscope, instead of in monochromator. The guidelines for use are available with each model.

Method of analysis

As in ultraviolet microspectrophotometry, in taking measurements, the data required are the intensity of the background light (I_o) which is taken from a blank area on the slide, and the reduced intensity of light after absorption by the objective (I_s), the *transmission* being calculated as I_s/I_o. The transmission indicates the fraction of light that remains after loss by absorption. In a completely transparent object, this value is 1 and there is a logarithmic decrease in transmission with increase in absorbing molecule, due to concentration or thickness. Therefore extinction or absorbance is:

$$E = \log_{10}\frac{1}{T} = \log_{10}\frac{I_o}{I_s}.$$

This law applies to all cytophotometric calculations. The law, developed for dilute solutions, holds equally well for all densitometric observations if reactions are carried out under strictly controlled conditions. From the optical density or absorbance data, the mass of material can be measured, on the basis of the fact that what is obtained in cytophotometry is transmittance (T). In conversion to mass (M), the following equation is adopted: $M = A\log(1/T)/K$ where A is the area πr^2, of which π may be omitted and K, the extinction coefficient. The K is a function of the wavelength and in relative mass determination, being constant, this value is not needed in calculation. This method holds good for homogeneous samples such as interphase. But in densitometry, the spatial heterogeneity of the specimen presents distribution error which is quite common to non-uniform microscope materials such as the distribution of chromosomes. More precisely, it may be attributed to unequal distribution or heterogeneity of DNA in the nucleus. In such cases, the total absorbance of a projected area is dependent on the total amount of such material present in the projected area. As such it should be the sum total of the absorbance of individual units within the area. To some extent, the distribution error is corrected by measuring the transmission at two different wavelengths (Patau, 1952) as outlined below.

The prerequisite of two-wavelength technique of absorption analysis is the uniform illumination of the area with a monochromatic source and

77

absence of light scatter through the use of proper mounting medium. In Feulgen-stained sections, it is desirable to carry out the analysis on the same slide, since acid hydrolysis—an essential feature of Feulgen staining affects the reaction to a significant extent. The method of analysis is as follows:

If the two wavelengths selected are λ_1 and λ_2, the extinction (E_1) at the former should be half the extinction (E_2) at the latter, so that $E_2 = 2E_1$. As mentioned above, $E_1 = \log I_o/I_s$ at λ_1, $E_2 = \log I_o/I_s$ and λ_2. Choice of the proper wavelengths is important. The total amount of absorbing material (M) in a measured area (A) is, $M = Kal_1D$ where K is constant; ($K = 1/e_1$, e_1 being the extinction coefficient in λ_1, K may be disregarded for relative determination in cell microspectrophotometry). L_1 is respective light loss and D is the correction factor for distributional error. Area (A) is measured as πr^2 where π may be omitted. From transmission T_1 and T_2, at λ_1 and λ_2, the degree of light loss may be worked out as follows:

$$L_1 = 1 - T_1 \text{ and } L_2 = 1 - T_2.$$

With the ratio L_2/L_1 at hand, the value of D can be worked out from the table given by Garcia (1962), with detailed principles. The following table taken from Pollister, Swift and Rasch (1969) gives the value of D corresponding to each L_2/L_1 ratio:

Values of D for different values of L_2/L_1

L_2/L_1	0.00	0.01	0.02	0.03	0.04	0.05	0.06	0.07	0.08	0.09	
1.0	—	4.033	3.461	3.134	2.907	2.734	2.595	2.479	2.380	2.294	1.0
1.1	2.218	2.150	2.089	2.033	1.982	1.935	1.892	1.851	1.813	1.227	1.1
1.2	1.744	1.712	1.683	1.655	1.628	1.602	1.578	1.555	1.533	1.511	1.2
1.3	1.491	1.471	1.453	1.435	1.418	1.400	1.384	1.368	1.353	1.339	1.3
1.4	1.324	1.310	1.297	1.284	1.271	1.259	1.247	1.235	1.224	1.213	1.4
1.5	1.202	1.191	1.181	1.171	1.162	1.152	1.143	1.133	1.124	1.116	1.5
1.6	1.107	1.098	1.091	1.083	1.075	1.067	1.059	1.053	1.045	1.038	1.6
1.7	1.031	1.024	1.017	1.011	1.004	0.998	0.991	0.985	0.979	0.973	1.7
1.8	0.968	0.962	0.956	0.950	0.945	0.940	0.934	0.928	0.923	0.918	1.8
1.9	0.194	0.909	0.903	0.899	0.894	0.890	0.884	0.880	0.876	0.871	1.9
2.0	0.867	—	—	—	—	—	—	—	—	—	—

The best method to avoid distributional error is to take integrated absorbance. As the means of the logarithms is not equivalent to the logarithms of the mean it is always desirable to scan absorbance from each individual point and to integrate the data.

Different types of microdensitometers are available, designed with the same objective but they differ with regard to their mechanism of operation. The difference mainly lies in the system of scanning. They can be classified under two categories: (a) in which the object is stationary and the scanning is done along the whole length of the object and in the other, (b) the object is moved with a motor-driven stage and the measuring area is fixed. The moving object densitometers are manufactured by Zeiss, Leitz, etc. while in the instruments of Vickers, Barr and Stroud, the specimens to be scanned remain stationary. In the Barr and Stroud type, the image of the object is enlarged and scanned by a mechanical device, whereas in the Vickers model the fixed object is scanned by moving the reduced image of an illuminated aperture with the aid of a pair of oscillating mirrors. In televised models, a television camera receives the image which is electronically scanned. The Vickers models are quite flexible in several respects and have the advantage of utilization for dry mass scanning through micro-interferometry with suitable adjustments. Gradual refinements in models have also resulted in high resolution scanning densitometry from photographic negatives of individual chromosomes.

Use in fluorometry
The principle of microscope photometry can be advantageously employed for fluorescence quantitation as well. Since the energy of fluorescent light is much lower than that of ordinary transmitted light, it is always desirable to use incident illumination (through suitable accessories); the illuminating lens being the objective itself. A xenon arc lamp may serve as the source of light. In order to secure large optical flux, it is desirable to use a fluorite or apochromat objective with a high numerical aperture. Automatic switchover device from low power transmitted light to fluorescence light for measurement is available. The illuminated field should be large but not exceed the area to be measured since illumination of other objects may cause change in fluorescence of the specimen to be measured. The immersion medium should also be free from any fluorescence effect.

For fluorescence analysis of chromosomes, it is necessary to get an integrated value of the intensity of the whole fluorescence spectrum. The transmission is obtained also from the barrier filter and the photocathode is so chosen as to be responsive to the entire spectrum. If necessary, spectral distribution of fluorescence can be measured by attaching a continuous spectrum monochromatizing device, and a suitable photocathode with a wide range of response.

Methods are also available for high speed quantitative karyotyping by flow microfluorometry. The technique involves isolation of metaphase chromosomes from cells, staining with a DNA specific fluorochrome and measuring for stain content at the rate of 10^5/min in a flow microfluorometer. The results tally well with scanning cytophotometry.

FLOW CYTOMETRY

Flow cytometry is an elegant and rapid method for sorting and quantifying chromosomal data at the microscopic level. It normally involves the isolation or preparation of chromosome suspensions, staining with DNA specific fluorescent dye and finally analysis through flow cytometer, such as Facstar plus Cytometer or Slit Scan Flow Cytometer (SSFCM). Approximately 10^6 chromosomes can be analysed per second and statistically large numbers of chromosomes can be processed within a few minutes. In SSFCM, the chromosomes are flown through a 1.3 mm thick laser beam (width—$1/e^2$ points), produced by an argon ion laser emitting 1.0 W at 488 μm. Fluoresence intensity along its length is detected photometrically and digitized every *ns* by means of a waveform recorder. The time varying fluorescence intensity is recorded as a measure of the distribution of the fluorescent dye along the length of the chromosomes.

Flow Cytometry, if coupled with sorting facilities, can provide details of chromosome aberrations such as translocations and deletions. Even marker chromosomes can be analysed in a flow karyotype. It can serve as the basic stock for preparation of DNA libraries of individual chromosomes. The sorted chromosomes can be utilized for mapping of genes through hybridization with probes on nitrocellulose filter. Lately, with the development of PCR technique, the need for a large number of sorted chromosomes for preparation of chromosome-specific DNA library has been much minimized and the technique for sorting is applied with convenience on suspensions.

However, chromosome sorting through flow cytometry depends to a great extent on the quality of the isolated chromosomes. Moreover, the proper adjustment of clean flow cytometer to give maximum power, good resolution of chromosomes and well defined sorting windows are essential (Fantes *et al*., 1994; Gray and Gram, 1990). For flow cytometry, the reader is referred to "Flow Cytogenetics" by Gray, J.W. (ed), 1989, Academic press, which gives a comprehensive treatment of the topic. However, Fantes *et al*., (1994) have also given a very lucid account of Flow cytometry with all technical details.

For the study of flow cytometry of plant chromosomes *Haplopappus gracilis* is the model system with two chromosomes (De Groot *et al*., 1986).

Before proceeding with the experiment, all the four steps, namely, cleaning, aligning, delivering and sorting, are to be fully understood and followed with meticulous care. The cleaning of the Flow cytometer at the initial stage is imperative. All the contaminants and residues are to be flushed out, before cleaning with the warm detergent for 30 min, followed by distilled water flushing for another 30 min. The cytometers need to be checked before addition of detergent. This process is to be followed by centering the beam properly so that two well aligned laser beams can be achieved, yielding intense fluorescence signalling with Hoechst 33258 and chromomycin azodyes.

Freshly prepared chromosome suspension is to be used for alignment. The centering and focussing of the UV laser beam should be carried out by minimizing width and maximizing fluorescence intensity. This procedure should be followed for both dyes. The flow rate is also to be properly controlled with a suspension dummy sample by adjustment of gas pressure. Finally, for sorting, it is essential that the droplet deflection system should be calibrated before the beginning of the experiment. The droplet delay needs to be measured in fractions of droplets, between signal detection and droplet charging. Collection of chromosomes into a tube and the amount of buffer to be used in the tube before sorting depend on the purpose for which sorting is desired. For example, for a PCR analysis, very few sorted chromosomes and the requisite amount of buffer will be needed as compared to the amount required for others. In the preparation of DNA libraries from specific chromosomes, millions of sorted chromosomes are needed, for which electronic sorting window is utilized.

Representative schedules

1. *For crop varieties*

Raise seedling for 1 or 2 weeks and isolate nuclei from different tissues according to the procedure described elsewhere. Chop segments of each tissue 2.5 cm in length into 2–3 mm discs and homogenize for 30 s in 10 ml of modified nuclear isolation buffer. The isolation buffer consists of 1 M hexylene glycol, 10 mM Tris (pH 8.0), 10 mM $MgCl_2$ and 0.5% Triton X-100. After grinding, filter the sample through nylon mesh having pore sizes of 250 μ and 53 μ. Place the sample in a 15 ml corex tube and centrifuge for 15 min at 500 × g at 4°C. After centrifugation, remove the supernatant and resuspend pellet in a staining solution consisting of 3% (w/v) polyethylene glycol, 50 μg per ml PI (propidium iodide), 180 units per ml RNase; 0.1% Triton X-100 in 4 mM citrate buffer (pH 7.2). After incubating at 37°C for 20 min, add an equal volume of salt solution consisting of the above components (excluding the RNase) in 0.4 M NaCl instead of citrate buffer. Keep the stained nuclei in dark at 4°C for one hour. Perform flow cytometric measurements with a PAS-III flow cytometer (Partec, Germany). A minimum of 5000 nuclei from each tissue sample is analyzed. The DNA distribution histograms for different tissues within a crop and among different crops can be compared. Different peaks representing diploid (2C), tetraploid (4C), octaploid (8C) and other higher ploidy level nuclei and percentage of nuclei, in different phases of cell cycle), from the 5000 nuclei are analysed. Statistical tests are performed to find out the significant differences among various ploidy levels of different genotypes.

2. For chromosomes and nucleus in pea, Pisum sativum
(after Gualberti et al., 1996)

Germinate seedlings at $25 \pm 1°C$ in dark. Incubate seedlings with 3 cm long roots for 18 h in Hoagland solution containing 1.25 mM hydroxyurea (HU). Wash roots in distilled water and incubate for 3 h in HU-free Hoagland solution. To block cells at metaphase, transfer seedlings into 10 μM amiprophos-methyl (APM) in Hoagland solution. Transfer to acetic acid–ethanol (1:3) mixture after 2 h incubation in APH. Fix overnight and stain according to Feulgen procedure.

To isolate nuclei, cut roots 1 cm from tip, rinse in distilled water and fix in 2% (v/v) formaldehyde in Tris buffer (10 mM Tris, 10 mM Na_2 EDTA, 100 mM NaCl, pH 7.5) with 0.1% Triton X-100, for 20 min at 5°C.

To isolate chromosomes, fix AP-treated roots in 4% (v/v) formaldehyde in Tris buffer for 30 min at 5°C. Wash roots three times (20 min each) in Tris buffer at 5°C. Homogenise tips in an automatic homogenizer for 15 s at 9500 rpm in 5 ml polystyrene tubes (Falcon 2054, containing 0.5 to 1 ml LBO1 lysis buffer (15 mM Tris, 2 mM Na_2 EDTA, 80 mM KCl), 20 mM NaCl, 0.5 mM spermine, 15 mM mercaptoethanol, 0.1% Triton X-100, pH 7.5). Channel suspension of nuclei and chromosomes through a 53-μm nylon mesh to remove large fragments. Then carefully syringe the suspensions once through a 22 G hypodermic needle and filter through a 21-μm nylon mesh.

To purify chromosome suspension from interphase nuclei and clumps, layer from the bottom, in a 10 ml glass tube, 1000 μl of 40% and 500 μl of 10% (w/v) sucrose in Tris buffer, followed by 500 μl of the chromosome suspension. Centrifuge at 40 g for 20 min. Remove 600 μl supernatant. Collect the next 500 μl with chromosomes and bring up the volume to 1 ml with LBO1. Store chromosome suspension at 5°C in 2BO1 supplemented with 1 mM NaN_3 for upto several weeks.

For flow cytometry, stain nuclei and chromosomes in suspension with DAPI, 5 μM and analyse with a "FAC-Star-PLUS" flow cytometer and sorter, equipped with an argon-ion laser tuned at $\lambda = 351$–363 nm at a 100 mW output power. The fluorescence is collected through a 400 nm long pass filter. Moniter fluorescence signals according to their height and area. Compare the analysis of pea flow karyotypes to the theoretical total DNA histogram generated with the software "KARYOSTAR" based on pea chromosome lengths, as measured on metaphase spreads.

In order to determine the chromosome content of flow-karyotype peaks, set sorting gates on fluorescence pulse area histograms. Sort chromosomes at a rate of 5 to $20\,s^{-1}$ directly into coverglasses and airdry.

In *Triticum monococcum*, chromosomes have been isolated from the cell lines following enzymic digestion of the cell wall and polyamine isolation buffer. The suspension was stained with Hoechst 33258 and chromomycin A_3 and analysed on a Facstar plus Cytometer (Leitch *et al.*, 1991).

3. *DNA preparation from sorted chromosomes* (after Fantes *et al.*, 1994)

Sort at least 5×10^5 chromosomes; if necessary by pooling sorted chromosomes from 2 d sorting. Dialyze against Ll of dialysis buffer for at least 16 h with at least one change of buffer. Carefully remove the chromosome suspension from the dialysis tubing. Wash tubing thoroughly with fresh dialysis buffer.

Take an aliquot of chromosomes to check on the purity. This can be a simple chromosome identification check after spinning the chromosomes down onto a slide either by banding analysis or by *in situ* hybridization with chromosome-specific centromeric sequences.

Pellet the chromosomes by centrifugation at $2000\,g$ for 15 min at 4°C. Remove as much residual buffer as possible. Store at -20°C.

Thaw the pellet, and thoroughly disperse the metaphase chromosomes. Add 10× extraction buffer to give 1 × final concentration.

Add 2 µg of rat tRNA from stock. This acts as carrier for the DNA through later stages. Make the suspension 0.5% in SDS. The vol should be $<100\,\mu l$. Add proteinase K to a final concentration of 1 mg/ml. Incubate for at least 5 h, or preferably overnight, at 37°C.

Extract the 100 µl with 100 µl of phenol. Extract the supernatant with phenol/chloroform and finally with chloroform/isoamyl alcohol. Each extraction should be back-extracted with 20–30 µl of TE. Dialyze against 1 litre of TE using a microdalysis filter for at least 5 h or preferably overnight. The dialysis filter should be preincubated with 1 ml of TE containing 5 µg tRNA, to block the membrane and prevent subsequent adsorption of the DNA sample. Remove DNA from dialysis thimble, and cut for 4 h with an appropriate restriction enzyme and buffer. Repeat the series of extractions. The volume at this stage should be 200–300 µl.

Repeat the dialysis step. Precipitate the DNA using 1/10 vol of 3 M sodium acetate, 100 µg/ml Dextran, and 2 vol of absolute ethanol. Incubate at -20°C for 2 h. Spin down at 30,000 g for 1 h. Wash the pellet in 70% ethanol, and spin again. Dry the DNA pellet briefly. Resuspend DNA in a small volume of TE or ligation buffer ready for ligation into chosen vector.

DNA CONTENTS—SCOPE AND ADVANCES

One of the important avenues of research opened up through quantitation of DNA following extraction, *in situ* estimation, as well as specific staining procedures, is the role of DNA amount and repeat DNA analysis in the plant kingdom. The basic fact that DNA is the genetic material presupposes the assumption that increase in complexity is associated with increase in DNA content during evolution. In nature, such a correlation has not been noticed.

Evidence so far gathered indicates that both increase and decrease in amount of DNA have taken place in evolution (vide Bennett and Leitch, 1995).

The DNA value of the entire haploid genome is termed as the "C" value. In the plant system the DNA values range from approximately 1 pg of DNA in 2C to almost 600 pgs in the 4 C nucleus.

It has been noted that in most cases, where large variations of DNA exist, this difference may be attributed to the repeated sequences which exist in large amounts (Flavell, 1980). Only a fraction of the total amount of DNA codes for structural proteins whereas the rest is repeats—non-coding, or coding for non-specific effects. Such a vast amount of non-coding sequences is present in several families of flowering plants and forms a significant portion of the total amount of DNA.

This huge amount, which has been conserved for millions of years, contributes to a great extent to the genotypic differences between the species and varieties as well (vide Rao and Sharma, 1987).

Precise significance of such sequences is not fully known (Charlesworth *et al.*, 1994). But their preponderance in some of the groups is distinct. The DNA content of temperate species in much higher than that of tropical species. In gymnosperms of the temperate regions, the amount is rather high. Similarly, temperate herbaceous perennials show high DNA.

In cereals in general, the DNA content is comparatively low, rice having very low amount of DNA (0.6 pg) while rye has the highest (8.60 pg). Despite low DNA content, the value of repetitive DNA is quite high in cereals which is almost 50% in rice, 68% in maize, 75% in rye and 50% in Triticale genomes.

There are certain parameters of growth and metabolism with which DNA content seems to be associated. High DNA is often associated with slow development. There is almost a positive correlation between nuclear dry mass, minimum cell cycle time, meiotic duration, pollen maturation time and mean generation time.

In extreme climatic situations such as alpine himalayas, there is almost complete absence of flowering plants with high DNA values. Species with heavy amount of DNA cannot be established from seed in the short growing season in such extreme conditions. In high altitudes, selection of species with low DNA value has thus been favoured because of the minimum generation time required for such environments.

The estimation of DNA, *vis-à-vis* repeated sequences, can be advantageously employed in determining the nature of the species. As high amount of DNA and slow mitotic cycle do not permit fast growth, these two parameters are used for screening of genotypes capable of fast growth—the essential requirements for energy plants. It has been extensively worked out in several species of *Acacia* that large amount of DNA is related to slow mitotic cycle and slow growth. Similarly, plants growing under stress conditions, of arid and semi-arid regions, do not have a high DNA content even after attaining polyploidy for increased tolerance (Mukherjee and Sharma, 1993).

Despite the continued evolution by amplification, the conserved nature of some of the repeated sequences is manifested in the structural features of genes as well as ribosomal RNA. The rRNA genes show considerable sequence homology between distantly related eukaryotes with certain regions within the genes showing virtually no divergence. Simultaneously, there are loci in rDNA which are polymorphic within a species.

In species of *Lathyrus* at the intraspecific level, the values reveal conservation of a different type. The strains of *Lathyrus sativus* do not show significant differences in DNA amount. The estimate of the total repeat DNA too does not reveal any marked difference. However, the analysis of repeats, on the basis of degree of reassociation kinetics, indicates certain genetic differences between different strains.

All the study of reassociation kinetics indicates that the strains have 5–10% of very fast repeats which undergo rapid renaturation (Chaudhury *et al.*, 1986). The above percentage of repeats does not show any interstrain differences in amount and thus appears to be highly conserved. The percentage of moderate and minor repeats on the other hand, varies between one strain and another, the total amount of repeats remaining constant in all. The constancy in a fraction of the repeats and variability in others may indicate the mechanism of maintaining hereditary stability of genotypes, keeping enough scope for augmenting genetic diversity. The analysis of repeated sequences provides with an important method for the detection of genetic polymorphism at the molecular level.

The repeated DNA sequences are additional sequences which have the property of amplification and a significant fraction of which are highly conserved. Such sequences are present throughout the chromosome, both at intra and intergenic sites. Their presence in identical sequences in different loci suggests that the dispersion is due to insertion. In a number of species, they have the property of mobility and are capable of insertion at different loci.

In view of their dynamic property, the term **"Dynamic DNA"** has also been attributed to such repeats (Sharma, 1983). As they do not necessarily contain structural genes responsible for qualitative characters and are present in multiple copies, the manipulation at such sites may not lead to any deleterious effect on the organism. In fact, these loci are sites for restriction enzyme operation and gene transfer. A clear understanding of the function and location of these conserved sequences, as well as their relationship with neighbouring genes, would lead to their better utilization in technology of gene transfer and unravelling of the rationale of the high DNA content in the plant system.

CHAPTER II.4

COMMON PROTOCOLS FOR ANALYSIS OF CHROMOSOMES AT STRUCTURAL LEVEL

Some sample schedules for the study of chromosomes in different tissues of plants are described. The principles and reagents have been given in an earlier chapter (II.1).

I. MITOTIC CHROMOSOMES

From root tips

Schedule A. From paraffin sections dehydrated through alcohol chloroform grades and stained in crystal violet

(i) *Fixation*: Cut fresh root tips e.g. of *Allium cepa* about 1 cm long. Wash away dirt particles. Transfer the root tips to a tube containing a mixture of 1% chromic acid and 10% formalin. Keep for 12–24 h.

(ii) *Washing*: Wash the roots in a porcelain thimble in running water for 3 h.

(iii) *Dehydration*: Transfer the roots to a glass phial containing 20% ethanol and keep for 1 h; then to 50% ethanol, keeping for 1 h; to 70% ethanol, treating overnight; through 80, 90 and 95% ethanol, keeping the roots for 1 h in each. Finally keep overnight in absolute ethanol.

(iv) *Clearing*: Keep in ethanol–chloroform mixtures (3:1, 1:1 and 1:3) successively for 1 h in each. Transfer to pure chloroform and add small shavings of paraffin.

(v) *Infiltration*: Keep overnight on a hot plate at 35°C. Remove the stopper, add a little more wax and keep the phial with contents at 45°C for 2 days and then transfer to 60°C. Change the wax with fresh molten wax at intervals of 30 min for 2 h.

(vi) *Embedding*: Pour the molten paraffin with roots into a paper tray and add some more molten wax, then orient the roots in groups of three with their tips pointing to the same side at the same level. After the wax has cooled slightly, plunge the block into cold water.

(vii) *Section-cutting*: Trim the block and cut transverse sections of the root tips 14 μm thick on the microtome. Cut the ribbons into suitable segments and mount serially in water on a slide previously coated with Mayer's adhesive. For coating a slide, put a tiny drop of Mayer's adhesive, on the slide and smear it three-quarters over the surface, then place the slide on a hot plate and help the ribbons to stretch with a pair of needles. Drain off water and keep the slide overnight on the hot plate to dry.

(viii) *Bringing to water*: Transfer the slide with sections to pure xylol grades I and II, keeping in each for 30 min; then transfer to a jar of ethanol : xylol (1 : 1) and keep for 15 min; pass through absolute ethanol, 95, 90, 80, 70, 50 and 30% ethanol, keeping in the first 3 for 10 min each, and 5 min in each of the rest, and then transfer to water.

(ix) *Pre-mordanting*: Keep the slide in 1% aqueous chromic acid solution overnight. Wash in running water for 3 h.

(x) *Staining*: Stain in 0.5% aqueous crystal violet solution for 20 min. Rinse in water.

(xi) *Mordanting*: Keep in 1% iodine and 1% KI mixture in 80% ethanol for 30–45 s, then dip in absolute ethanol for 2 s.

(xii) *Dehydration*: Pass through three successive grades of absolute ethanol, keeping in each for 2 s.

(xiii) *Differentiation*: Transfer to clove oil I, then differentiate under the microscope after keeping in clove oil II for 2 min.

(xiv) *Clearing*: Transfer to xylol grade I and keep for 30 min; pass through pure xylol II and III, keeping for 1 h and 30 min respectively.

(xv) *Mounting*: From xylol III, mount in Canada balsam under a coverglass. Allow to dry overnight on the hot plate.

Alternatives Any one mixture of the list of fixatives given in the chapter on fixatives (II.1) can be used for fixation depending on the material. *n*-Butyl alcohol can be used instead of chloroform in the clearing process.

For materials which stain easily, pre-mordanting in 1% chromic acid solution can be omitted; for materials difficult to stain, Navashin's fluid A can be used in pre-mordanting.

For *mordanting* Gram's aqueous solution of potassium iodide and iodine can be used instead of KI and I_2 in 80% ethanol. For materials in which the background retains stain even after mordanting in KI and I_2 solution, keep the slide in 1% chromic acid solution after step (xi) and before step (xii) for 15 s.

For high accuracy in staining, after dehydration in absolute ethanol, transfer the slides to very thin terpineol, keep for 1–2 min, then rinse in xylol and transfer to clove oil. Differentiate as usual.

Schedule B. From paraffin preparations dehydrated by rapid dioxane method and stained in crystal violet, suitable only for root tips
Fixation and washing are as in the previous schedule.

(iii) *Dehydration*: Transfer the roots through aqueous solutions of dioxane, 25 and 75%, keeping 2 h in each. Keep in pure dioxane overnight.

(v) *Infiltration*: Transfer to 60°C, adding paraffin of low melting point at intervals of 30 min for 4 h, then add pure molten wax and keep the roots in it for 2 h before embedding.

The remaining steps are the same as the previous schedule.

Schedule C. From paraffin sections stained in haematoxylin
The initial stages, from fixation to block preparation and bringing the sections down to water, are the same as followed in Schedule A, (i) to (viii).

In an older method Mordant the slide in alum (3% ferri-ammonium sulphate) solution for at least 3 h. Rinse thoroughly in water. Stain for 24 h in 0.5% aqueous haematoxylin solution. Rinse in water. Differentiate and de-stain in alum for 5 min or more. Rinse for 15 min in running water. Dehydrate by passing through ethanol series 50, 70, 90, 95% and absolute, keeping for 5 min in each. Pass through ethanol–xylol (1:1) and pure xylol I, keeping in the former for 15 min and in the latter for 1 h. Mount in Canada balsam.

In a more rapid technique Mordant in 1% alum solution for 10–20 min. Rinse in running water for 10 min. Stain in 0.5% haematoxylin (ripened) solution for 5–15 min. Wash in water and de-stain for 5–20 min in saturated aqueous picric acid. Keep for 1 min in a jar containing water with 1 or 2 drops of 0.88% ammonia. Rinse in running water for 30 min. Dehydrate through ethanol grades 20, 60, 80% and absolute. The remaining steps are similar to the previous method.

Schedule D. From paraffin sections stained in basic fuchsin
Reagent a: Basic fuchsin in 10% ethanol 0.05 ml; absolute ethanol 5 ml; phenol crystals 3 g; distilled water 95 ml. Dissolve the phenol in distilled water and add fuchsin solution and ethanol.

Reagent b: 10% aq. potassium metabisulphite soln 2 vol; 1 N aq. sulphuric acid soln 1–2 vol.

Bring the sections (originally fixed in alcoholic or formalin-mixed fixatives) down to distilled water. Hydrolyse in 1 N HCl for 5 min at room temperature and then in 1 N HCl at 60°C for 15 min. Immerse in 1 N HCl at room temperature. Wash in water. Stain for 2–3 min in reagent *a*. Transfer immediately, without washing, to reagent *b* and keep for 5 min. Immerse in a second lot of reagent *b* for another 5 min. Rinse in water. Immerse in 1 N H_2SO_4 in 96% ethanol, for 3–5 min. Wash thoroughly in water. Dehydrate through ethanol and xylol grades and mount in Canada balsam. DNA takes up magenta colour.

Schedule E. From pre-treated squash preparations

(*a*) Stained in Feulgen
Pre-treat fresh root tips (e.g. *Hemerocallis fulva*) in 0.02 oxyquinoline solution at 10–12°C for 3 h.

Fix in acetic–ethanol (1 : 1) mixture for 1 h.

Washing: Rinse in distilled water.

Hydrolysis: Hydrolyse the root tips in N HCl at 60°C for 12 min.

Washing: Rinse in water.

Staining: Transfer root tips to leuco-basic fuchsin solution and keep in it for 30 min to 1 h till the tips are magenta-coloured.

Squashing: Transfer each tip to a drop of 45% acetic acid on a slide, cut out the tip region and discard the other tissue. Place a coverglass over the tip and squash, applying uniform pressure with the thumb on a piece of blotting paper placed on the whole.

Ring with paraffin wax and observe.

Making permanent: Invert the slide in a closed tray containing glacial acetic acid–ethanol (3 : 1) mixture. Following detachment of coverglass, pass both slide with material and coverglass through acetic–ethanol (1 : 1, 1 : 3) mixtures, pure ethanol, ethanol–xylol (1 : 1) mixture and xylol I and II, keeping 5 min in each. Mount in Canada Balsam.

Other pre-treatment chemicals can be used, like *p*-dichlorobenzene, acenaphthene, coumarin, aesculine, isopsoralene, etc., for different materials. For difficult materials like root tips of sugarcane, pre-treat in sat. α-bromonaphthalene solution in 0.05% saponin for 3 h in cold, fix in acetic–ethanol (1 : 3) for 48–72 h, hydrolyse in N HCl at 60°C for 7–8 min, treat with 3% pectinase solution in pH 2.6 acetate buffer for 60–90 min and stain according to the Feulgen technique.

For cereals, soak root tips for 1–2 h at 18–20°C in aqueous α-bromonaphthalene; 1–2 h in water, and 0.5–1 h at 10–14°C in a mixture of 1% chromic acid 5 ml, 2% osmic acid 1 ml, and aqueous 0.002 M OQ 1 ml. Treat successively in water for 1–2 min, 1% sulphuric acid soln for 10–15 min, water for 1–2 min, 1% chromic acid solution for 30 min or 1 h, and squash in acetic–carmine.

Alternative method Treat in aqueous *a*-bromonaphthalene followed by successive washing in water and 22% acetic acid, fix in N HCl : 22% acetic acid mixture (1 : 12), prior to usual hydrolysis and Feulgen staining for members of Triticinae.

For grasses like *Briza*, pre-treat in sat. *p*-dichlorobenzene solution for 18 to 20 h at 4°C. Fix in acetic–ethanol (1 : 3), treat with aqueous pectinase for 2 h at room temperature; hydrolyse at 60°C in N HCl for 8 min and stain in Feulgen for 2 h. For grasses, use conc. HCl to separate cells, stain in haematoxylin, squash in 0.5% aceto-carmine.

(b) Stained in acetic–orcein solution

(i) *Pre-treatment* Treat fresh root tips (e.g. *Aloe vera*) in saturated aqueous solution of *p*-dichlorobenzene for 2.5–3 h at 12–14°C.

(ii) *Fixation* Transfer to glacial acetic acid–ethanol mixture (1:2) and keep for 30 min to 2 h, followed by treatment in 45% acetic acid for 15 min.

(iii) *Staining* Transfer the root tips to 2% acetic–orcein solution and N HCl mixture in the proportion of 9:1, and heat gently over a flame for 5–10 s, taking care that the liquid does not boil.

(iv) *Squashing* Lift a root tip from the mixture, place it in a drop of 1% acetic–orcein solution on a slide. Cut out the tip, place a coverglass on it and squash by applying uniform pressure with the thumb through a piece of blotting paper.

(v) *Ring* the coverglass with paraffin and observe.

(vi) *For making permanent*, invert the slide with coverglass after squashing in a covered tray containing tertiary butyl alcohol and keep till coverglass is detached. Mount slide and coverglass separately in euparal.

Alternative pre-treatment chemicals can be used like OQ, coumarin, aesculine, α-bromonaphthalene, etc.

2% propionic–orcein, 1% propionic acid and propionic–ethanol can be used instead of 2% acetic–orcein, 1% acetic–orcein and acetic–ethanol solutions in the respective steps. Treatment in 45% acetic acid in step (ii) is optional.

To intensify stain add a drop of aqueous ferric chloride solution to 2% acetic–orcein–N HCl mixture or keep the tissue in the staining mixture for a period extending up to 12 h and then mount in 45% acetic acid.

(*c*) *Stained in acetic–lacmoid solution* Steps (i) and (ii) are same as for the acetic–orcein schedule.

(iii) Stain the tissue by transferring the root tips to a glass phial containing 10 ml standard acetic–lacmoid solution and 1 ml N HCl, and heat for 5–10 s over a flame, taking care not to boil the fluid, then leave for 10 min.

(iv) Transfer a root tip to a drop of standard acetic–lacmoid solution on a slide and squash as usual as in previous schedules. The next steps are similar to the acetic–orcein schedule.

Schedule F. From macerated preparations
The steps for pretreatment and fixation are similar to earlier schedules, the fixative used being acetic–methanol (1:3). Digest the fixed root tips in a mixture of 2.5% cellulase and 2.5% pectinase (1:1) at 25°C for 5–10 h. Wash and keep in distilled water for hypotonic treatment. Keep root tip on a slide, add a few drops of fresh fixative and smash with tweezers into bits. Prepare cell suspension in fresh fixative. Add a few drops to a clean chilled slide, allow to flow down the slide and then flame dry. Dry in air.

From leaf tips

Fixation for paraffin blocks is not effective. Usually squashes made after pre-treatment yield the best results. A sample schedule is given below (Sharma and Mookerjea, 1955):

(i) Dissect out very young leaf tips of *Cestrum nocturnum*, wash in water and immerse in saturated aqueous solution of aesculine and keep at 12–14°C for 15 min to 24 h.

(ii) Fix in acetic–ethanol (1:1) mixture for at least 3 h, the period being extended, if necessary, upto 24 h.

(iii) Transfer the tips to a mixture of 2% acetic–orcein solution and N HCl (9:1), heat over a flame for 3–4 s, then leave in the mixture at 30°C for 30 min.

(iv) Squash the tips on a dry slide in a drop of 1% acetic–orcein solution with a coverglass applying uniform pressure with the help of blotting paper.

(v) Making permanent is similar to root tip smears.

Alternative pre-treatment chemicals for root tips can be used here. The period of fixation in acetic–ethanol should be increased, if necessary, to remove chlorophyll completely. Acetic–ethanol mixture (1:2) or (1:3), chilled 80% ethanol or acidulated ethanol (conc. HCl–95% ethanol 1:3) can be used instead of acetic–ethanol (1:1) mixture.

Modifications

(i) Pre-treat in saturated solution of α-bromonaphthalene in 0.05% saponin for 3 h in cold for good results with grass leaf chromosomes. Expose young leaf, sever top of the shoot and cut slits into the remaining tube for penetration.

(ii) For *Saccharum* leaf, pre-treat with 0.2% colchicine at room temperature for 2 h.

(iii) For tea leaf, pre-treat in aqueous saturated *p*-dichlorobenzene (2–3 h) at 4–10°C, fix for 6–12 h in a mixture of propionic acid, chloroform and ethanol (1:3:6), stain with 2% propionic–orcein at 80°C and squash in 1% propionic-orcein.

(iv) For grass leaf, soak longitudinal sections of leaf shoots for 2–4 h in aqueous 0.002 M OQ at 25°C, blot and fix in ethanol, chloroform and acetic acid mixture (3:4:1). Macerate at 45°C for 30 min in pectinase solution before staining and squashing.

(v) Rice chromosomes need longer fixation, elimination of hydrolysis and phase contrast microscopy.

(vi) Keep for 2–3 h at 23–27°C in 2% aceto-orcein:N HCl mixture (9:1), followed by heating to 90°C in the same mixture for 8 s for better preparations in *Crinum*.

From pollen grains

The pollen grains of Angiosperms are utilised for studying (a) mitotic division in pollen grains; (b) fertility of the grains, and (c) mitotic division in the generative cell inside the growing pollen tube.

For studying the *mitotic division* in pollen grains

(i) Dissect out an anther from a flower bud of suitable size (e.g. *Nothoscordum fragrans*), put in a drop of 1% acetic–carmine solution on a dry slide and smear the anther with a clean scalpel; cover with a coverglass and observe. In flower buds of a suitable size, mitotic division is observed in the pollen grains.

(ii) Dissect out the remaining anthers from the flower bud in which pollen grain division was observed. Place the anthers on a clean dry slide, cut off the edges with a clean scalpel, squeeze out the inner fluid and reject the empty lobes, then smear the fluid with a clean scalpel.

(iii) Immediately invert the slides in a covered tray containing Navashin's A and B fixatives, mixed in the proportion 1 : 1 and keep overnight.

(iv) Wash the slides in running water for 3 h, then stain in 0.5% aqueous crystal violet solution for 30 min and rinse in water.

(v) The subsequent steps, namely *mordanting*, *dehydration*, *differentiation*, *clearing* and *mounting* are similar to the corresponding steps (xi, xii, xiii, xiv and xv) of the technique followed in staining root tip sections cut from paraffin blocks (see Schedule A).

Modifications

(i) For wheat pollen, treat anthers at 18–20°C in 0.5% aqueous colchicine or 0.002 M aq. oxyquinoline solutions at 10–14°C for 1 h, fix in Carnoy's fluid for 6 h, wash, hydrolyse in N HCl, stain in leucobasic fuchsin and smear in 1% acetic-carmine. For *Saccharum* and related genera, pre-treat in 0.5% aqueous colchicine for 1 h, wash, treat in 0.002 M aqueous OQ for 1 h, wash, fix in a mixture of methanol 60 ml; chloroform 30 ml; water 20 ml; picric acid 1 g and mercuric chloride 1 g, for 24 h. The remaining schedule is similar to that adopted for wheat.

(ii) For studying chromosomes from herbarium sheets of *Impatiens*, soak anthers overnight in a saturated solution of iron acetate in 45% acetic acid, rinse and smear in dilute acetic–carmine. Heat several times to boiling and seal.

The slides can be made permanent by ethanol vapour or tertiary butyl alcohol techniques. If the stain taken by the chromosomes is not satisfactory, pre-mordanting in 1% chromic acid solution, as in the case of root tip sections, described before, is followed.

In order to *study the apparent fertility of pollen grains* two sets of methods are available: based on staining the contents of the grains and on the germination of the pollen tube.

The first method involves *staining the grains with different dyes* and counting the percentage of empty and coloured grains. A number of staining media are available. A very convenient method is to stain the pollen grains in Müntzing's mixture of glycerol and 1% acetic–carmine solution for some hours. The filled grains take up the stain while the empty ones do not. The acetic acid evaporates, but glycerol remains. By ringing with paraffin, the preparation can be stored for a long time. Another method is staining with Owzarzak's methyl green phloxine. The medium preserves the slides for several months. The walls of all the grains take colour but only the filled grains take up both colours.

These methods show the frequency of filled pollen grains, but only suggest their fertility percentage, since while the empty grains are certainly sterile, the filled grains are not necessarily able to germinate.

In generative cells inside pollen tubes:
For the study of pollen tube mitosis the practice is to study the mitotic division of the generative nucleus into the two sperm nuclei, which usually takes place between 2 and 48 h after germination of the pollen tube. The methods generally include two main stages, (a) germination of the pollen tube, and (b) study of the chromosome structure. The chief factors necessary for the artificial germination of pollen tubes are temperature, humidity and culture media.

Hanging drop culture method in a moist chamber:

(i) *Pollen grain culture*: fit a ring on a slide and smear both rims of the ring with vaseline so that it is attached to the slide at one end. Place a drop of 3% sugar solution on a clean coverglass, then dust pollen grains from an opened flower e.g. of *Papaver* sp. into the solution. Invert the coverglass with the drop of sugar solution on it and attach it to the other vaselined rim of the ring, so that the drop with pollen grains hangs in the closed chamber enclosed by the ring. Note growth of the pollen tubes in the ringed slide under the microscope. After about 3 h remove a few pollen tubes for observation at intervals of 1 h till the optimum time is reached.

(ii) *Treatment for chromosome study*: Lift out the pollen tubes with a brush, put in a drop of 1% acetic–orcein solution on a clean slide, warm slightly and squash as usual under a coverglass, applying uniform pressure.

(iii) *Permanent preparation*: The slide can be made permanent by the alcohol vapour technique.

For *accumulating metaphase*, add 0.05% colchicine to the sugar solution. For controlling humidity of the hanging drop chamber, add a drop of water on the slide or place a drop of sugar solution beside the hanging drop.

Coated slide technique

(i) Weight out 12 g of lactone, 1.5 g of agar and 0.01 g of colchicine. Heat the lactone and agar with 100 ml distilled water in a double boiler until the agar is dissolved, then add colchicine when the medium has cooled to 80–60°C. Keep the medium at about 60–70°C.

(ii) Coat slides cleaned in ethanol with a thin layer of egg white. Dip in a beaker containing the medium at 60°C until it warms up, then withdraw the slide, drain off the medium, and wipe the back with a piece of clean cloth.

(iii) When the medium has set, but is not dry, pick up the pollen from a mature flower (e.g. of *Tradescantia virginiana*) with a brush and dust a thin film on the medium. Immediately place the slide sown with pollen in a moist growing box, or a horizontal glass staining dish lined with damp blotting paper on the two sides and top, and keep the temperature between 20 and 25°C. Observe at intervals till the optimum period for maximum division is found.

(iv) Add a drop of 1% acetic–carmine solution to the pollen tube, squash under a coverglass and observe.

(v) Make permanent by the ethanol vapour technique.

Modifications

(i) Stain the slides by the Feulgen schedule after 12 min hydrolysis in N HCl at 60°C. Handle carefully, as the medium may be washed off.

(ii) For species with binucleate pollen, germinate in a medium containing H_3BO_3, 0.01 g; $Ca(NO_3)_2 \cdot 4H_2O$, 0.03 g; $MgSO_4 \cdot 7H_2O$, 0.02 g; KNO_3, 0.01 g; sucrose, 10 g; water, 100 ml, on a slide resting on moist filter paper in a closed petri dish for 24 h. Place crystals of acenaphthene on filter paper. After 24 h, squash pollen tubes in propionic–orcein solution.

(iii) For convenient handling of germinated pollen, grow pollen on an autoclaved membrane filter (Millipore AA WP 025 00) in contact with a sterilized medium which contains agar 0.5–1%, sucrose 0.1–0.5% and boric acid 0.01% for 2 h to overnight at 2–4°C on a filter paper with a mixture of OsO_4, 1 g; CrO_3, 1.66 g and water, 233 ml. For *Persica* pollen add 10% acetic acid. Wash in water, bleach in a mixture of 3% hydrogen peroxide and saturated aqueous ammonium oxalate solution on filter paper. Hydrolyse with 4 N HCl for 18 min at room temperature, stain in Feulgen, wash in three changes of 2% $K_2S_2O_5$ at pH 2.3 (KH_2PO_4, 1.4 g; conc. HCl, 0.35 ml; distilled water, 100 ml) by placing membrane on filter paper wet with the respective fluids. Transfer pollen to glacial acetic acid, squash and process.

(iv) Sow *Tradescantia* pollen on lactose–agar medium at 38–39°C for 16 h. Fix slides in acetic–ethanol (1 : 3) for 1–3 h, hydrolyse in N HCl at 60°C, treat with water at 65°C. Delaminate upper layer of medium in cold water. Flatten and fix remaining single layer of pollen tubes to the slide by pressing under a coverglass by quick freeze technique, stain in Feulgen and mount.

(v) For palm chromosomes, sow pollen in a medium containing H_2BO_3, 100 ppm; colchicine, 0.04%; lactose, 12%; gelatin, 5% and egg albumen, 1 drop in 10 ml.

Collodion membrane technique

Mix 1 part of collodion (necol collodion solution from BDH, England) with three parts acetone. Prepare 3% aqueous sugar solution in a petri dish and warm and put a drop of collodion–acetone mixture on it. This drop rapidly spreads out in a thin film over the surface and the film hardens into a thin membrane as acetone evaporates. Cover the open dishes lightly by filter paper and leave in a warm dry place for 3 h till acetone evaporates completely.

Dust the pollen with a brush or directly from the anther onto the smooth areas of the floating membrane towards the centre, replace the lid on the petri dish and transfer it to an incubator.

After the pollen tubes have germinated, with a pair of scissors cut out a piece of the membrane, about 1 cm in diameter, with the pollen tubes. Lift with a needle onto a clean dry slide, add a drop of 1% acetic–orcein solution and squash under an albuminized coverglass, applying uniform pressure. Ring with paraffin.

Invert the slide, after wiping off the paraffin with xylol, in a covered tray containing acetic–ethanol mixture (1 : 3). After the coverglass is detached with the membrane run it through two changes of absolute ethanol–xylol (1 : 1) mixture and pure xylol, keeping for 10 min in each. Mount the coverglass, pollen side down, in a drop of Canada balsam on a clean slide.

During Feulgen staining as an alternative method the earlier steps are similar, up to the germination of pollen.

Float the cut membrane with pollen tubes on a slide and blot off excess solution. Add a drop of acetic–ethanol mixture (1 : 3) and cover with an albuminized coverglass. Press gently and invert in a tray containing acetic–ethanol mixture. After 1.5 h, transfer coverglass with pollen tubes to

95% ethanol and keep overnight. Bring down to water through 80, 70, and 30% ethanol; keep 2 min in each, rinse and hydrolyse for 12 min at 60°C in N HCl. Rinse in water, stain in leuco-basic fuchsin solution for 30 min and squash in 45% acetic acid on a clean slide.

For making permanent, the usual techniques described before can be followed.

Metaphase arrest technique completely excludes the use of nutrient medium in the germination of the pollen tube. Line both bottom and cover of a pair of petri dishes with well-moistened filter paper and place a clean slide in the petri dish and dust on it pollen grains from a newly opened flower (e.g. *Tradescantia virginiana*). Spread 50–100 mg of fine acenaphthene crystals on the filter paper close to the slide. The fumes disturb the spindle to give contracted chromosomes. Cover and keep at 20–22°C for 24 h. Add a drop of 1% acetic-carmine solution to the pollen tubes on the slide and squash under a coverglass.

Floating cellophane method is based on growing pollen grains on a square of cellophane paper (about 2 cm × 2 cm) which is floated in sugar solution in a petri dish. The pollen is dusted on to the upper surface of cellophane paper. The petri dish is covered and kept at 20°C.

The germinated tubes can be stained in a staining-cum-fixing mixture and observed; otherwise they can be fixed in acetic–ethanol mixture for 2–24 h, hydrolyzed in N HCl for 6 min and stained in fuchsin sulphurous acid solution.

From endosperm

For the study of endosperm chromosomes, different methods of pre-treatment and staining are available which are more or less similar to those followed for other mitotic chromosomes. Methods are available for the study of endosperm chromosomes in the living state, clarification of their birefringent property, the effect of chemicals as well as their cinemicrographic analysis.

Feulgen squash method

(i) *Fixation*: dissect out very young developing seeds and fix in acetic–ethanol mixture (1:2) for 1–2 h. Keep overnight in 95% ethanol.
(ii) *Washing*: run the seeds through 70, 50 and 30% ethanol, keeping for 10 min in each, then wash in running water in a porcelain thimble for 10 min.
(iii) *Hydrolysis*: hydrolyse in N HCl at 60°C for 8–12 min.
(iv) *Staining*: rinse in water and stain in leucobasic fuchsin solution for 2 h, then wash for 10 min in two changes of tap water.
(v) *Mounting*: dissect out the endosperm on a clean dry slide in a drop of 45% acetic acid solution, under a dissecting microscope using tungsten

needles pointed in molten $NaNO_2$. Using Mayer's adhesive, film a coverglass and dry it by passing over a flame, then squash the dissected endosperm under the coverglass, exerting strong but uniform pressure under a piece of blotting paper.

(vi) *Make permanent* following any one of the schedules described earlier.

Modifications: for intensifying the stain, 1% acetic–orcein solution can be used instead of 45% acetic acid as the mounting medium.

Acetic–orcein squash method

Dissect out the very young developing seeds of *Cestrum nocturnum* under a dissecting microscope and place immediately in a saturated solution of aesculine; keep at 10–12°C for 3 h. Fix in acetic–ethanol mixture (1:1) at room temperature for 2 h.

Heat in a mixture of 2% acetic–orcein solution and N HCl (9:1) over a flame for 9 or 10 s, removing the tube at intervals so that the fluid does not boil; keep for 30 min.

Transfer each seed to a clean slide in a drop of 1% acetic–orcein solution, cut it into two or three pieces with a scalpel and keep the pieces at a little distance from each other. Squash the whole under a long coverglass, exerting strong and uniform pressure and ring with paraffin.

Make permanent following any one of the methods described previously.

The seeds selected should be as young as possible, preferably taken within a week of fertilization. If very small, the entire seed should be squashed, but if the seeds are comparatively large, the endosperm is dissected out and then squashed. The endosperm can be excised, placed directly in acetic–carmine, macerated and brought to boiling and the nuclei and mitotic figures separated by centrifugation.

Permanent preparation of endosperm cells flattened in the living stage

Select a suitable ovule (*Haemanthus katherinae*). Take out the contents of the embryo sac. Press the contents on to a coverglass smeared with a thin layer of a mixture of 0.5% agar, 0.5% gelatin and 3.5% glucose. Surface tension aids in flattening the cells. Keep in a moist chamber so as to avoid drying before fixation.

Fix coverglass with the material in chrome–acetic–formalin fixative, diluted in equal parts with distilled water for at least 12 h. Rinse in water and treat in a mixture of decinormal potassium cyanide and 2% magnesium sulphate mixed in equal parts for 1 h. Wash off the cyanide by several changes in water. Hydrolyse the coverglass with material for 4 h at room temperature in a mixture of rectified spirit and concentrated hydrochloric acid (3:1). Dip in water and stain in Feulgen solution for 12 h. Wash in sulphur dioxide water, dehydrate as usual, pass through xylol and mount in balsam.

Study of birefringence in endosperm mitosis
With the same material follow the first step as above without using gelatin and use coverglass originally ringed with vaseline paraffin. Place the cover-glass with the material on another glucose agar coated coverglass and seal with ringed vaseline-paraffin. After flattening of cells, tilt the preparation to allow excess liquid to drain onto the lower coverglass. Blot.

By proper adjustment, flatten the endosperm cells to obtain the chromosomes in one plane. Break the liquid contact between the top and bottom coverglasses and stabilise the preparation.

Study birefringence in a special polarising microscope with non-rectified coated stain-free 25×0.65 n.a. Leitz oil immersion objective in conjunction with 10×0.25 n.a. American optical coated objective as condenser. Pure green light (546 nm) from a high-pressure mercury arc lamp can be isolated with a multilayer high transmission interference filter.

Polytene chromosomes from suspensor cells

Endomitotic replication of chromosome is common in certain organs of plants. In plant systems, polytenic constitution of chromosomes has been reported in suspensor cells of certain legumes like *Psophocarpus tetragonolobus* L. DC.

Dissect out the ovules under a dissecting microscope. Halve each ovule longitudinally and take out the suspensor stalk attached to the embryo very carefully. Fix the tissue in acetic acid–ethanol (1 : 2) overnight and then keep in 45% acetic acid at room temperature for 2–3 h to separate the suspensor from surrounding tissue. Take out the suspensor, squash in 2% acetic-carmine, ring with paraffin and observe.

Alternatively, hydrolyse the fixed tissues in N HCl and stain following Feulgen reaction schedule.

In the polytenic chromosomes, the presence of clear bands and interbands as well as coiling and uncoiling cycles can be observed.

II. MEIOTIC CHROMOSOMES

Meiotic chromosomes are studied usually from pollen mother cells and occasionally from embryosac mother cells.

From pollen mother cells

Temporary squash technique

(i) Take flower buds (e.g. of *Solanum torvum*) serially from an inflorescence, starting from the smallest and working up to the largest, until the correct bud having divisional stages is found.

(ii) Dissect out a single anther from a bud with a needle. Place it on a clean slide.
(iii) With a clean scalpel, smear the entire anther on the slide and add a drop of 1% iron–acetic–carmine solution to it immediately. Remove the debris.
(iv) Heat slightly over a flame. Cover with a coverglass and ring with paraffin.
(v) Instead of fresh anthers, anthers fixed in acetic–ethanol (1 : 1) mixture or in Carnoy's fluid, and later stored in 70% ethanol can also be observed. If stored in 70% ethanol, keep 1 h in each of acetic–ethanol (1 : 1) and 45% acetic acid solutions before smearing in 1% acetic–carmine solution. For materials taking bright stain, treatment in acetic–ethanol can be omitted.

Instead of iron-acetic-carmine solution, 1% acetic–carmine solution can be used and a trace of iron added by rubbing a rusty needle in the drop of stain on the slide. 1% acetic–carmine solution can be replaced by acetic–orcein, acetic–lacmoid, nigrosine solutions.

The slides can be kept as such in a refrigerator for a few weeks and then made permanent.

or

Fix very small buds in tight inflorescences in acetic–ethanol mixture or in Carnoy's fluid for 1 h, hydrolyse at 60°C for 5–10 min in N HCl, rinse in water and stain for 1–3 h in leuco-basic fuchsin solution. Rinse in two or three changes of 45% acetic acid solution. Dissect out single anther; squash in 45% acetic acid and make permanent as usual.

or

Macerate in a mixture of 15% chromic acid, 10% nitric acid, 5% HCl (2 : 1 : 1) for 5–7 min and harden in ethanol–propionic acid (1 : 1), between fixation and staining, for *Gossypium* microspores. Restore deteriorated temporary acetic–carmine preparations by replacing the acetic–carmine under the coverglass first with 2 N HCl and then with 1% acetic–carmine.

Permanent smear technique
Take flower buds (e.g. *Datura fastuosa*) of different sizes. Dissect out a single anther from each bud and observe by squashing in 1% iron–acetic–carmine solution as given in the last technique until meiotic divisional figures are observed.

Dissect out the remaining anthers of the bud showing division and place each on a clean slide, cut off one end with a clean scalpel and squeeze out the contents by pressing with the left thumb. Discard the empty anther lobe. Quickly draw the fluid into a thin smear on the slide with a clean scalpel and immerse immediately in a tray containing Navashin's fluids A and B, freshly mixed in equal proportion. Keep in the fixative for 3–12 h.

Wash the slide in running water for 1 h. Stain in 0.5% aqueous crystal violet solution for 20 min or more and rinse in water.

Mordant in 1% solution of iodine and KI in 80% ethanol for 45 s. Dehydrate by passing the slide through absolute ethanol grades I, II and III, keeping about 2–3 s in each.

Transfer the slide to clove oil I, keep for 2–3 min. Take out the slide and observe the staining under a microscope. If satisfactory, transfer the slide to clove oil II and keep for 2–3 min. Pass the slide through xylol grades I, II and III for 30 min, 1 h and 30 min respectively. Mount in Canada balsam or clarite X.

Premordant materials difficult to stain overnight in 1% chromic acid solution after fixing and washing. Wash in running water for 3 h before staining.

In an alternative method, after smearing on a clean slide, immerse the material in acetic–ethanol mixture and keep for 1 h. Treat the slide with material in 45% acetic acid for 15 min and hydrolyse in N HCl at 60°C for 15 min. Finally, rinse in water, stain in leuco-basic fuchsin solution for 1–2 h and mount with 45% acetic acid.

Permanent paraffin section technique

(i) *Determination of size*: test single anthers of individual flower buds (e.g. *Allium cepa*) until a bud of suitable size with divisional figures is obtained. Collect several buds of approximately the same size.

(ii) *Fixation*: dip each bud, holding it with a pair of fine forceps, in Carnoy's fluid and keep for 2–3 s. Rinse in water thoroughly, then transfer to a phial containing a mixture of Navashin's A and B fluids (1:1) and keep for 24 h.

(iii–vii) *Washing, dehydration, clearing, infiltration and embedding* are to be carried out in the same way as described under the corresponding steps in the paraffin section technique for root tips (Schedule A). While embedding, orient the flower buds in groups, depending on their size, about 1 cm away from each other.

(viii) *Section-cutting* is similar to that followed for root tips already described. Cut longitudinal sections 12 µm thick. Mount the ribbons as given before.

(ix–xvi) *Bringing down to water, pre-mordanting in 1% chromic acid solution, staining in crystal violet, mordanting, dehydration, differentiation, clearing and mounting* are similar to the methods followed in the case of root tips.

(xvii) Haematoxylin staining can be done after step (ix), bringing the slides down to water, following the schedule given for root tips. Pre-mordanting in 1% per cent chromic acid can be omitted. In plants with very small flowers, like members of Araceae, the entire inflorescence is cut into equal segments and fixed. Transverse sections of the inflorescence are cut serially for study. In plants with very large flowers, the anthers are dissected out and fixed.

For scattering chromosomes, 0.002 M oxyquinoline solution can be mixed with the fixative (1 : 1). Instead of Navashin's A and B fluids, other regular fixatives can also be used.

From embryosac mother cells

Meiotic division can also be studied from the embryosac mother cells. It usually takes place after the meiotic division in anthers, sometimes as long as three weeks afterwards.

The preliminary step in the process is to expose the ovules to the treatment fluids, having first dissected them out. For the very large ovules it is necessary to remove the ovary wall. In a number of monocotyledonous plants, the ovules can be dissected out on entire strings. Small ovules are more difficult to handle, as they tend to get lost if dissected out. Ovule squash technique by detaching the small ovules from the placenta just before squashing, involves maceration in N HCl and staining with Feulgen reaction or by the acetic–orcein, acetic–carmine or acetic–lacmoid schedule. The former technique is more useful for larger ovules while the latter can be used for both small and large ones. The unwanted tissue is dissected out before final squashing. For squash preparations, the most effective fixative is a modification of Carnoy's fluid having four parts chloroform, three parts absolute ethanol and one part glacial acetic acid. The high proportion of chloroform is necessary to keep the pliability of cell structures. The period of fixation may vary from two days to three weeks. Whole mounts can be prepared by fixing in bulk for 24 h followed by hydrolysis and staining in Feulgen solution. For the study of small ovules and difficult materials paraffin blocks are prepared; the fixative used should be strong and should penetrate rapidly the covering tissues. La Cour's 2BX is effective.

Squash technique for young embryosac
Dissect out ovules from the ovary and fix in Carnoy's fluid for 1 day. Keep in 95% ethanol for 1–2 days. Run through 90, 80, 70, 50 and 30% ethanol, keeping in each for 5–10 min. Rinse in water. Two separate staining schedules can be followed after this stage:

(a) *Staining in Feulgen solution* Hydrolyse for 8–10 min at 60°C in N HCl. Rinse in water and stain in leuco-basic fuchsin solution for 2 h. Intensify the stain by keeping in water for 15 min. Transfer to a drop of 45% acetic acid on a clean slide and squash under coverglass, applying uniform pressure. Dehydrate by inverting in tertiary butyl alcohol and mount in euparal.

(b) *Staining in acetic–orcein solution*　Transfer the ovules from water to a mixture of 2% acetic–orcein solution and N HCl (9 : 1). Heat gently for 5–10 s without boiling the fluid. Keep for 20 min in the mixture. Transfer to a drop of 1% acetic–orcein solution on a clean slide and squash under a coverglass, exerting uniform pressure. Make the slide permanent by any of the schedules described in the chapter on mounting. In plants with very small ovules, the entire ovary can be cut into tiny pieces and treated. Acetic–lacmoid solution can be used instead of acetic–orcein solution and acetic–ethanol mixture (1 : 1) instead of Carnoy's fluid.

Squash technique for mature embryosac
Fix the ovary in Carnoy's fluid for 2 days. Transfer to a mixture containing 10 drops of saturated solution of iron acetate in 45% acetic acid and 10 ml of 4% iron alum solution. Keep the phial containing the material in the mixture in a water bath at 75°C for 3 min. Give two changes with distilled water heated to 75°C, keeping for 2 min in each. Transfer to cold water and keep for 2–3 min. Hydrolyse in 50% HCl for 10 min. Rinse in several changes of distilled water for 20 min. Transfer the ovary to a drop of 1% iron–acetic–carmine solution on a clean slide. Dissect out the ovules with needles into the stain and remove the rest of the ovary. Tap the ovules with a flat-bladed scalpel until the cells are separated. Apply a coverglass and heat the slide gently. Squash and seal.

Paraffin block preparation for embryosac mother cells
Fix the dissected ovules overnight in La Cour's 2BX fixative (2% chromic acid, 2% potassium dichromate, 2% osmic acid, 10% acetic acid, 1% saponin, distilled water in ratio 10 : 10 : 12 : 6 : 1 : 5). Wash overnight in running water.

The later steps, namely, dehydration, clearing, infiltration, embedding and section cutting are similar to the corresponding steps in paraffin preparations of root tips. Sections are cut 15–20 μm thick. Run the slides with sections through xylol I and II grades, keeping in each for 30 min. Pass through ethanol–xylol and absolute ethanol grades (15 min in each). Keep in 95% ethanol for 10 min. Keep in a mixture of 80% ethanol and hydrogen peroxide (3 : 1) for 24 h. Observe, and if the background is not clear, keep for another 24 h. Transfer to 70% ethanol, then 50 and 30% ethanol, keeping 5 min in each. Rinse thoroughly in water.

The next steps in the schedule, pre-mordanting in 1% chromic acid, staining in crystal violet, mordanting, dehydration, differentiation, clearing and mounting are done as in the case of paraffin preparations of root tips described in earlier schedules.

Pachytene chromosomes

Due to their configuration during the pachytene stage in pollen mother cells, the chromosomes of a large number of plants present very good material for the study of their individual structure and the nature of pairing.

Fix a complete spike (e.g. barley) with awns cut close in acetic–ethanol mixture (1:4) for 3–4 h. Keep overnight in 95% ethanol and 70% ethanol. Dissect out the three anthers of each flower.

Transfer one anther to a drop of iron–acetic–carmine solution (two drops of iron acetate in 10 ml acetic–carmine solution). Under a wide-field binocular microscope, cut the anther into two pieces, press each half gently to squeeze out the pollen mother cells, and remove all pieces of anther wall and tapetum with a needle, leaving only the pollen mother cells.

Observe under the microscope if the nuclei are in the pachytene stage. If so, add another drop of stain, place a coverglass on the material and heat gently over a flame three or four times.

Apply uniform pressure vertically with a U-shaped needle-point, blotting off excess stain from the slide; check under a microscope. If greater pressure is wanted, add more stain, heat again and press. Careless application of pressure may break up the chromosomes.

Heat again and invert the slide whilst still warm in 10% aqueous acetic acid and after the coverglass falls off, pass both slide and coverglass through acetic–ethanol and ethanol–xylol grades and mount in Canada balsam.

III. RESTAINING SCHEDULES FOR MITOTIC AND MEIOTIC PREPARATIONS

Permanent slides kept for a long period usually lose the brightness of their stain. The principal factors responsible for this fading are: (a) progressive acidity of the mounting medium, chiefly Canada balsam—this defect can be remedied by using neutral mounting media; (b) exposure to ultraviolet light by leaving the slides lying about carelessly or by exposure to arc lamp projectors, and (c) failure to remove all extraneous chemicals. If dehydration and later clearing in xylol are insufficient and traces of any solvent of the stain are carried over into the mounting medium, the stain fades quite rapidly.

Preparations, in which the stain has faded, can be re-stained in some cases, For re-staining, the original stain itself or some other stain which suits the fixative in which the tissue was originally fixed is used. The process consists of the following steps:

Removal of the mounting medium: place the slides in a solvent of the mounting medium, usually xylol, till the coverglass is detached. Give a change in xylol to remove the mounting medium completely.

Bring the slides down to water through ethanol–xylol mixture (1:1), absolute ethanol, 95, 90, 80, 70, 60, 50 and 30% ethanol grades, keeping 10 min in each.

Staining: the original schedule is followed. Both pre-mordanting and post-mordanting in 1% chromic acid are done for slides being stained by crystal violet schedule. Slides originally fixed in Flemming's fluid and stained in crystal violet can be restained following Feulgen schedule.

Make permanent following the usual procedure.

The restaining procedure is also applied to slides which have been rejected during differentiation. For example, during the crystal violet schedule, if on differentiation in clove oil the tissue is seen to have taken insufficient stain, the slide is transferred directly to down grade xylol I and allowed to remain overnight. Afterwards it is brought down to water, pre-mordanted and stained as usual.

The exceptions to the re-staining process are Feulgen stain and fluorochromes. Slides with insufficient stain cannot be hydrolysed again and restained in leucobasic fuchsin solution. However, certain other alternatives suggested are:

For tissues originally fixed in acetic fixatives, use acetic stains, like acetic–orcein or acetic–lacmoid solution.

Tissues fixed in aqueous fixatives can be stained following the crystal violet schedule.

After fixation in alcohol fixatives, use haematoxylin staining.

For slides with insufficient Feulgen stain, the stain can be brightened by immersing in 1% acetic–carmine or acetic–orcein solution for 2–5 min.

IV. STUDY OF NUCLEOLUS

Some of the schedules developed are outlined below:

(1) *Staining of nucleolonema*: (a) Silver impregnation procedures and iron–pyrogallic stain for fixed material, and (b) phase contrast microscopy, dark-field illumination and oblique transillumination for fresh material have been used for studying the filamentous structures within the nucleolus.

(2) *For staining nucleoli*, treat tissues with 95% ethanol 2 parts; formalin, 1; 5% glacial acetic acid, 1; hydrolyse with N. HCl at 60°C for fixed periods and squash in 1% acetic-carmine solution.

(3) *Feulgen-light green schedule* (Semmens and Bhaduri, 1941): Bring down sections fixed in a fixative without acetic acid to water. Treat for 2–3 h in 75% ethanol. Wash in distilled water. Hydrolyse for 10 min in N HCl at 60°C. Stain for 2 h in leuco-basic fuchsin solution. Wash in two changes of SO$_2$ water, keep for 10 min in each. Rinse successively in distilled water, 50 and 70% ethanol. Mordant for 1 h in 80% ethanol saturated with Na$_2$CO$_3$.

Dip in 80 and 95% ethanol. Stain for 20–25 min in filtered saturated alcoholic solution of light green with 2–3 drops of aniline oil. Drain. Rinse in a mixture of saturated solution of Na_2CO_3 in 80% ethanol, 10 ml and 90 ml 80% ethanol. Differentiate in 95% ethanol. Dehydrate through absolute ethanol, ethanol–xylol and xylol grades and mount in balsam. For smears and squashes, the initial treatment in 75% ethanol is omitted. In squashes, squash cells after Feulgen staining in 45% acetic acid. Separate coverglass in 40% ethanol and proceed as usual.

(4) *Toluidine blue molybdate method* is based on the principle of gradual blocking of -NH_2 group of nucleoprotein and unmasking of phosphate groups of nucleic acids binding cationic dyes. The dyes giving satisfactory results are: Coleman Bell CU-3, National Aniline NU-2 and NU-17, Harlcuco NU-14, Matheson-Coleman Bell CU-9 and Biological Stain Commission NU-19. The dyes are dissolved in McIlvaine's buffer, pH 3.0 at 20°C.

Wash tissue in 0.85% saline for 10 s before fixation. Treat two slides in 5% aqueous trichloracetic acid for 10 min. Wash in distilled water. Fix one slide for 5 min and another for 10 min in formal sublimate containing 40% formaldehyde, 1 part, and 6% aq. mercuric chloride, 9 parts. Rinse in tap water and treat with Ingol's iodine for 5 min. Immerse slides in 5% sodium thiosulphate for 5 min and rinse in water. Stain for 30 min in toluidine blue. Treat with 4% aqueous ammonium molybdate solution for 15 min. Wash in tap water. Dehydrate in tertiary butyl alcohol, clear in xylol and mount in a synthetic resin. The *pars amorpha* of the nucleolus takes up bluish-green stain while the nucleolini are bright purple.

V. STUDY OF CHROMOSOMES FROM LOWER GROUPS

Algae

Chromosome structure in algae is of particular interest since the group presents a variable pattern. In most forms, including a majority of Chlorophyceae, chromosome morphology is comparable to that of higher plants, whereas in others, like Conjugales, Euglenophyceae, and especially Dinophyceae, the structure demonstrates a number of unusual characteristics. In Conjugales, the chromosomes are devoid of localised centromeres, comparable to some extent with the structure reported in *Luzula* of angiosperms. Euglenophyceae shows certain features of special interest like the absence of a typical equatorial plate and centromere and the persistence of RNA-containing endosome throughout mitosis, as well as a different type of chromosome aggregation and movement from higher plants. In Dinoflagellates the cytochemical and ultrastructural data have shown the chromosomes to be sausage-shaped structures, composed of continuous fine fibrils of DNA only, without any basic protein—a structure finding a parallel with the genophore of prokaryotes, that is bacteria and blue green algae, but differing fundamentally from the latter in having a nucleus.

The methods of chromosome study in algae principally include two separate stages: culture of the algae to obtain suitable divisional figures, and processing to observe the chromosomes.

Culture is usually necessary, for acquiring a sufficient number of metaphase plates in polar view for chromosome analysis. Algae growing in the wild condition show both somatic and germinal divisional figures, as seen in marine *Cladophora*, and *Ulva* and most forms of red algae. In the simplest form, the alga is grown in its natural medium, but progressively complex media with different proportions have been devised for obtaining better growth and synchronization of mitosis to collect the highest number of mitotic stages. The other major factor is light period. In most cases, artificial light and dark photoperiods are provided.

For successful culture of any alga, and green algae in particular, the bulk material after collection is teased out under dissecting/compound microscopes to separate pure filaments from mixed ones. Media ranging from water of the habitat to different types of synthetic media have been devised.

In Chlorophyceae and certain other algae, synchrony in division can be obtained even in culture without any special treatment. However, synchronization of division maybe achieved through temperature control in continuous light with bubbled air enriched with 4% carbon dioxide and by continuous renewal of culture medium. Smaller members of marine Dinophyceae can be grown in sea water, to which nitrate, phosphate and soil extracts have been added (50 ml soil extract; 0.2 g sodium nitrate and 0.03 sodium phosphate, both in 10 ml of water added per litre of sea water). Both soil extract and sea water can be replaced by artificial components. The cultures are kept under fluorescent lighting between 1076–21521 lx at a temperature of 15–25°C for 12–18 h.

Culture of gametophytes and sporophytes of the Laminariales is influenced by nutrient, temperature and light. Sea water enriched by various nutrients has been used. The most suitable temperature is between 10–16°C and an indirect light intensity of 538–3228 lx, but never direct sunlight. Two commonly used nutrients are: (A) 0.01 M KH_2PO_4, 0.5 ml and 0.01 M KNO_3, 0.5 ml, added to 25 ml of sea water every 14 days and (B) 10^{-5} M KH_2PO_4 (1/20th of nutrient A stock); 5×10^{-5} M KNO_3 (1/4th of nutrient A stock); 0.5 ml of each added to 25 ml sea water every 14 days.

In a number of Phaeophyceae, large pieces of the fruiting region are collected, washed in jets of boiled sea water and immersed in boiled sea water. After the zoospores are released, the suspension is poured in a flat dish containing sterilized slides or coverglass. After 30 min, the coverglasses are removed and placed in a culture chamber. Rapid growth is ensured by long hours of day-light or by permanent illumination with three 80 W fluorescent tubes at about 15°C. In Rhodophyta, culture methods have only a limited application. The habitat medium, with frequent changes, is sufficient in most cases.

As in the higher plants, schedules for *studying the chromosomes of algae* involve mainly fixation, staining and mounting. Pre-treatment is given in specific cases, as are also different maceration techniques. There is considerable variation in response to fixatives and stains, not only between the different groups, but also between different genera and species, and in some cases between different parts of the same plant. Decolorization of the pigment is an important factor in the choice of fixative.

Different fixatives used include:

(i) *Formalin–ethanol* in different proportions. In Rhodophyceae, it is suitable for haematoxylin or brazilin stains but useless for Feulgen or acetic-carmine. For green algae, however, fixatives containing formalin or mercuric chloride are unsatisfactory.

(ii) *Formalin–acetic–ethanol* in different formulae. A fresh combination of glacial acetic acid, 2.5 ml; 40% formaldehyde, 6.5 ml, and 50% ethanol, 100 ml is recommended for Rhodophyceae.

(iii) *Amongst the metallic fixatives*, chromic acid–acetic acid mixtures have been used in a large range of variations, a very common one being the modification of Karpechenko's fluid. It includes: Solution A containing chromic acid, 1 g; glacial acetic acid, 5 ml; sea water, 65 ml and Solution B with 40% formaldehyde, 40 ml and 35 ml sea water. The two solutions are mixed immediately before use. Sea water is replaced by distilled water for fresh-water forms and the use of formaldehyde is optional. Chrom-acetic–formalin has been used as fixative for large parenchymatous Phaeophyceae, followed by softening in sodium carbonate and occasional bleaching with H_2O_2. Osmic acid vapour, nitric acid vapour, 2% osmic acid solution, Belling's Navashin solution enriched with osmic acid, iodine water, bromine water and chromic acid solution have all been successfully used on different members of Chlorophyceae. Fixation in osmic acid must be brief and the osmium must be washed out before it forms a black deposit.

(iv) *Acetic acid : ethanol and acetic acid : methanol mixtures* in different concentrations and proportions. Mixtures of glacial acetic acid and 95% ethanol (1:1, 1:2, 1:3) are effective for Rhodophyceae, the two former giving better results. However, they can cause shrinkage in large-celled delicate forms and disintegration of calcified materials. Smaller thalli need only a few minutes treatment, but 1–6 h immersion is recommended for cartilaginous forms. The materials should be processed immediately after fixation and storage avoided.

Within Phaeophyceae, filamentous forms can be fixed in acetic acid–ethanol mixture (1:3) for up to 24 h, preferably changing the fixative after 1 h. Lithium chloride has been used for large parenchymatous plants after fixation as a softening agent. Decolorization is usually satisfactory after 24 h.

Acetic–ethanol (1:3) gives good results with members of the Dinophyceae which are large and easy to handle. In different proportions it gives very good results. Fixation is carried out in centrifuge tubes and the cells collected by centrifugation. Addition of a few drops of saturated ferric acetate (in acetic acid) to the fixative 1 h before staining is beneficial for certain species.

Mixtures of glacial acetic acid and 95% ethanol (1:1, 1:2, 1:3) and also ethanol or methanol alone, gave good results in different forms of Chlorophyceae.

Processing can involve

(i) Whole mounts for uniseriate filamentous thalli in the Rhodophyceae and for various members of the Dinophyceae and Chlorophyceae.
(ii) Serial sections of paraffin-embedded material only for certain Rhodophyceae following the usual block preparation schedule.
(iii) Squash preparations are most commonly used in studying algal chromosomes. The fixed material can be squashed directly or it may require a softening process, depending upon the hardness of the material. The softening agent and the period and conditions of treatment depend mainly upon the material but also on the fixature and stain used. Dilute solutions of acid or alkali in water or ethanol are used for softening, the more common one being HCl or NaOH in concentrations ranging from 1 to 50% and also dilute solutions of sodium carbonate.

Since in the Rhodophyceae, the carpogonium or young carposporophyte is encased by a large amount of unwanted tissue, sections of the fertile axis (50–100 μm) are cut on a freezing microtome with dilute gelatin as the supporting medium and then the sections are squashed. Since the brown algae are rather tough and resilient, they require special softening pre-treatments with sodium carbonate, lithium chloride or a mixture of ammonium oxalate and hydrogen peroxide.

Stains commonly applied to higher plants have also been used for algae, like acetic–carmine, Feulgen, haematoxylin, brazilin and methyl green-pyronin. *Acetic–carmine* stain is most widely applied following the iron alum-acetic–carmine method.

In Chlorophyceae, fix and then mordant in aqueous iron alum solution, used in different dilutions, depending on the material, for a period not exceeding 30 s. Wash repeatedly. Add super-saturated carmine solution in 45% acetic acid and boil the preparation to dissolve the starch. Hold filamentous forms with forceps and pass through the solutions. Handle spores as settled on slides or coverglasses, and unicellular cultures in centrifuge tubes. Pass carmine-stained preparations as usual through acetic acid–ethanol (1:3), 95% and absolute ethanol grades before mounting in euparal. Acetic–orcein has been used instead of carmine but gives a paler stain. In another method a few drops of super-saturated solution of ferric acetate in 45% acetic acid is added to the fixative (acetic–ethanol 1:3). This addition of iron salt has also been found to be very effective in Charophyceae. The material is transferred directly from the fixative to carmine and back to the fixative, followed by 95% euparal essence and euparal.

Members of the Dinophyceae are stained after fixation in a drop of acetic–carmine on a slide. Acetic–orcein is equally effective. For Laminariales, add acetic–carmine, containing 1 drop of saturated ferric acetate solution per 25 ml, to the material *after* squashing; followed by alternate heating and cooling. Large parenchymatous brown algae require fixation in acetic–ethanol (1:3), enriched by a few drops of ferric acetate. Wash the material in 70% ethanol, cut hand sections and mount in 6% Na_2CO_3. After squashing, add distilled water and later the stain and boil gently. This schedule has also been used successfully for the Rhodophyceae after slight modifications.

Feulgen staining schedule has been applied in almost all forms of algae. The method is similar to that for aceto-carmine staining, both for Chlorophyceae and Dinophyceae. Gametophyte materials of Laminariales take up bright stain. For large parenchymatous brown algae, fixation in Karpechenko's fluid is followed by washing in running water and bleaching for 3–4 h in 20% aqueous H_2O_2 solution and again washing in running water. The material is heated to 60°C in distilled water, hydrolyzed in N HCl at 60°C for 7–10 min. Without bleaching, hydrolysis has to be done for 15–30 min. The material is transferred to cold distilled water, washed in running water for 10 min, hand sections are cut and squashed in SO_2 water (N HCl, 5 ml; 30% $K_2S_2O_5$, 5 ml; water, 100 ml). Feulgen staining, however, does not yield consistent results with the red algae.

Heidenhain's haematoxylin schedule and its modifications have been used in the red algae. Haematoxylin in acetic acid stains the nuclei in green algae but it is not permanent.

Brazilin was extensively used in red algae. Both the stain and its mordant require a long period of ripening in the dark to give satisfactory preparations. Treat sections or squash preparations in a 2% ferric ammonium sulphate solution in 70% ethanol for 1 h, wash in 70% ethanol and stain for 12–16 h in

0.5% brazilin in 70% ethanol. Finally wash twice in 70% ethanol and dehydrate through ethanol grades. This stain avoids swelling of chromosomes due to aqueous solutions and the procedure is much shorter than haematoxylin.

Methyl green pyronin staining schedule has been applied in the green algae only, using a BDH dye mixture. A very small amount is dissolved in water and the material, after acetic fixation, is mounted in it.

Preparation of permanent slides from temporary ones can be made by any of the schedules followed for higher plants.

In addition to the general methods, as outlined above, a few *sample schedules used specially for members of the Phaeophyta* are given below since in this group a special method for softening is necessary.

(i) Acetic–carmine preparations as used in Laminariales:
Fix in acetic–ethanol (1:3) for 12–18 h, wash in running water; immerse in 1 M lithium chloride solution for 15 min; keep in water for 15 min; dissect out material and squash under a coverglass; insert a few drops of acetic–carmine solution with a trace of ferric acetate at the side of the coverglass, heat at intervals for 30 s without boiling, squash again and blot excess stain. If over-stained, de-stain by heating in a mixture of acetic–carmine solution and glacial acetic acid (1:5) for 15 s.

In filamentous Phaeophyta, fix for 24 h and add a few drops of saturated ferric acetate in 45% acetic acid to the fixative. After washing in acetic–ethanol (1:3), add stain, cover, boil, squash.

(ii) Feulgen schedule for Phaeophyta:
Fix coverglass with growing gametophyte in acetic–ethanol (1:3), wash, hydrolyse in N HCl at 60°C for 8–10 min, transfer to cold water, and then keep in decolorised Schiff's reagent for 8 h at room temperature; bleach in three changes of SO_2 water; squash in SO_2 water and later dehydrate through ethanol grades and mount in euparal. Variations in hydrolysis and staining periods are adopted for different materials.

(iii) For softening and isolating female conceptacles, split the receptacles longitudinally and fix for 12 h, followed by treatment for 20 min in a mixture of saturated ammonium oxalate solution and 20 vol H_2O_2 (1:1); wash for 15 min before staining. Alternatively, for male material, macerate 30 g for 4 min in 150 ml of fixative in a Waring Blendor; transfer to a graduated cylinder; pipette off middle layer with sex organs and stain as usual.

Fungi

A group of simple organisms living as parasites or saprophytes and categorized as fungi, holds a unique position in the plant kingdom. Even though belonging to Eukaryota in its nuclear constitution this group is distinguished by its extremely simple thalloid constitution and absence of such pigments that are universal for higher plants. The meagre cytological data so far available have also indicated the occurrence of unusual features in certain groups such as the existence of nuclear membrane during division and the controversial double reduction division during meiosis. From the standpoint of the evolution of the structure and behaviour of chromosomes, there is immense potentiality in the cytological analysis of this assemblage.

For chromosome studies, Helly's modified fluid (mercuric chloride, 5 g; potassium dichromate 3 g; distilled water, 100 ml; add 5 ml formalin just before use) is found to be very successful as a fixative in a wide range of fungi. Other fixatives used are osmium tetroxide vapour and acetic–ethanol but their effects are less desirable.

Different chromosomal stains have been tried out.

Giemsa gives effective results for quite a large number of filamentous fungi and yeasts. The preparations can be stained directly or after hydrolysis in N HCl. A stain used for yeast chromosomes contains 16 drops of Gurr's Giemsa R66 dissolved in 10–12 ml of Gurr's Giemsa buffer at pH 6.9. If hydrolysis is required, fixed cells are usually extracted for $1\frac{1}{2}$ h with 1% NaCl at 60°C, the time, concentration and temperatures depending on the material. Then treat for 10 min with N HCl at 60°C, rinse with tap water and keep in the stain for several hours.

Haematoxylin staining has been advocated for squash preparations, following fixation in BAC fixative.

Feulgen staining is utilized according to the usual schedule of acid hydrolysis followed by staining. Schiff reagent prepared with Diamant Fuchsin is used for yeast chromosomes. The smears can be mounted in water or acetic–carmine.

Schedules

For yeast Place a loopful of slimy growth from a 2–5 day-old plate culture on a No. 1 cover glass. Place another coverglass on top with corners turned away at an angle of 5 degrees to the former. Allow the drop to spread, pull apart the coverglass so that a smear is formed on each and immerse in Helly's modified fixative for 10 min. Rinse thoroughly in 70% ethanol, transfer to Newcomer's fixative for preservation in cold. The film must not be permitted to dry prior to fixation. The yeast culture may be exposed briefly to formalin vapour before the smear is drawn for better preparation. In Giemsa staining, extract fixed cells for 1.5 h with 1% NaCl at 60°C, treat for 10 min in N HCl at 60°C, rinse in tap water and stain for several hours in Giemsa solution. Differentiate by moving the smears repeatedly for 10–12 s at a time in 40 ml distilled water to which a few drops of acetic acid have been added (pH 4.2). Observe under an immersion lens to control extraction of excess stain. When the nuclei are brightly differentiated, mount the coverglass on a slide in a drop of buffer containing 2–3 drops of Giemsa per 10 ml. Blot excess medium, squash with uniform pressure, seal.

In Feulgen schedule, hydrolyse the smear for 10 min, rinse with tap water, stain in Schiff's reagent for 3.5 h, rinse quickly in 10 changes of SO_2 water (tap water 90 ml; N HCl 5 ml; 10% sodium metabisulphite 5 ml), keep in running water for 20 min and mount in water or acetic-carmine.

For mitosis in basidiomycetes Centrifuge the culture and suspend in distilled water. Homogenise, spread in a thin film on slide, air dry, fix in acetic–lactic–ethanol (6:1:1) for 10 min, pass through 95% and 70% ethanol and rinse in water. Hydrolyse in N HCl at room temperature for 5 min, then at 60°C for 6 min, wash thoroughly in distilled water and suspend for 5 min in a phosphate buffer (pH 7.2). Stain for 25 min in Giemsa's stain, rinse successively in water and buffer, drain, flood with Abopon and cover with a coverglass.

For general mitotic preparations Pour a mixture of BAC fixative (*n*-butanol, acetic acid and 10% aqueous chromic acid solution, 9:6:2) over the culture in a petri dish and store under partial vacuum at 0–6°C for 1–5 days. Dissect out fruit bodies in a mixture of conc. HCl and 95% ethanol (1:1), heat gently, wash with Carnoy's fixative for 1–2 min, heat in propionic–carmine solution with a trace of iron, apply coverglass, allow to stand, heat to boiling and press the coverglass. Add a drop of 45% acetic acid to all corners, apply a drop of glycerin–acetic acid mixture to one corner, blot and seal.

For members of the Pezizales Fix in Carnoy's fluid for 24–72 h followed by either direct squashing in acetic–carmine or hydrolysis in N HCl. Stain in Feulgen and squash in acetic–carmine.

For basidia of agarics Fix bits of hymenial tissue in Newcomer's fixative for 1–12 h, hydrolyse and mordant in aqueous N HCl containing 2% aluminium alum, 2% chrome alum and 2% iodic acid for 5 min at room temperature, then at 60°C for 10–15 min. Wash in three changes of distilled water and keep in Wittmann's acetic–iron–haematoxylin for 2 h. Mount in a drop of stain, cover with a coverglass, press, heat to just below boiling and seal.

For chromosomes of Neurospora Dip strips of agar containing perithecia, after four days of crossing, in ethanol : acetic acid : 85% lactic acid mixture (6:1:1) in closed vials and store in deep freeze. Prepare stain by mixing a stock solution of 47 ml acetic acid and 20 ml lactic acid solution (1 ml 85% lactic acid in 24 ml distilled water) with 5 ml N HCl and 28 ml distilled water at room temperature. Mix 5 ml of this fluid with 100 mg Natural Green (Gurr) and reflux for 4 min after boiling over a low flame, in a small beaker covered by a petri plate bottom containing two ice cubes. Tease out the asci in a drop of stain, mount after separation in another drop under a coverglass, heat slightly and seal with dental wax.

Bryophyta

Culture living material in moistened glass pots, keep in refrigerator with glass windows at 10–15°C.

Cut out tips of new developing shoots. Fix in modified Carnoy's fluid (1:1:1) for 3 h at 8°C. Transfer to 45% acetic acid and keep for 5–10 min. Stain in 2% acetic–orcein solution for 10 h at 15°C. Dissect out stained tips in 0.5% acetic–orcein with needles under dissecting microscope. Cover with a coverglass and keep on a hot plate at 100°C for a few minutes.

Squash, applying uniform pressure, remove excess stain by blotting with filter paper, seal.

Pteridophyta

Diploid mitotic division can be studied from leaf-tips and haploid division from growing apex of young prothallus.

Pretreat extreme inside tip within the circinnate vernation dissected out, with colchicine or p-dichlorobenzene or aesculine in cold for durations similar to angiosperm leaf-tip. Fixation in acidulated ethanol, heating in 2% acetic–orcein mixture and squashing in 1% acetic–orcein are similar to leaf-tips of angiosperms.

For meiotic studies, dissect out sori, and process as for angiosperm anthers.

Gymnosperms

Mitotic division is studied from leaf-tips. The procedure is similar to that of angiosperm leaf-tip. The extreme tip is cut out, brushed to remove hairy growth and pretreated and squashed as usual. The durations of pretreatment and fixation are longer.

Meiosis is studied from microspore and megaspore mother cells with a procedure modified from angiosperms.

VI. STUDY OF CHROMOSOME COMPONENTS

Nucleic acids

Pyronin-methyl green technique
Of all the methods employed for the differential localization of RNA and DNA, the universally accepted standard technique is that of Brachet based on Pappenheim and Unna's methyl green-pyronin G mixture. In this

schedule, methyl green imparts a green colour to chromatin or, more precisely, DNA, whereas RNA present in the cytoplasm, nucleolus, etc., appears pinkish red with pyronin. In both cases the reaction involves the phosphoric groups of the nucleic acid moiety. Methyl green is correctly represented as 'Methyl green OO', a basic dye of the triphenyl methane series. Kurnick (1947) claimed that selectivity of DNA staining by methyl green depends on the polymerized nature of the DNA molecule, and depolymerized DNA could not be satisfactorily stained with methyl green, and also that histones competed with the dye for nucleic acid. Kurnick and Mirsky (1950) asserted that one dye molecule combines with ten phosphoric groups of DNA, basing their observation on the stoichiometry of the reaction by dialysis, precipitation of stain-nucleic acid mixtures and staining of nuclei of known DNA content. This ratio of 1:10 was given by heptamethyl pararosaniline (CI 684) and the ratio of 1:13 by hexamethyl pararosalinine (CI 685) (Pearse, 1972). Kurnick (1955) and Errera (1951) claimed that stable binding of methyl green should involve at least two amino groups of the dye. The stainability of methyl green is therefore controlled largely by different agents which under certain conditions may bring about depolymerization. Goldstein (1961) claimed that dye cations, having a combined atomic weight between 350 and 500, can penetrate and adhere to nuclear DNA whereas smaller cations adhere to denser molecules such as cytoplasmic RNA. Methyl green, having a cationic weight of 387, is specific for DNA. Heat or other agents may alter the structure of DNA in such a way that it becomes denser and, as such, stainable with pyronin.

Pyronin is a basic dye of the xanthene group and is available in three different forms, 'pyronin B', 'pyronin Y' and 'pyronin G'. Pyronin Y (Michrome No. 339, Gurr) is a tetramethyl whereas pyronin B (Michrome No. 44, Gurr) is a tetraethyl compound. Pyronin preferentially stains low polymers of nucleic acid, but stoichiometric studies did not reveal constancy in the binding of depolymerized DNA by pyronin. To secure selective staining, pyronin G or Y is more suitable since pyronin B results in non-specific cytoplasmic staining. Evidently, methylation may have some connection with the staining property of pyronin.

Schedule (*Kurnick 1955*) The recommended fixative is Carnoy's fluid or freeze-drying. Dissolve 2 g of pyronin Y in 100 ml of distilled water. Add chloroform and shake the mixture in a separating funnel till the layer of chloroform becomes colourless. Separate the dissolved dye. Similarly prepare a 2% solution of methyl green and extract it with chloroform. The solutions can be kept as stock. For use, mix together 12.5 ml of pyronin Y solution and 7.5 ml of methyl green solution and add 30 ml of distilled water. Alternatively, prepare staining mixture by adding together: 2% aq. pyronin solution 50 ml; 2% aq. methyl green solution 30 ml; dist. water 10 ml; chloroform 10 ml.

Bring paraffin sections of tissues, previously fixed in Carnoy's fluid, down to distilled water. Shake the staining mixture before use. Then immerse the slides in it for 15 min at 20°C. Blot excess stain with a filter paper and dip the slide in 50% ethanol and pure *n*-butanol mixture (1:1) for 2 s. Immerse in *n*-butanol for 10 min. Again immerse in *n*-butanol for 5 min. Treat in toluol for 1 h. Mount in neutral balsam. Chromosomes and nuclear chromatin stain bright green while nucleolar and cytoplasmic RNA stain pink to deep rosy red.

Alkaline phosphatase

The technique for the demonstration of alkaline phosphatase, a term loosely applied to phosphomonoesterase, is one of the most important ones for localization of specific enzyme sites on chromosomes. The demonstration of the enzyme activity in the cell, and especially in the chromosome, can be carried out principally through two different methods: (a) calcium phosphate deposition; and (b) azo dye methods.

The calcium phosphate precipitation technique, developed independently by Gomori (1939) and Takamatsu (1939), was based on the principle that if a tissue containing the enzyme is incubated at 37°C in a medium containing phosphate salt as the principal constituent at an alkaline pH (9.4), the liberated phosphoric acid can be deposited at the site of the enzyme as an insoluble phosphate precipitate if a calcium salt is present as one of the components. The visualization of calcium phosphate can be done by converting it into silver sulphate and then to metallic silver or into cobalt phosphate through a cobalt salt and finally to a black precipitate of cobalt sulphide through ammonium sulphide.

Several phosphate salts have been used as substrates, such as α,β-glycerophosphate, fructose diphosphate, adenosine triphosphate, sodium dihydrogen phosphate, etc., but the most commonly used one is sodium-β-glycerophosphate. Since the enzyme is effective in an alkaline medium, the maintenance of proper pH (9.2–9.8) is necessary.

Several methods for the quantitative estimation of alkaline phosphatase activity, with the aid of spectrophotometry, radioactivity as well as interferometry, have been developed.

The azo dye technique is based on the principle of precipitation of the alcoholic part of the phosphate ester instead of the phosphoric acid. The method, as originally developed by Menten, Junge and Green (1944), used β-naphthyl phosphate as the substrate. The liberated β-napthol, after hydrolysis, is acted upon *in situ* by diazotized β-naphthyl amine at pH 9.4 when a red precipitate is obtained. It can be applied to both fresh and formalin-fixed tissues. Danielli (1947) used phenyl phosphate and β-naphthyl phosphate to obtain the reaction. Mannheimer and Seligman (1948) used magnesium ion as activator and α-naphthyl diazonium naphthalene-1, 5-disulphonate for coupling. The purplish red dye, which precipitates out, has been found to be insoluble. This method has the added advantage over the original technique in that the coupling reagent used here is stable as compared to the diazonium salt. In Gomori's modification of the original method, sodium α-naphthyl phosphate has been used instead of calcium β-napthyl phosphate, as the sodium salt is relatively less soluble.

In calcium cobalt method, incubation mixture contains 3% aq. sodium-β-glycerophosphate soln. 10 ml; 2% aq. sodium diethylbarbiturate 10 ml; distilled water 5 ml; 2% aq. calcium chloride solution 20 ml; 5% aq. magnesium sulphate solution 1 ml; 2% aqueous cobalt nitrate or acetate solution. Aqueous dilute yellow ammonium sulphide solution contains about 50 drops of concentrated liquid in 20 ml water. Fix in cold acetone at 4°C with three changes for 24 h. Transfer through progressive ethanol grades to absolute ethanol, keeping 30 min in each grade. Treat with ethanol–ether-mixture (1:1) for 1 h and transfer to 1% celloidin. Decant excess celloidin and treat successively in chloroform and benzene, keeping 1 h in each. Embed in paraffin. Cut sections 5 μm thick and mount on albuminized slides. Dry at 37°C and store at 4°C. Bring down the slides in distilled water through immersion successively in petroleum ether and absolute acetone. Incubate in the incubating mixture at 37°C for 30 min to 16 h, 4 h being the optimum period. Wash thoroughly in distilled water after bringing down the slides to room temperature. Immerse in 2% cobalt nitrate solution for 3–5 min. Wash in distilled water. Immerse in yellow ammonium sulphide solution for 1–2 min. Rinse in distilled water. Stain with 1% eosin solution for 5 min if required. Dehydrate and clear through ethanol and xylol grades and mount in neutral balsam. Deposition of a black precipitate at the sites of alkaline phosphatase activity is seen.

Alternatives: For frozen sections, cut very thin sections and mount on slides with out albumin. Dry at room temperature for 1–2 h. The remaining steps are similar to paraffin sections. In a later schedule by Gomori, the incubation mixture contains: 3% aq. sodium glycerophosphate solution 20 ml; 2% aq. sodium diethylbarbiturate solution 30 ml; 2% aq. calcium chloride solution 4 ml; 2% aq. magnesium sulphate solution 2 ml; distilled water 30 ml. Cobalt chloride can be used instead of cobalt nitrate or acetate. Non-metallic fixatives give quite good results. The preparation of paraffin blocks can be done through the usual alcohol–chloroform grades. In no step, from fixation to mounting, should the temperature exceed 56°C. The incubation mixture should be freshly prepared before use.

α-Napthyl phosphate method   Incubation mixture contains: Sodium-α-naphthyl phosphate 0.05 g: 5% aq. borax solution 10 ml; 10% aq. magnesium chloride or sulphate solution 0.5 ml; cold distilled water (20°C) 100 ml;

a stabilized diazonium salt 0.25 mg. Either tetrazotised-*o*-dianisidine (michrome blue salt 250), 3-nitroanisole-4-diazonium chloride (michrome red salt 612) or 3-nitrotoluene-4-diazonium naphthalene-1,5-disulphonate (michrome scarlet salt 618) can be used. Also needed are haematoxylin solution in water, 70% ethanol 99 ml; glacial acetic acid 1 ml. Mix together.

For paraffin block preparations, fix thin cold slices of tissue at 4°C for 24 h in three or four changes of acetone. Clear in xylol. Embed in paraffin rapidly. Cut sections and attach to albuminized slides. Deparaffinize as usual. Rinse three times in pure acetone and then in three changes of distilled water. Incubate in incubation mixture at 37°C for 10–30 min or more. Remove a slide at intervals and observe under the microscope until the correct brightness of colour is attained. Stir the mixture mechanically during incubation. Rinse thoroughly in distilled water after bringing the slides to room temperature. Immerse in haematoxylin solution for counterstaining. Treat with acetic–ethanol mixture for 5–10 min. Rinse thoroughly in water. Mount in glycerine jelly or dehydrate, clear and mount in neutral balsam.

The sites of alkaline phosphate activity stain purplish black with michrome blue salt 250; purplish brown with michrome red salt 606 and reddish brown with michrome red salts 612 and 618.

Alternatively, sodium β-naphthylphosphate may be used instead of sodium-α-naphthyl-phosphate but α-salts yield a more specific and non-diffusible precipitate.

A modified coupling azo dye method for alkaline phosphatase For frozen sections: Fix thin slices of tissue in 10% neutral formalin in the cold for 10–16 h, or use fresh frozen cold microtome sections, mounted on coverglass. Cut frozen sections 10–15 μm and mount on slides without adhesive. Dry in air for 1–3 h. Dissolve 10–20 mg sodium-naphthyl phosphate in 20 ml 0.1 M stock tris buffer (pH 10). Add to it, with stirring, 20 mg of the stable diazotate of 5-chloro-*o*-toluidine. Cover the sections on the slides with the filtered solution and incubate at room temperature for 15–60 min. Rinse in running water for 1–3 min. Counterstain for 1–2 min in Mayer's haemalum. Rinse in running water for 30–60 min and mount in glycerine jelly. The sites of alkaline phosphatase activity appear brown with Fast Red TR and Fast Violet B or black with Fast Black B. The nuclei are dark blue.

For paraffin sections of material fixed in cold acetone: Bring down the sections to water after passing successively through light petroleum and absolute acetone. Cover with freshly prepared filtrate of substrate: diazonium salt mixture as in previous schedule. Incubate for 30 min to 4 h for salt Fast Blue RR or for upto 2 h for salt Fast Red RC or for upto 12 h for salt Fast Red TR. Rinse in water. Counterstain with haemalum as given in previous schedule, wash in running water and mount in glycerine jelly. Salt Fast Red TR gives the best results, the sites of alkaline phosphatase activity appearing reddish brown and the nuclei blue.

CHAPTER II.5

LOCALIZATION OF CHROMOSOME SEGMENTS THROUGH BANDING PATTERNS OF CHROMOSOMES AND SISTER CHROMATID EXCHANGE

I. BANDING PATTERNS: PRINCIPLES, SCOPE AND ADVANCES

The study of linear *banding patterns* in the chromosomes has emerged as a powerful tool in identification of chromosome segments. Chromosomes of an individual genotype show specific patterns of longitudinal differentiation, which, in absence of gross structural differentiation, serve as identifying criteria of a species. In certain cases, even functionally differentiated regions of chromosome, such as ribosomal or nucleolar regions, have specific banding patterns.

Initially, Caspersson and his colleagues noted that treatment with Quina-crine dyes revealed strongly fluorescent chromosome segments when viewed through ultraviolet light (Caspersson *et al.*, 1971). They further recorded differentially fluorescent segments, following staining with various fluoro-chromes under ultraviolet light. The bands were termed as Q bands. A concomitant discovery showed that such fluorescent segments correspon-ded to Giemsa-stained bands, obtained following a short saline treatment. Such bands at intercalary regions were referred to as G bands (Drets and Shaw, 1971). Since then, G-banding pattern has been obtained following treatments with a variety of chemicals, including saline, NaOH/HCl prior to SSC, and trypsin (vide Sumner, 1994).

The protocol for molecular hybridization has also been utilized in banding technique. Molecular hybridization is based in the principle that single strands of RNA or DNA are able to recognize and pair with their comple-mentary base sequences. Through renaturation of DNA complex, originally denatured by various treatments, the eukaryote genome has been shown to contain repetitive sequences of DNA in addition to having unique or a few copy sequences of DNA (Pardue and Gall, 1970). The degree of reannealing and the time taken, or reassociation kinetics, is an index of the degree of repeats present in the chromosome. Repetitive DNA sequences are found to be present in high amounts in certain segments of chromosomes, specially the centromeres. When this protocol for molecular hybridization, that is denaturation and reannealing, is followed by Giemsa staining, intense positive bands are revealed in similar segments of chromosomes. Such Giemsa-stained bands are termed as C bands which indicate the presence of highly repeated sequences. Such bands invariably represent heterochromatic segments. For nucleolar regions, a special banding technique has been developed termed as N banding, in which acid extraction is primarily involved. Differentiation of heterochromatin by N banding from C banded centromeric sites in plants is also possible (Kakeda *et al.*, 1991).

Another banding technique, the R banding, gives bands opposite to that of Q and G and is termed *Reverse* banding. It was initially obtained through controlled heating of euchromatic regions. The O banding technique (Lavania and Sharma, 1979) utilizes trypsin and acid treatment for plant systems, revealing intercalary bands through orcein staining. In general, fluorescent banding, if enhanced with counterstain, is much more effective than normal cytochemical staining with absorption.

In the plant system, G, C and Q banding have been applied by various authors (Kakeda *et al.*, 1991; Marks and Schweizer, 1974; Vosa, 1973). Success is comparatively sporadic in view of the inherent disadvantages of securing air-dried preparations of solid tissue, which is an essential protocol for Giemsa staining. Plant tissue as a whole is less responsive to Giemsa staining. Feulgen staining has been found to be successful in certain genera.

Several DNA binding specific antibiotics such as chromomycin and mithramycin are used as fluorescent dyes. It is possible to stain chromosomes with either dye, followed by DAPI (4-6-diamidino-2-phenylindole). The latter is an AT-specific fluorochrome. Counterstaining with nonfluorescent compounds such as methyl green with chromomycin, and actinomycin D with DAPI, can be carried out. The C bands appear bright with the former and pale with the latter. Similarly, ethidium bromide removes or replaces quinacrine from pale staining regions and QM staining in AT rich sequences becomes distinct. With acridine orange, the picture is reversed specially with GC rich telomeres. Alcoholic extracts of several alkaloids of Papaveraceae also show quinacrine-like fluorescence. As both AT and GC specific bands are available, the combined application permits study of sequence complexity in chromosomes.

The chemical basis of Giemsa banding is derived primarily from the fact that G bands are comparable to Q bands after treatment with fluorescent compounds and observation under UV light. It has been shown that with quinacrine mustard, fluorescent amino-acridine nucleus becomes intercalated within the double helix of DNA. The basic N_2 atoms form ionic bands with DNA phosphate, and alkylating side groups form covalent bonds with guanine DNA (Modest and Sengupta, 1973). The antimalarial drug, quinacrine, though fluorescent, does not form covalent links as no alkylating side group is present. Therefore, it has been suggested that primary binding of both compounds is through intercalation of the acridine nucleus into the double helix whereas in case of quinacrine mustard, the selective alkylation of guanine is a secondary effect. The similarity between Giemsa and fluorescent quinacrine reaction is also borne out by the guanine specificity of fluorescent analogue of actinomycin D (Hecht *et al.*, 1974). The brightly fluorescent Q and G bands are AT rich as evidenced by fluorescence with anti-adenosine antibodies which yield patterns similar to Q and G bands.

Notwithstanding the above evidences that preferential renaturation of repetitive sequences in DNA is responsible for C banding, which is due to disruption and subsequent reformation of chromatin, Lisanti and Stockert (1973) have demonstrated the manifestation of C bands to be equal, in both single and double stranded DNA.

Though dye intercalation and secondary alkylation have been suggested to be responsible for G banding, the involvement of acid protein is also indicated by the induction of bands through the use of trypsin and compounds rich in SH group of proteins. Similarly, chelating agents and other proteolytic enzymes alter the banding pattern (Lavania and Sharma, 1979; Sharma and Sharma, 1973). G and C bands can be related in a sequential manner by the progressive extraction of protein components associated with DNA (Chattopadhyaya and Sharma, 1988). Equally relevant is the report of Comings (1974) showing alkali and/or saline treatment to be a necessity for C banding. It removes 50% of DNA. Repetitive sequences of centromeric DNA must therefore be packaged in a structure much more resistant to extraction than the arms.

Mild treatment with trypsin produces G bands whereas prolonged treatment produces C bands instead of G. Comings has claimed that differences in the proteins associated with DNA may explain the two types of banding. Since acid extraction does not produce bands, it has been suggested that non-histone protein binding is the crucial factor. Under normal conditions, the entire chromosome stains with Giemsa, whereas removal of proteins from certain sites i.e., non-band regions, results in staining in areas (bands) where protein is still left.

The importance of DNA-protein linkage, as in G banding, is also noted in Orcein banding, where pattern is the same. Orcein is an amphoteric dye which stains both DNA and protein, the primary reaction involving tertiary amines of chromosomal polypeptide (Sen, 1965). Orcein stain is very suitable for somatic chromosomes because of its capacity for staining proteins as well. Treatment with mixture of sodium chloride and sodium citrate involves removal of proteins from certain sites. In such sites, due to the nature of the DNA, possibly because of the unique sequences, binding to the dye may be comparatively weak. Dye is retained as shown by bands at sites where the binding is comparatively strong, due to the compact and homogeneous nature of repetitive DNA sequences. Increase in the duration of treatment leads to stepwise disappearance of the band, except C bands which disappear only after prolonged treatment. The removal of chromosome proteins by NaCl (Mirsky, 1947) and trypsin treatment suggests that non-histone proteins are gradually removed for *O*-banding. *C*-bands, observed after both Giemsa and orcein staining in centromeric regions, contain highly repetitive DNA. The fact that short treatment results in banding in intercalary segments while prolonged treatment permits only *C*-bands to be retained, may be considered as an index of the relative difference in binding of chromosomal protein to highly repeated sequences and to less repeated sequences of DNA. Apparently, binding of proteins is governed by the nature of the repeats. The more homogeneous and extended the repeats are, the stronger is the link between DNA and protein. The distribution of proteins, as indicated by orcein banding, varies in the segments of chromosomes, depending on the type of DNA to which they are attached, or more precisely, to DNA of different functions.

The application of *restriction enzyme* on chromosome has also resulted in differential staining of segments, with consequent manifestation of bands. The *restriction RE bands*, observed following fluorescence, may indicate base composition in certain areas of the chromosomes. Such bands may identify different classes of proteins as well as their interaction with each other and DNA, if it is followed by Giemsa staining. The recognition sites of the restriction enzymes are known, such as Hae II for GGCC, Hind III for AAGCTT and *Eco*R for GAATTC. The bands may therefore represent *G*-enriched or *A*-enriched intercalary regions. Such RE bands, induced by restriction endonucleases, can be correlated with distribution frequencies of recognition sites of the corresponding restriction endonucleases in DNA, even in highly organized chromatin (Kamisugi *et al*., 1992). There are also evidences to show that organization of chromatin may be a factor in the activity of these enzymes (vide Sharma and Sharma, 1994). Restriction endonucleases have been demonstrated to modify the action of the buffer solution. The presence of Mg ion in the buffer solution may therefore affect markedly the manifestation of bands (Kamisugi *et al*., 1992).

Theoretically, the stained areas, after digestion with endonuclease and staining with Giemsa or ethidium bromide, represent areas which are not affected by digestion because of the absence of recognition sites. Light stains may also suggest enzyme cuts followed by extraction of DNA. Such RE banding permits the distinction of different types of heterochromatin within a heterochromatic block. Most of the bands after endonuclease cuts manifest a *C* band pattern.

In the *restriction endonuclease nick translation* (RE/NT) banding, the procedure includes three steps, firstly, the digestion of chromosome preparations by endonucleases recognizing specific targets of DNA. This is followed by polymerization with biotinylated nucleotides at the cuts generated by endonucleases. Lastly, the incorporated biotinylated nucleotides are detected through different immunocytochemical methods. However, the pattern obtained through endonuclease treatment, Giemsa staining or nick translation, depends to a great extent on the access of the enzymes to the sites and the extent of DNA extraction. It does not merely depend on the enzyme recognition sites. Despite the limitations of the accessibility of targets to enzymes, the *RE/NT* bands are reliable methods for locating restriction sites on chromosomes. Alternatively, treatment with drugs like 5-azacytidine, to disturb complex chromosome architecture and facilitate accessibility, has also been suggested (de la Torre and Sumner, 1994).

The development of banding pattern technology has also led to the improvement of techniques for demonstration of sister chromatid exchanges in somatic cells. *Sister Chromatid Exchange technique*, otherwise termed as SCE technique, is initially based on the principle of semiconservative replication of chromatids through thymidine uptake and autoradiography (Taylor *et al.*, 1957). In the first cycle, only one chromatid shows incorporation. In SCE technique, bromodeoxyuridine (BrdU) is incorporated through one or two cycles followed by fluorochrome and fluorochrome-cum-Giemsa staining (Latt, 1974; vide Sharma and Sharma, 1980, 1994), or by indirect immunofluorescent method (Yanagisawa *et al.*, 1993). Because of the problem of incorporating BrdU in plant DNA, such reports in plant system are very few (Vosa, 1977). To some extent, this limitation has been overcome by fluorodeoxyuridine (FdU), which inhibits thymidine synthetase and thus thymidilic acid. Simultaneously, additional uridine needs to be added to keep RNA synthesis unhampered. The SCE technique, with Giemsa staining, can clearly differentiate substituted chromatids from non-substituted ones. Similarly BrdU-Giemsa staining can also differentiate late replicating DNA by pale colour or dot formation. Certain authors however suggest that protein modification is responsible for differential staining with Giemsa. UV irradiation and trypsin treatment can also induce differential disintegration of BrdU-substituted chromatids. However, the technique is extremely useful in the study of spontaneous exchanges as well as the effect of environmental mutagens on the plant system and is regarded as the most sensitive cytological method for detecting potential mutagenic and carcinogenic agents. Moreover, the use of indirect immunofluorescent method permits. the study of cell cycle and synchronization.

REPRESENTATIVE SCHEDULES

Differential banding patterns of chromosomes were initially developed for the analysis of human chromosome segments as low and high intensity regions under the fluorescence microscope or as differentially stained areas under the light microscope. The methods were then extended first to different animals and later to plant chromosomes.

The protocol for molecular hybridization, that is, denaturation at the cytological level, if followed by renaturation and staining with different dyes, particularly Giemsa, gives intensely positive reaction at similar segments of chromosomes, which otherwise show repetitive DNA. This banding, following denaturation–renaturation and Giemsa staining, is termed C-banding.

Saline treatment, followed by Giemsa staining, gave a set of patterns, similar in relative staining intensities to the Q-bands. Such bands, known as G-bands, did not require any denaturation-renaturation. Since then G-banding has been obtained following treatments in a variety of chemicals and even mere heating.

R banding or reverse banding, where the pattern is opposite to that of Q and G-bands, was obtained following controlled heating. Acid extraction has been employed for N-banding for the nucleolar-organising region. A CT-banding procedure produces bands in the telomeric and centromeric regions alone. Orcein-staining after incubation in XSSC results in O-banding in plants which is similar in general to the G-band pattern.

Fluorescent banding or Q-banding

Quinacrine mustard binds to chromatin by intercalation of the three planar rings with the large group at position nine lying in the small groove of DNA. Most pale staining regions are caused by a decreased binding of Q, predominantly due to non-histone proteins. DNA-base composition influences the fluorescence by correlation with G-C bonding at lower Q:DNA ratios and by conversion of dyes bound near G-C bases into energy sinks at higher ratios. Chromosomal proteins possibly have a much less pronounced effect.

With Quinacrine mustard (QM) or Quinacrine dihydrochloride (Q)
Pre-treat root tips in aqueous 0.05% colchicine, fix in acetic–ethanol (1:3) overnight, macerate following enzyme treatment, squash directly in 45% acetic acid and stain in 0.5% aqueous quinacrine dihydrochloride for 10–15 min. Rinse in water and mount in distilled water. Alternative stains are H33258 (0.02%) and ethidium bromide.

With Hoechst 33258
Hoechst 33258 (2-[2-(4-hydroxyphenyl)-6-benzimidazolyl]-6-(1-methyl-4-piperazyl)-benzimidazole), shows enhanced fluorescence with both AT and GC-rich DNA. It can be used as a probe for identifying all types of AT-rich regions in chromosomes, including those which are not demonstrable with

quinacrine. Its interaction with DNA and chromatin is characterised by changes in absorption and circular dichroism measurements.

Fix root tips, after pre-treatment in colchicine (4 h), in acetic–ethanol (1 : 3). Treat 1-day old slides in barium hydroxide/saline mixture and stain in H33258 (40 μg/ml) in McIlvaine's buffer, pH 4.1, for 10 s. Rinse and mount slides in same buffer and seal. Heat the sealed slides for about 6 s on a hot plate (120°C) and cool rapidly by placing them, coverglass down, between two slabs of dry ice. A combination with Giemsa banding has been employed in rye.

Reverse fluorescent banding with chromomycin and DAPI

Two DNA binding guanine specific antibiotics, chromomycin A_3 (CMA) and closely related mithramycin (MM), are used as fluorescent dyes. Metaphase chromosomes in roots can be sequentially stained with CMA or MM and the DNA-binding AT-specific fluorochrome 4′-6-diamidino-2-phenylindole (DAPI). Non-fluorescent counterstain may be used, methyl green with CMA and actinomycin D (AMD) with DAPI. In general, C-bands which are bright with CMA and MM, are pale with DAPI and *vice versa.*

Treat root tips with colchicine (0.05%, 3–6 h), fix in acetic–ethanol (1 : 3, overnight), squash in 45% acetic acid and air-dry after removing coverglass by dry ice.

For CMA staining Pre-incubate air-dried slides in McIlvaine citric acid-Na_2HPO_4 buffer (pH 6.9–7.0), containing 10 mM $MgCl_2$ for 10–15 min. Stain slides in buffer containing 10 mM $MgCl_2$ and 0.12 mg/ml CMA ('reinst', Serva, Heidelberg) or 0.11 mg/ml MM for 5–10 min, wash and mount in McIlvaine's buffer (pH 6.9–7.0) and seal with rubber solution.

For counterstaining, pre-incubate slides in McIlvaine's buffer (pH 4.9) for 10–25 min, stain for 5–15 min in a buffered solution (pH 4.9) of 0.1% methyl green GA (Chroma, Stuttgart), from which methyl violet has been removed with chloroform extraction. Rinse in buffer (pH 4.9) and in neutral buffer containing 10 mM $MgCl_2$, stain in CMA as described before.

For sequential staining with CMA and DAPI Observe and photograph CMA or MM fluorescence as for Q-banding. Remove coverglass by dry ice or rinsing in buffer and then acetic–methanol and methanol successively. Remove chromosome-bound CMA or MM, extracting with pyridine for 2–3 days. Stain with DAPI solution (1 μg/ml in McIlvaine's buffer, pH 7.0).

For sequential fluorochrome staining with CMA and Hoechst 33258 Immerse slides in freshly prepared 0.02% Hoechst 33258 (Sigma) in ethanol for 3 min; drain and transfer to ethanol for 2 min, air-dry. Place about 30–50 μl of a 0.5 mg/ml solution of chromomycin A3 (Sigma) in McIlvaine's buffer (pH 7.0) plus 5 mM $MgCl_2$ over each preparation and cover with a coverglass. Incubate the slides for 1 h in a moist box at room temperature in dark. Rinse off coverglass with distilled water and briefly air-dry. Mount each slide in a small droplet of 50% Citifluor **PBS** (Agar Scientific) in McIlvaine's buffer plus $MgCl_2$ (pH 7.0).

Observe chromosome fluorescence in a Zeiss Axiophot microscope with mercury vapour lamp. Use Zeiss filter set 1 (excitation 365 nm; emission 395–297 nm) for *Hoechst fluorescence* and Zeiss filter set 6 (excitation 436/ 8 nm; emission 460–470 nm) for *chromomycin staining*.

Bands that are bright with Hoechst 33258 will be dim with CMA and *vice versa*. All positive *G* bands could be differentiated into two types with these two stains, indicating the presence of AT-rich or GC-rich heterochromatin.

Ethidium bromide as counterstain for Quinacrine
Fix cellular smears in 95% ethanol. Keep in 95 and 70% ethanol for 3 min each; distilled water, 3 min; citric acid phosphate buffer (0.01 M, pH 5.6) 3 min; staining solution (50 μg/ml QM in citric–phosphate buffer, 0.01 M, pH 5.6, containing 0.5% zinc sulphate), 10 min; two changes of 6 min each in pH 5.6 buffer; 5% zinc sulphate solution, 5 min; two changes in distilled water of 4 min each; ethidium bromide solution (2 μg/ml EB in 7.0 pH phosphate buffer, 0.01 M), 5 min; 7.0 pH phosphate buffer (0.01 M), two changes of 2 min each; 7.6 pH buffer (0.01 M) two changes of 4 min each; mount in same buffer and seal. With this counterstain, cytoplasm fluoresces pale green, nuclei pale orange and interphase fluorescent bodies appear as bright yellow spots, as also the brightly fluorescing bands in metaphase chromosomes.

The mechanism of the counterstaining by EB is not yet clear. Presumably the brilliant areas contain AT-rich DNA binding firmly with QM, and EB removes, replaces or obscures the quinacrine in weakly stained areas, so that densely stained chromatin, by contrast, seems more prominent.

Other chemicals observed to give quinacrine-like fluorescence include alcoholic extracts of the alkaloids from fresh roots of eight genera of Papaveraceae and Fumariaceae as also from *Chelidonicum majus*, *Macleaya cordata* and *Glacium flavum*. Of limited use are sarcolysinoacridin, berberine sulphate, and 2,7-di-*t*-butyl proflavine DBP.

Giemsa or *G*-banding

Giemsa is a complex mixture of thiazine dyes and eosin. Of them, methylene blue and Azure A, B and C alone give good banding. Thionin, with no methyl groups, gives poor banding while eosin has no effect.

Strong banding is favoured by presence in stain of high concentrations of methylene violet and its immediate homologues. Chromatography shows that only small amounts of these dyes occur in dry Leishman and Giemsa powders. However, additional active dye may be generated during preparing solution from dry powder.

According to Matsui and Sasaki (1975) during G-banding, macromolecules like DNA and proteins are lost, leading to an uneven distribution of chromatin. Non-histone proteins of relatively larger molecular sizes are removed and the G-positive bands represent relatively thermostable chromatin consisting of smaller protein molecules.

Several chemicals have been used for inducing G-bands in addition to Ba(OH)$_2$, like trypsin, urea, SDS and NaOH. The chromosomes may be prepared by pretreatment, fixation, degradation, hypotonic treatment, cell suspension and air-drying or by conventional squash technique.

G-banding for root tips
Grow roots at 10°C; pre-treat in 0.05% colchicine in the dark for 4 h at room temperature, fix in acetic–ethanol (1:3) for 48 h and store in 70% ethanol at 4°C. Hydrolyse root tips for 8 min in 1 N HCl at room temperature, squash in 45% acetic acid, remove coverglass by dry ice method, dip in absolute ethanol, air-dry and store in desiccator for 5 days. Treat with aqueous 0.064 M Ba(OH)$_2$8H$_2$O for 50 min at room temperature, wash in two changes of deionized water for 5 min, incubate in 2× SSC (pH 7.0) at 60°C for 40 min, rinse in two changes of deionized water and air-dry. Stain in fresh Giemsa stock solution, diluted 50 times with M/15 Sörensen phosphate buffer (pH 6.8) for 5–8 min at room temperature, rinse in deionized water and mount in Permount.

Alternative schedule Pre-treat root tips with colchicine (0.1% for 1–2 h or 0.05% for 2–3 h 30 m); fix in acetic–ethanol (1:3) for 3 h, store overnight in 90% ethanol; squash in 45% acetic acid on slides 'subbed' with Haupt's adhesive. Remove coverglass with dry ice, rinse in ethanol, air-dry and incubate in hot Ba(OH)$_2$, varying between 15 min at 50°C in 5% Ba(OH)$_2$ to 20 min at 60°C in 6.5%. Rinse in distilled water, incubate in 2× SSC at 65°C for 1–2 h, rinse, stain in Giemsa (Gurr's R66 improved stock solution diluted with 50 times M/15 Sörensen buffer pH 6.9) for 3–24 h. Wash in distilled water, air-dry and mount in DPX. Modifications of the BSG techniques have been employed with varying degrees of success in several angiosperms.

G-banding after maceration
Schedule 1: Pre-treat root-tips in suitable pretreatment agent and fix in acetic–methanol (1:3) at room temperature. The periods for pretreatment and fixation will depend on the material used. Macerate root-tips in 2.5% cellulase and 2.5% pectinase at 27°C for 3.5 h. Squash on a slide in a few drops of fixative with forceps and flame dry.

Treat preparations on slides in an oven at 90°C for 50 min, then incubate in 2× SSC (pH 7.4) at 60°C for 40 min; rinse in distilled water. Stain in Giemsa diluted 50-fold with Sörensen's buffer, M/15, pH 6.8 at 25°C for 40 min.

Schedule 2: Treat preparations on slides with either (i) 0.01% trypsin (made up with Difco 250 : 1, 0.02% EDTA-CMF solution) for several sec at 4°C; or (ii) 8 M urea (made up with 0.02% EDTA) and 1/15 M phosphate buffer (pH 7.4) in proportion of 2 : 1 at 20–25°C.

Then wash with shaking with 0.85% NaCl and stain with 2.5% Giemsa (phosphate buffer, pH 7.4 and Giemsa 40 : 1) for 5–10 min.

In an *alternative trypsin-urea* method, treat the preparation as in (i) with trypsin, wash with 0.85% NaCl; keep in urea-phosphate buffer (as in (ii) above) for several sec at 4–10°C. Wash in 0.85% NaCl and stain with Giemsa as given above.

G-banding for flower-buds
Fix buds for 24 h in Pienar's fixative (6 : 3 : 2, methanol, chloroform, propionic acid). Store at 4°C for 1 week in 90% ethanol. Smear anthers in 45% acetic acid, remove coverglass with dry ice and air-dry. Immerse slides in 45% acetic acid for 20 min at 60°C, wash for 15 min in tap water. Rinse finally in distilled water.

Place in fresh saturated aqueous $Ba(OH)_2$ for 5 min at room temperature, wash in tap water for 1 h. Incubate in 2× SSC at 60°C for 1 h, rinse in distilled water, stain in 2% Giemsa (GT Gurr's R66 improved stock solution diluted 50 times with M/18 Sörenson phosphate buffer pH 6.9) for about 1 × h. Rinse rapidly in distilled water, air-dry, rinse in euparal essence and mount in euparal.

Schedule for kinetochore staining in meiosis
Immerse air-dried slides for 5 min in 2× SSC (pH 7.0) at 90°C, 1 h in 2× SSC at 60°C, stain in Giemsa.

Or, keep air-dried slides for 10 min in potassium phosphate buffer (pH 6.8, 0.12 M) at 90–94°C; 15 s in ice-cold phosphate buffer quinch; 1–24 h in 60°C phosphate buffer; 15 s in ice-cold phosphate buffer quinch and stain in Giemsa. Kinetochore takes up stain. These methods were tried in *Triticale*, wheat-rye hybrids and *Secale*.

Stains other than Giemsa may produce *G*-banding after the usual banding schedule. Methylene blue results in preferential staining pattern of *Vicia* chromosomes similar to *G*-staining. It binds to nucleic acids and interacts weakly, mainly as outside binding with chromatin.

Feulgen-bands
Feulgen staining has been employed in lieu of Giemsa in producing *G*-bands. Store pre-treated fixed root tips in 70% ethanol at 4°C for at least four days. Hydrolyse in 10% HCl at 60°C, wash thoroughly, stain in leucobasic fuchsin

solution, wash and stain in acetic–carmine for 5 min. Squash in a drop of acetic–carmine and heat overa steambath for 1 to 2 min. This process has, however, limited application.

High resolution G bands in maize

(i) Germinate maize seeds. Remove root-tips about 1 cm long and pretreat with 0.05% colchicine solution containing either 10 ppm actinomycin D or ethidium bromide for 2 h at 25°C for *G*-banding. Transfer to the Ohnuki's hypotonic solution (55 mM KCl, 55 mM NaNO₃, 55 mM CH₃COONa, 10:5:2) for 30 min to 1 h at 25°C. Fix in acetic acid–methanol (1:3) for 1–4 days at −20°C. The period can be reduced to 2 h at 25°C for root-tips pretreated with actinomycin D. Macerate fixed root-tips using a tweezer either in fresh fixative or in enzymatic mixture. For enzymatic maceration, wash fixed root-tips for 10 min, and then macerate in a mixture (pH 4.2), containing 2% Cellulase RS and 2% Macerozyme R 200 for 20–60 min at 37°C in a 1.5 ml Eppendorf tube. Rinse in water 2–3 times. Transfer each macerated root-tip to a glass slide with the help of a pasteur pipette, add fresh fixative and cut into small pieces with a sharp pointed tweezer. Observe under phase contrast. Airdry good preparations for 2 days in an incubator at 37°C.

(ii) For banding, directly stain samples pretreated with actinomycin D in 10% Wright's solution diluted with 1/15 M phosphate buffer (pH 6.8) for 10 min at 25°C. Wash and airdry. Again fix enzymatically macerated samples in 2% glutaraldehyde solution, diluted with the phosphate buffer, for 10 min at 25°C. Wash. Immerse the post-fixed slides either in 2% trypsin, dissolved in PBS (pH 7.2) for 10 min at 25°C or in 0.02% SDS dissolved in Tris-HCl buffer (20 mM, pH 8.0) for 2–5 min at 25°C. Briefly wash and airdry the slides. Stain in 5% Wright's solution in 1/30 M phosphate buffer (pH 6.8) for 5 min.

To identify Knob positions, a slight modification can be used. Airdry slides after preparation at room temperature for a week, immerse in 5% Ba(OH)₂ for 6 min at room temperature and quickly transfer to 0.1 N NaOH for 30 s at room temperature. Dip slides in 2× SSC for 1 h at 60°C. Wash. Stain with 2% Wright's solution for 30 min to 1 h. Photograph well banded plates in black and white film.

Analyse enlarged microphotographs by the CHIAS method (Fukui, 1985, 1986).

(iii) The typical high resolution *G*-bands can be seen at prometaphase and become coarser in metaphase. Hypotonic treatment with Ohnuki's solution gave clearer bands than control. Trypsin and SDS were more suitable than urea and Triton X-100 and Wright staining solution than Giemsa solution.

C-banding

C-banding techniques were discovered as a byproduct of the *in situ* RNA/DNA hybridization procedure. The methods were initially devised for mammalian chromosomes but have been found to be effective in other systems as well.

Giemsa C-banding

C-banding should be considered as a technique that stains all constitutive heterochromatin whereas N-banding and modified C-banding are specialized staining techniques.

C-banding schedules

(A) Pre-treat root tips in 0.05% colchicine for 4 h, fix in acetic–ethanol for 12–15 h, hydrolyse in 1 N HCl at 60°C for 25 s, squash in 45% acetic acid with albuminized coverglass. Separate coverglass in absolute ethanol or by CO_2 freezing; dry at 60°C; denature in $Ba(OH)_2$ for 5 min at room temperature. Rinse in running distilled water, dry, incubate in 2× SSC for 1 h at 60°C, rinse and dry again. Stain in 0.5% Giemsa at pH 6.8 for 5–15 min, wash in distilled water, dry, dip in xylol and mount in canada balsam or DPX. Modifications involve changes in time of incubation and temperature.

(B) Hydrolyse pretreated and fixed root-tips in 45% acetic acid for 25 min, incubate for 5 min in 5% barium hydroxide at 18°C. Stain in 2% Giemsa (Gurr's R66) in Difco buffer, pH 6.8 for 5–10 min. Rapidly airdry slides by shaking. Mount in euparal. Alternatively, dip dry slide in xylol and mount in Canada balsam or DPX.

(C) For *Pinus* ovules, fix in acetic acid–ethanol (1:3) for 24 h; keep for 1 week in 90% ethanol at 4°C; squash in 45% acetic acid after gentle heating. Remove coverglass by dry ice and airdry. Incubate in 45% acetic acid for 20 min at 60°C. Wash thoroughly. Incubate in 5% barium hydroxide at 54–56°C for 15 min; wash in water for 1 h. Treat in 2%× SSC (pH 7.0) at 60°C for 1–2 h. Wash. Stain in 2% Giemsa for 1 h, wash, airdry and mount.

N-banding

Treat freshly excised root-tips with ice-cold water for 22 to 24 h. Fix in glacial acetic acid–99% ethanol (1:3) for upto 3 days in room temperature or even upto one month in cold. Stain root-tips in 1% acetic–carmine solution for 1–2 h in room temperature and squash in 45% acetic acid. Remove the coverglass by freezing. Treat the preparation with 45% hot acetic acid at 55 to 60°C for 5–10 min. Air dry overnight. Incubate the preparation in hot phosphate buffer (1 M NaH_2PO_4) at 94 ± 1°C for 2 min, rinse briefly in tap water and airdry. Next, stain preparation in Banco Giemsa stain (1 drop per ml Sörenson's phosphate buffer) for 30–40 min at room temperature, rinse briefly in tap water and airdry. This method has been used in Gramineae, specially wheat (Gerlach, 1977; Gill *et al.*, 1991).

Modified C-banding (MC)

MC combines parts of the protocols of N-banding and C-banding procedures.

Treat freshly excised roottips in cold water for 24 h and then fix in glacial acetic acid–ethanol (99%) 1:3 for one day. Stain in 1% acetic–carmine solution for 1–2 h at room temperature and squash in 45% acetic acid. Remove coverglass by freezing, treat with 45% acetic acid at 50°C for 5–10 min and airdry overnight. Next treat preparations in a saturated barium hydroxide solution at 50°C for 2.5 min, rinse in tap water, immerse in 2× SSC at 50°C for 3–10 min. Rinse in tap water. Stain wet slides in Wright or Leishman Banco Giemsa staining solution (one drop per ml Sörenson's phosphate buffer) for about 20 min, rinse in water and airdry.

Alternatively, treat freshly excised roottips with 0.05% colchicine for 3 h. Fix in glacial acetic acid–ethanol (99%) 1:3 for upto 3 days at room temperature. Keep in 45% acetic acid for 2–3 min. Squash. Remove coverglass by freezing. Keep in 99% ethanol overnight and air dry for a few min. Incubate the preparations in 0.2 N HCl at 60°C in a water bath for 2 min. Wash in distilled water, incubate in saturated barium hydroxide solution at room temperature for 7 min, rinse in distilled water and incubate in 2× SSC for 1 h at 60°C. Transfer slides directly to 1–5% Giemsa staining solution (Fisher) in phosphate buffer and keep for upto 30 min. Wash and airdry. Place airdried slides in xylene, mount in Permount.

Combined C and N banding for plants

Preparation of slides: Excise fresh root-tips from germinating seedlings. Pretreat with distilled water at 0°C for 16 h. Fix in freshly prepared acetic acid–methanol (1:3) and keep at −20°C for about a week. Wash the roots with distilled water and then macerate with an enzymatic mixture (4% cellulase Onozuka RS; 1% pectolyase Y-23; 75 mM KCl, 7.5 mM Na$_2$EDTA, pH 4.0) at 37°C for 45–60 min. Rinse in distilled water. Place roottip in a few drops of fixative on a glass slide, tap carefully with the tip of a pair of forceps and airdry.

For C banding: Incubate the air-dried slides after ageing for a week in 0.2 N HCl at 55°C for 3 min, in 5% barium hydroxide solution at 25°C for 5 min and then in 2× SSC at 55°C for 20 min. Rinse in distilled water between each step.

For N-banding: Incubate air-dried aged slides in 1 M sodium dihydrogen phosphate solution at 90°C for 3 min. Rinse in distilled water. When staining for either C or N-bands stain the treated slides with 8% Wright solution (Merck) diluted with 1/30 M phosphate buffer (Na$_2$HPO$_4$ or KH$_2$PO$_4$, pH 6.8) for 1 h at room temperature. Rinse and airdry. Adjust the concentration of Wright's solution and duration of staining within the extents of 5–10% and 30 min–2 h respectively, according to quality of slides and Wright's solution.

For combined staining: Photograph after staining for either *C* and *N* band and then subject to the procedure for the other banding without destaining.

The two patterns are in general similar. However, in barley, every centromeric site showed an *N*-band positive and *C*-band negative heterochromatin at the centromeric site. It is suggested that the differential alteration of chromosome structure rather than the differential extraction of chromatin, may be responsible for such *N* + *C*-heterochromatin (Kakeda *et al*., 1991).

Alternatively, pretreat freshly cut roottips with 0.05% colchicine solution containing either 10 ppm actinomycin D or ethidium bromide, for 2 h at 25°C. Transfer to Ohnuki's hypotonic solution (55 mM KCl, 55 mM NaNO$_3$, 55 mM CH$_3$COONa, 10:5:2) for 30 min to 1 h at 25°C. Fix in acetic acid–methanol (1:3) for 1 to 4 days at −20°C in a freezer. For actinomycin D-treated roottips, the fixation can be reduced to 2 h at 25°C. Smear or squash roottips either with a tweezer in fresh fixative or macerate enzymatically. For enzymatic maceration, wash fixed roottips for about 10 min and macerate in an enzymatic mixture (pH 4.2) containing 2% cellulase RS and 2% macerozyme R-200, both of Yakult Honsha Co. Ltd. Tokyo for 20 to 60 min at 37°C in a 1.5 ml Eppendorf tube. Rinse with water 2 or 3 times. Pick up macerated roottip onto a glass slide using a pasteur pipette, add fresh fixative and cut into small pieces. Squash, air dry for 2 days in an incubator at 37°C. Stain samples prepared by actinomycin pretreatment in 10% Wright solution diluted with 1/15 M phosphate buffer (pH 6.8) for 10 min at 25°C. Wash and air dry.

Fix enzymatically macerated samples again in a 2% glutaraldehyde solution diluted with the phosphate buffer for 10 min at 25°C and wash. Immerse post-fixed slides either in 2% trypsin (Merck) dissolved in PBS (pH 7.2) for 10 min at 25°C, or in 0.02% SDS dissolved in tris-HCl buffer (20 mM, pH 8.0) for 2–5 min at 25°C. Briefly wash and airdry. Stain in 5% Wright solution in 1/30 M phosphate buffer (pH 6.8) for 5 min (Kakeda *et al*., 1990).

Three types of *G* bands are obtained: by actinomycin, trypsin and SDS. Pretreatment with colchicine or colchicine and ethidium bromide aid in the accumulation of prometaphases in which the fine patterns can be observed. Image analysis can be done from enlarged microphotographs through the digitally manipulated chromosome image analysing system CHIAS (Fukui, 1986).

CT-banding

CT-banding is a further amplification of *C*-banding techniques. Treat air-dried preparations with Ba(OH)$_2$ solution at 60°C, incubate in 2× SSC at

60°C and stain in cationic dye 'Stains All'. The bands are of C and R types, located mainly at telomeric regions.

Cd-banding

Cd-banding reveals two identical dots (centromeric dots Cd) at the centromeric region, one for each chromatid. Store airdried preparations for one week at room temperature. Incubate in Earle's BSS medium (pH 8.5 to 9.0) at 85°C for 45 min. Stain in 4% Giemsa in 1/30 M phosphate buffer (pH 6.5).

Orcean banding or O-banding

It was developed primarily for use in plant chromosomes (Sharma, 1975, 1978). The technique principally involves an elimination of the denaturation step and consists of pre-treatment of the tissue, fixation in a strong concentration of × SSC, washing, staining in acid-orcein and mounting in 45% acetic acid. Orcein-positive bands appear in different segments of chromosomes, including the centromeric and intercalary ones. The mechanism of reaction possibly involves the DNA-protein linkage, since orcein is an amphoteric dye, capable of staining both DNA and protein. The gradual removal of non-histone protein through SSC treatment is principally responsible for O-banding at the sites of stronger DNA-protein linkage. As the removal of protein by SSC application is gradual, mild treatment results in intercalary bands comparable to G-banding whereas strong prolonged treatment ultimately shows only C-bands where the linkage is strong due to highly homogeneous repeats. The method therefore allows localization of major and minor reiterated sequences in chromosomes.

Fix pre-treated tissue in acetic acid–ethanol (1 : 2) for 2–12 h. Treat in 45% acetic acid for 5 min. Wash in water. Treat in a mixture of 1 M sodium chloride and 0.1 M sodium citrate (1 : 1) at 27–28°C for 2–3 h. Wash in water. Warm for a few seconds at 90°C in a mixture of 2% acetic–orcein solution and N HCl (9 : 1). Keep in the mixture at 27–28°C for 1 h. Squash under a coverglass in 45% acetic acid and seal.

Reverse banding or R-banding

The pattern is the exact reverse of G-banding, indicating a similarity of the mechanism involved. The pattern is produced by incubation at high temperature or a suitable pH.

NOR-banding

NOR-banding has been employed principally in the localization of nucleolar organising regions.

By Giemsa staining

Incubate air-dried slides at $96 \pm 1°C$ for 15 min in 1 M NaH_2PO_4 solution (pH 4.2 ± 0.2 adjusted with 1 N NaOH). Rinse in distilled water, stain in Giemsa (diluted 1 in 25 in 1/15 M phosphate buffer, pH 7.0), rinse in tap water and air-dry.

By Ag-NOR staining

Solutions used   Silver nitrate: Dissolve $AgNO_3$ (4g) in 8 ml distilled water; keep in dark. Discard if blackening occurs.

Colloidal developer: Add gelatin (2 g) to 100 ml distilled water. Stir continuously with gentle warming to dissolve. Add 1 ml pure formic acid. Discard after 2 weeks.

Giemsa: 5% Gurr's Improved R66 (BDH) in Gurr's buffer (pH 6.8).

Method (after Sumner, 1994)   Use slides 2–3 days after preparation for chromosomes. Mix two drops of the colloidal developer and four drops of $AgNO_3$ solution in an Eppendorf tube. Pipette the mixture onto the preparation and cover with a coverglass. Place on a hotplate preheated to approx. 70°C for 1–2 min. Remove when the solution turns golden-yellow. Remove slide from hot plate. Wash off coverglass with a stream of distilled water. Wash thoroughly with distilled water. Counterstain with Giemsa for about 5 min. Rinse with distilled water, blot, dry thoroughly, mount with synthetic neutral mountant like DPX. The time for silver staining should be as short as necessary in order to prevent other structures from taking up stain.

Silver staining for nucleolar organizing regions (*NOR*)   Pretreat root-tips in aq. 0.05% colchicine solution for 2.5–3 h at 20°C. Fix in acetic–ethanol (1 : 3) for 2 h or overnight at 4°C. Dip successively in 50% and 30% ethanol and then wash with distilled water for 5–10 min. Macerate in a mixture of 95% ethanol–acetic acid–1 N HCl (3 : 2 : 5) for 5–10 min at 60°C. Wash with distilled water 3 times for 10 min each and then soak in distilled water. Squash in a drop of 45% acetic acid under a coverglass. Remove coverglass with ice-freezing method and then dip in 95% ethanol for 5 min. Air-dry at room temperature for 2 h or overnight. Take a clean slide, add a drop of 1% aqueous solution of gelatin (add 1% formic acid) and 2 or 3 drops of fresh $AgNO_3$ (dissolved in deionized water) by pipette. Mix and cover with coverglass.

Incubate slide in a moist chamber at 60–65°C for 5–15 min. When the staining solution turns golden-brown, examine under the microscope. Rinse with distilled water to remove the coverglass. Fix in sodium thiosulphate solution for 5–10 min, Rinse in distilled water. Counterstain with an aqueous solution of 0.001% methylene blue for 1–3 min. Rinse in distilled water 2–3 times. Airdry and mount in neutral balsam.

The nuclei and chromosomes stain light green, the nucleoli and NOR are brown to black while the cytoplasm is yellow.

Alternative method   After fixation, macerate root-tips in a mixture of cellulase and pectinase (2% each) at 37°C for 3 h and refix for 3 h. Smear the tip cells by flame-drying method. Store overnight at 37°C. The subsequent procedure for silver staining is similar to that given above.

HKG banding

Pretreat germinated seeds directly with 0.05% 8-hydroxyquinoline solution for 2 h 30 min at 22°C. Dissect out the embryos, wash with distilled water and fix in three changes of chilled acetic acid–methanol (1 : 3), keeping 15 min in each change. Store at −20°C for one to several days. Wash in water thoroughly. Dissect out root-tips, mince in tiny fragments in distilled water. Centrifuge at 150 g for 2 min. Resuspend pellet in 2 ml freshly prepared enzyme (0.2 ml flaxzyme NOVO and 1.6 ml distilled water). Incubate at 35°C for 2 h 30 min. Remove supernatant after centrifuging for 1 min at 150 g. Gently resuspend pellet in 2 ml distilled water. Filter through nylon mesh into another centrifuge tube and make up volume to 5 ml with distilled water, centrifuge for 1 min at 150 g. Wash pellet with cold acetic acid–methanol (1 : 3) and resuspend in 0.5 ml of same fixative. Drop 4 or 5 drops of cell suspension on a slide covered with 45% acetic acid, air dry, place slide on a hot plate at 50°C for 10 min. Age at 35°C for one to several days.

Treat two-weeks old slides with 5% barium hydroxide solution at 56°C with continuous agitation for 10–15 s, wash in two changes of 70% ethanol, then absolute ethanol and then transfer to acetic acid–methanol (1 : 8). Dry on a hot plate for a few minutes, stain with 3% Giemsa (Merck) in phosphate buffer, pH 6.8 for 8–10 min. Wash twice in distilled water and air dry. Observe for *C banding*. For *HKG banding*, take one to five day old slides. Hydrolyse at 60°C in N HCl at 4–6 min, wash four times in distilled water for a total of 10 min. Briefly immerse the slides in 0.9% NaCl, dip 10 times in 70% ethanol, dry on a hot plate at 50°C for a few seconds. Immerse in 0.06 N KOH solution for 8–12 s in room temperature with continuous agitation. Wash in two changes of 70% ethanol, absolute ethanol and transfer to acetic acid–methanol (1 : 3). Dry on a hot plate and stain with Giemsa as described earlier for *C*-banding. Dark distinct bands are produced which are different from *C* bands. The method has been applied in both wheat and maize (De Carvalho and Saraiva, 1993; Shang *et al.*, 1988).

HY-banding

HY-banding was initially applied to somatic chromosomes of some members of the Liliiflorae. Root tips, fixed in acetic–ethanol (1 : 3), on treatment with

0.1 or 0.2 N HCl between 60 and 80°C and staining with aceto-carmine, gave banded chromosomes. Heterochromatic regions stain differentially, mentioned as Hy^+ and Hy^- bands but do not always coincide with *G*-bands.

High resolution (*HR*) banding (Zhu and Wei, 1987)

Schedule 1 (AMD technique)
Pretreat root-tips in a mixture of actinomycin D and colchicine at 25°C for 2 h. Wash. Treat in Ohnuki's hypotonic solution (0.055 M each of KCl, $NaNO_3$ and CH_3COONa as 10:5:2) at 25°C for 0.5–15 s. Wash. Treat in 2.5% enzyme solution (pectinase and cellulase) at 25°C for 2 h. Wash. Fix in acetic–methanol (1:3) for 2 h. Transfer to a few drops of fixative on a clean slide, smash and smear. Stain in Wright–Giemsa solution.

Schedule 2 (Trypsin technique)
Pretreat root-tips in 0.5% colchicine for 2 h. The steps of enzymolysis and fixation are similar to the earlier schedule. Stain root-tips in 1% acetic–carmine, squash in 45% acetic acid. Remove coverglass by freezing. Immerse in 45% acetic acid at 60°C for 15 min and 95% ethanol at 25°C for 10 min. Transfer to an oven at 60°C for 2 h and then airdry for 24 h. Treat the dried preparations with 0.025% trypsin (pH 8.0) at 35°C for 0.5–2 min. Wash with 0.85% NaCl solution and then distilled water. Stain in Wright–Giemsa solution.

Restriction endonuclease banding

Two categories of chromosome bands are observed,

(a) Related to DNA base composition in certain areas, by simply staining fixed chromosomes with quinacrine or Hoechst 33258 or chromomycin $A_3(CMA_3)$.

(b) Depending on the different classes of chromosomal proteins, in their interaction with each other and/or with DNA by physical or chemical treatment of chromosomes, followed by staining in dyes like Giemsa. Restriction endonucleases selectively digest fixed chromosomes and subsequent staining with Giemsa or Ethidium bromide gives differential staining.

RE bands
Prepare working solution of the restriction endonuclease by dissolving the enzyme in the assay buffer suggested by the manufacturer to give a final concentration of units/ml according to enzyme.

Allium cepa, Hordeum vulgare, Vicia faba Pretreat root-tips suitably (in 0.05% aq. colchicine or 0.025% in a saturated solution of alphabromonaphthalene) for 3 h. Fix in acetic acid–ethanol (1 : 3). Digest roottips at 37°C in 1% cellulase for 90 min and 2% pectinase for 2 h; dilute in 0.01 M sodium citrate buffer, pH 4.4 to 4.8. Squash roottips in 45% acetic acid. Remove coverglass with dry ice. Wash in 96% ethanol and airdry. Slides can be stored in glycerol for several months.

Before enzyme treatment, wash out glycerol by 2× SSC for 10 min. Air dry. Equilibrate the airdried slides in the incubation buffer for 1 h. Put 20 µl drops of buffer and buffer-containing enzyme respectively on the slides. Incubate in moist chamber at 35°C to 45°C (according to enzyme) for 2–24 h. Remove the drops by pipettes. Briefly rinse slides in distilled water and stain in 4% Giemsa (Merck) for 2–5 min, airdry and mount in entellan (Merck).

RE banding in *Vicia faba* gave almost the same patterns as conventional Giemsa banding but the method is still rather variable (Schubert, 1990).

Barley (after Kamisugi *et al.*, 1992) Germinate barley seeds in a Petri dish at 25°C. Excise primary roots when they are 1–1.5 cm long and dip in 0°C chilled distilled water for 14–20 h. Fix in acetic acid–methanol (1 : 3) overnight. Wash thoroughly. Macerate with the enzyme cocktail containing 2% cellulase, 1.5% macerozyme R200, 0.3% pectolyase Y-23; 1 mM EDTA, pH 4.2, at 37°C for 20 min on the glass slide (Kamisugi and Fukui, 1990). Rinse and tap the root tips briefly by a fine forceps with the fresh fixative. Airdry the slides.

Enzymes used were: *Mbo*II, GAAGA8/7; *Rsa*I, GT↓ AC; *Hae* III, GG↓CC; *Hinf*I, G↓ANTC; and *Dra*II, PuG↓GNCCPy.

Working solution for each: 200 units in 100 µl of incubation buffer with suitable salt concentration. Buffer solution was 10 mM Tris-HCl (50 mM for H buffer); 10 mM $MgCl_2$, 1 mM dithiothreitol (DTT), pH 7.5 with different NaCl concentrations. (NaCl concentration for L buffer O mM; M buffer 50 mM; H buffer 100 mM.)

Drop 20 µl of the working solution on chromosome spreads on the slides. Apply a clean coverglass. Incubate at 35°C for 2–6 h in a humid chamber. Rinse, stain with a 2% Giemsa solution (1/15 M phosphate buffer, pH 6.8) for 15–30 min. Specific bands were seen with the different enzymes, while buffer solutions could alter these bands.

II. SISTER CHROMATID EXCHANGE

The initiation of banding methodology led to improvements in differentiating sister chromatids. It was initiated by incorporating bromodeoxyuridine (BrdU) during one or two successive replications and staining with fluorochromes afterwards. Alternatively the slides were stained with Giemsa after treatment with a fluorochrome, exposure to light or storing and heating.

This technique permits the identification of two chromatids of each chromosome on the basis of differential staining intensities. The chromosomes of *somatic tissues* show the formation of sister chromatid exchanges (SCEs).

Another set of methods utilizes BrdU-Giemsa staining to identify late DNA-replicating sites by pale colour or dot formation. Dotted chromosomes may also be produced with sodium phosphate solution supersaturated with $NaHCO_3$. A combination of *G*-banding and autoradiography has been employed also to map sister chromatid exchanges.

BrdU-dye techniques provide a new approach to study both structural and functional properties of metaphase chromosome bands. Incorporation of BrdU for an entire cycle followed by 33258 Hoechst staining does not give marked bands but its incorporation for only part of one S phase differentiates between early and late replicating chromosome regions, the latter corresponding to the *Q* and non-centromeric *C*-band positive chromosome segments. After two cycles of BrdU incorporation, sister chromatids can be identified. The random assortment of sister chromatids of homologues may be observed after a third cycle Baseline sister chromatid exchanges (SCEs), spontaneous or BrdU-induced, are more frequent in $Q+$ bands. They are sensitive indices of DNA damage by alkylating agents and light.

Schedule

Expose lateral roots to aqueous solution containing $100\,\mu M$ 5-BrdUrd, $0.1\,\mu M$ 5-FdUrd and $5\,\mu M$ Urd for 22 h. An alternative solution is $10^{-7}\,M$ FdU, $10^{-4}\,M$ BrdU and $10^{-6}\,M$ Urd Transfer to aqueous solution containing $100\,\mu M$ thymidine (dThd) and $5\,\mu M$ Urd for 21 h. Treat in 0.05% colchicine for 3 h, fix overnight in acetic–methanol cold (1 : 3) in dark at 20°C. Rinse in 0.01 M citric acid–sodium citrate buffer (pH 4.7), incubate for 75 min at 27°C with 0.5% pectinase, dissolved in the same buffer. Squash in 45% acetic acid, coat with a mixture of 10 : 1 gelatin and chrome alum. Remove coverglass by dry ice and bring preparation to water through descending ethanol grades. Incubate in moist chamber for 60 min at 27°C with RNase (1 mg RNase in 10 ml 0.5× SSC) $200\,\mu l$. Cover. Rinse in 0.5× SSC, stain 20 min in H33258 (1 mg dissolved in 100 ml ethanol; 0.1 ml of this solution is added to 200 ml 0.5× SSC). Rinse and mount in 0.5× SSC.

Store over distilled water for four days at 4°C, incubate for 60 min at 55°C in 0.5× SSC, Rinse in 0.017 M phosphate buffer, pH 6.8, stain for 6–7 min in 3% Giemsa (R66) solution in same buffer. Rinse in phosphate buffer, then water; air dry, pass through xylene and mount in Canada balsam (Kihlman and Kronborg, 1975).

Alternative method   Omit RNase and H 33258. Incubate the BrdU-labelled chromosomes in 2× SSC at 60–80°C for 40 mm under UV (40W) irradiation.

Stain with Giemsa. It was effective in barley and rye (Zhang and Yang, 1986).

or

After fixation of BrdU-labelled chromosomes, hydrolyse root-tips at 28°C in 5N HCl for 40–70 min and then stain in Feulgen solution (Tempelaar *et al.*, 1982).

or

Dry labelled chromosome preparations for 7 days, treat with 1N NaH_2PO_4 (pH 8.0) at 87°C for 8 min, wash in water and stain with Giemsa (Zhang *et al.*, 1991).

or

Stain BrdU-substituted chromosome directly in an EDTA-Giemsa solution. (5% Giemsa solution. in 2% EDTA, pH 11.5) at 26°C for 5–15 min, rinse in tap water and dry (Yi and Zhang, 1992).

Detection of SCE sites by sequential staining and imaging methods (after Nakayama and Fukui, 1995)

(i) Germinate barley seeds. When the roots are 2–3 cm long, treat germinating seeds with a solution containing 100 μM 5-bromo-2-deoxyuridine and 200 μM deoxycytidine for 12 h. Rinse with distilled water for 10 min. Keep in a solution containing 100 μM thymidine and 200 μM deoxycytidine for 12 h. Excise roottips. Pretreat with 0.01% colchicine solution for 3 h. All treatments should be carried out in dark at 25°C.

(ii) For chromosome preparation, fix the roottips immediately in freshly prepared acetic acid–methanol (1:3) and store at −20°C for about a week. Afterwards, rinse roottips in distilled water and macerate with an enzymatic mixture (4% w/v cellulase Onazuka RS; 2% w/v macerozyme R200; 1% w/v pectolyase Y-23; 7.5 mM KCl; 7.5 mM Na_2EDTA; pH 4.0) for 13 min at 37°C. Rinse. Chop roottips into small pieces, after adding fresh fixative; flame dry and air-dry overnight.

(iii) For fluorescence plus Giemsa staining, stain chromosome preparation with H33258 fluorochrome (1%) solution diluted with McIlvaine's buffer (6.6 mM citric acid; 88 mM Na_2HPO_4; pH 7.0) at 4°C for 15 min. Rinse with 2× SSC (0.3 M NaCl; 0.03 M sodium citrate; pH 7.4) twice at 25°C. Briefly dip samples 10 times in 2× SSC at 60°C and then mount in 2× SSC under coverglass at 60°C. Subsequently irradiate samples with UV-light at a distance of 3–5 min for 5–10 min at 60°C using a transilluminator (HP-6L, peak wavelength 365 nm, 700 μW/cm²). Remove coverglass. Dip the samples 5 times in 2 × SSC at 60°C. Rinse with water for 5 min at 25°C. Stain slides with Wright solution (4%) diluted with Sörensen phosphate buffer (67 mM Na_2HPO_4; 67 mM KH_2PO_4, pH 7.0) at 4°C for 10–20 min. Rinse with running water. Photograph good metaphase plates (SCD) with black and white negative film.

(iv) For C-banding treatment, destain photographed slides through ethanol series (100%, 70%, 50%, 30%), keeping 5 min in each. Wash in distilled water. Treat slides with barium hydroxide (1%) solution at 37°C for 1 min, rinse with distilled water after each step of treatment; and stain with Wright solution (8%) diluted with Sörensen buffer at 4°C for 20 min.

(v) For image analysis of photographic chromosome images, enlarge images on black and white printing paper magnifying ×1000 times. The CHIAS method was followed (Fukui, 1985). Freeze the C-banded chromosome images in the image frame memories of the CHIAS through a TV camera and extract contour lines of the C-banded chromosomes. Record the SCD chromosome images in the other image memories, adjusting the chromosome position along the C-banded chromosome regions, along the extracted contour lines. Superimpose the banded regions, discriminated from the C-banded chromosome images, onto the SCD (sister chromatid differentiation) chromosome images so as to define the SCE sites.

(vi) This method is a simplified one, demonstrating sister chromatid differentiation (SCD), combined with C-banding and imaging methods, to determine the exact sites of SCE.

Simplified method for SCE

Immerse seeds in tap water for 3 h. Then keep on moistened filter paper for germination at 25 ± 0.5°C for about 24 h. When the roots are about 0.3 to 1 cm long, transfer the germinated seeds to an aqueous solution containing BrdU $(10^{-3}\,M)$; Uridine (Urd, $10^{-6}\,M$); and 5-fluorodeoxyuridine (FdU, $10^{-8}\,M$). Grow for one cell cycle (12 h), rinse in tap water, incubate for another 12 h at 25°C in an aqueous solution containing uridine $(10^{-6}\,M)$ and thymidine $(10^{-4}\,M)$. Wash the seedlings, immerse in tap water and chill (0°C, 36 h) for pretreatment. Fix in acetic acid–methanol (1:3) at 0°C for 24 h. The procedure upto fixation is carried out in the dark to prevent photolysis of the BrdU-substituted DNA.

To prepare slides, excise the meristematic regions, digest in a mixture of cellulase and pectinase (2.5% each) at 25°C for about 5 h; wash and keep for hypotonic treatment in distilled water for 1 h. Add fresh fixative to prepare cell suspension. Drop a few drops of the suspension on a clean chilled slide. Flame-dry the fixative to scatter the cells. Dry the slide in air. Stain BrdU-substituted chromosomes directly in a staining solution of sodium-EDTA (2%, pH 11.5) and giemsa stock solution at about 26°C for 5–15 min; rinse in distilled water and air dry. This part has been described earlier in this chapter (Yi and Zhang, 1992) for *Hordeum vulgare* and *Secale cereale*.

CHAPTER II.6

CYTOCHEMICAL ANALYSIS INCLUDING ISOLATION AND EXTRACTION

ISOLATION OF NUCLEI

The primary requisite for the extraction of nuclei is to isolate them in normal condition in adequate amounts and to keep the chromatin in an undamaged state. The procedure followed may be direct or indirect. In the former, the nucleus is isolated directly from the cytoplasm with the aid of a micro-manipulator and observed under the microscope. In the indirect mass isolation technique, the general stages include the disintegration of the cytoplasm through mechanical or chemical means, keeping the nucleus intact, followed by filtration through mesh or cheesecloth, differential centrifugation in suitable liquids and sedimentation. The general specific gravity of the liquid lies between 1.35 and 1.45 and centrifugation at 1000–6000 rpm is needed.

Protocol

Wash fresh, young roots or leaves in chilled water, remove the midribs of leaves, weigh 20 g and remove to 4°C chamber for next stages.

Transfer the tissue to a Waring blender containing 120 ml extraction buffer (0.25 M sucrose, 20 mM Tris HCl, pH 7.8; 10 mM NaCl, 1 mM $MgCl_2$, 2.5% Ficoll and 5% dextran 40) and continue blending for 15 s (at Variac Setting 50 volts). Moisten a flannelette of two layers with buffer. Filter homogenate through it and adjust the final volume to 120 ml. Centrifuge for 10 min at $2500 \times g$. Separate the pellet and suspend in 10 ml extraction buffer. Slowly layer 5 ml of the suspension in a Corex tube (30 ml) on the upper surface of discontinuous sucrose gradient solution (7 ml 60%, 7 ml 50% and 7 ml 25% sucrose). Centrifuge for 30 min at 8000 rpm. The pellet contains the nuclei only. Suspend it in 4 ml extraction buffer. It can be precipitated with 2 vol 95% ethanol and stored at $-20°C$. The method is also applicable to algae and other organs of angiosperms.

Alternative method Extract 6 g wheat flour with petrol ether. Suspend in 300 ml fresh ether and grind for 46 h in a ball mill. Suspend in 500 ml of mixture of cyclohexane and carbon tetrachloride, adjusted to a sp. gr. of 1.395. Centrifuge. Separate the supernatant. Add to it one-third its volume of ether and centrifuge. Suspend the earlier sediment in cyclohexane and carbon tetrachloride mixture (sp. gr. 1.447), centrifuge at 6000 rpm for 6 min.

Resuspend the supernatant in petrol ether and again centrifuge. Purify supernatant again by centrifuging at 6000 rpm in cyclohexane and carbon tetrachloride mixture (sp. gr. 1.416) and then again in petrol ether.

ISOLATION OF CHROMOSOMES AND CHROMOSOME COMPONENTS

For the analysis and quantitation of chromosomes and chromosome components, chemical isolation is the other alternative, in addition to *in situ* technique as described elsewhere (II.3). The basic necessity in such technique is to keep the nuclei and chromosomes in an undamaged state as far as practicable, despite limitations of chemical extraction, as compared to conditions *in situ*. However, there are methods of direct isolation through micrurgical techniques which have been dealt with in Section V. In the indirect mass isolation technique, the different steps include the disintegration of the cytoplasm through mechanical or chemical means while keeping the nucleus intact, followed by filtration through mesh or cheese cloth; differential centrifugation in suitable liquids; or discontinuous density gradient separation. In addition to mineral cations, polyamines, such as spermidine or spermine, help in stabilizing the chromosome structure. Moreover, addition of Triton is desired to ward off other cytoplasmic constituents from nuclei. After separation, if not directly utilized, storage at −20°C in 70% ethanol is preferred.

ISOLATION OF CHROMOSOMES

Chromosome isolation often becomes necessary for identification of functional segments and mapping of gene loci. Individual chromosome isolation through micromanipulation is well suited for preparation of chromosome-specific DNA library through cloning and amplification. Isolated chromosomes can be subjected to scanning electron microscopy as well. The principle of the method is to treat metaphase cell populations with colcemid and cool at 4°C in fresh medium to inactivate trypsin, dissolve mitotic apparatus and remove residual colcemid. Incubate the pellet in cold buffer after centrifugation at 37°C for 10–15 min to allow the cells to be broken easily and to equilibrate the chromosomes with the buffer.

Hexylene glycol in the buffer prevents instability and disintegration The latter is checked by raising calcium concentration. The material is syringed through a 22 gauge needle for rupturing the cell membrane. All steps are continuously checked through phase contrast microscopy and the temperature not allowed to drop below 37°C before cell breakage. After the chromosomes are liberated, the remaining process is carried out at 4°C. Isolated chromosomes can be stored in a stable condition without disintegration at 4°C for several months. Detailed schedules are given at the end of this chapter.

Separation of DNA and protein after chromosome isolation

After chromosome isolation, the initial step for the analysis of components is the separation of DNA and protein, and then histone and nonhistone moieties. The methods earlier tried for separation of the protein components involved (a) use of dilute mineral acids; (b) quantitation with polyacrylamide gel electrophoresis; and quantitation after initially fractionating chromatin proteins through chromatography on Bio Rex-70; different media, like hydroxyapatite; Sephadex and cellulose. Detailed representative schedules are given at the end of the chapter.

To eliminate limitations in these techniques, Sonnebichler *et al.* (1977) suggested a method for analysis of chromosomal proteins after complete separation of DNA and protein from chromatin through high speed centrifugation with salt and urea.

The outline of this method is as follows:

Separation of DNA from protein
Wash isolated chromosomes once with 0.024 M EDTA, 0.075 M NaCl (pH 7.0) and twice with 0.15 M NaCl. Add 2 M NaCl, 5 M urea and 0.01 M NaHSO$_4$ (for checking protein degradation); transfer the chromosomes to centrifuge tubes, and centrifuge in an angle rotor at 100,000× g for 35 h. The DNA forms a pellet and protein remains in the supernatant. Subject the pellet after separation to the same procedure to free it completely from protein. This pellet in now pure DNA. (For small amount of nucleoprotein, separation can be achieved by density gradient centrifugation with 2 M CsCl and 5 M urea for 70 h at 100,000× g).

Dialyse the supernatant for 3 h, reducing the salt concentration to 0.6 M and precipitate protein with 6 vol of acetone. Wash the protein with pure acetone. Dry the purified protein in vacuum.

Separation of protein components
Dissolve proteins in 1 mg/100 ml 1% acetic acid, 0.01 M beta mercaptoethanol, and 8 M urea. For electrophoresis use refrigerated teflon surface in a moist chamber.

Prepare 0.6 M ammonium borate buffer with 6 M urea, 0.01 M EDTA, 0.01 M mercaptoethanol (pH 10). Strong urea checks histone-nonhistone overlapping during electrophoresis. Adjust the buffer with 25% ammonia. Equilibrate cellogel strips (4 × 17 cm) with buffer, place in the chamber, and blot with filter paper to prevent formation of air bubbles. Apply 1 mm of protein in 1 to 2 µl.

Use electric power of 60 v/cm for 1 h. Histones migrate towards the cathode and nonhistones move towards the anode or remain stationary. Slight non-histones may move towards the cathode, but for their slow rate can easily be differentiated. Stain cellogel strips with 0.5% amidoblack in 45% methanol, 45% water and 10% glacial acetic acid. Remove excess dye by successive washing with solvent for 15 min. For quantitation, cut the coloured areas from blank cellogel strips; dissolve in glacial acetic acid. Measure the intensities at 630 nm on the basis of the standards prepared with known quantities of protein on cellogel strips.

This method does not allow any loss of protein and the analysis gives an accurate assessment of protein composition. With low resolution electrophoresis for short duration, a series of samples can be analysed within a short period.

Analysis of chromosome proteins using polyacrylamide gel
Dissolve chromosomes in 3% SDS (sodium dodecyl sulphate), 0.062 M Tris (pH 6.8) at 100°C for 10 min and observe under spectrophotometer. Analyse chromosome protein in 9% polyacrylamide gel using Tris-glycine-buffered SDS system. Stain the gels with 0.05% Coomassie blue in methanol–acetic acid-water (40 : 7.5 : 52.5). Destain by diffusion in 7.5% acetic acid. Analyse the protein bands in spectrophotometer. For photography, develop in D-11 (Kodak) for 5 min at 20°C. Scan in a scanner.

Separation of heterochromatic and euchromatic parts of chromosomes
The disruption of nuclei is best obtained through removal of the outer membrane. The method involves repeated suspension of nuclei in 0.01 M Tris buffer (pH 7.1), containing divalent calcium and magnesium, through stirring and centrifuging at $500 \times g$ for 5 min. The quantity of buffer used is 100 ml/ml of nuclei. The treatment varies within different tissues, depending on the type and concentration of cations used in the initial homogenizing medium and Tris buffer. The suspension should be prepared at least thrice.

If only calcium or both calcium and magnesium are used, repeated suspensions with heavy stirring for 30 s in each are necessary for removal of the outer membrane. It is always desirable to use in the buffer the same cations originally used in homogenizing medium.

In order to disrupt the nuclei, and secure chromosomes and chromosome fragments for further separation through density gradient centrifugation, two methods are in vogue—sonication and passage through a French pressure cell. In both, nuclei treated with Tris buffer are suspended on 0.25 M sucrose (25 vol) and stirred with Potter-Elvehjem pestle for 30 s in plastic tube (clearance 1 mm, 20%). To obtain nuclear swelling, the optical density at 420 nm is adjusted to 1–3 with 0.25 M sucrose and the mixture stirred gently for 20 min in cold. Fibrous materials are removed by passing through two layers of flannelette. The nuclei are periodically observed to detect optimal swelling which is indicated by a typical spherical shape, hyaline nature and size almost twice that of the original one.

In sonication method, a sonifier is used such as Branson sonifier operating at 7 to 11 A, generating sonic waves at 20 kc/s. The sample is kept at 0–4°C. Nuclear suspensions in 15 to 18 ml aliquots are sonicated for 5 s bursts and the sonicate is periodically examined for disruption of all nuclei.

French pressure cell method is employed to disrupt nuclei by processing through a pressure cell.

Homogenize the tissue in 10 vol of 0.25 M sucrose, with 5 mM NaCl and filter through several layers of cheesecloth. Stir the filtrate gently with a magnetic stirrer. Take 40 ml of aliquot in a French pressure cell and apply pressure through the hydraulic press up to 7000 psi. Slowly open the needle valve of the cell to expel the homogenate, without allowing the pressure to reach below 5000 psi. Take 20 ml of the pressate in 30 ml tubes (Spinco 25.1 rotor), follow a two layer system with different concentrations of sucrose and separate the nucleoli as pellet at 25,000 rpm. Pipette out the upper layers and separate chromatin through high speed centrifugation. Carry out the entire procedure at 2–4°C.

For separation of euchromatin and heterochromatin, filter the chromatin suspension through two layers of flannelette and subject to differential centrifugation or separation of chromatin fractions of different densities, referred to as heterochromatin, intermediate chromatin and euchromatin. Since the compaction and amount of heterochromatin vary from species to species, the degree of centrifugation also differs. Different velocities, namely $500 \times g$, $1000 \times g$, $4000 \times g$, $6000 \times g$, $12,000 \times g$, $20,000 \times g$ and $78,000 \times g$ per 30 min may be used and sediments collected for each fraction. The euchromatin part of very low density is obtained after the final centrifugation from the supernatant by making it 0.15 M with NaCl and precipitating with 2 vol of cold ethanol. Fix preparations periodically with acetic–methanol and stain with Wright's or Feulgen stain.

In order to estimate the amount of heterochromatin, highly homogeneous repeated fractions of DNA, specially satellite DNA, are separated through CsCl density gradient centrifugation. These fractions are species-specific. Even within the species, the genotypic diversity can be measured through an analysis of the repeat DNA fractions.

Representative schedules for isolation of chromosomes

Purification of chromosomes from plants may involve isolation of a single morphologically distinct chromosome through micromanipulation or the preparation of pure chromosome suspension. The latter type may omit enzymatic digestion of cell walls and use formaldehyde fixation before the rupture of cells to prevent stickiness of cytoplasm and aggregation of the chromosomes (Schubert *et al.*, 1993). Micromanipulation techniques have been included under Section V.

Isolation from root-tips

Expose root-tips at room temperature to a solution of 0.05% colchicine, 2% cellulysin (Calbiochem); 1% macerase (Calbiochem); 0.25% pectinase (Sigma); 0.25% rhozyme (Rohm & Haas) and 13% mannitol at pH 5.7. Microsporocytes can also be used. The period of incubation ranges from 30 min for the meiotic tissue to 18 h for the mitotic tissue. Tease apart the digested material, pass gently through a pasteur pipette and incubate for another hour. Filter through several layers of cheesecloth to remove large debris. Collect the protoplasts by centrifuging at $200 \times g$ for 15 min. Wash twice with 20 vol. each time of 5 mM 2 (*n*-morpholine) ethane sulphonic acid (MES) and 13% mannitol at pH 6.0. Resuspend in chromosome lysis buffer consisting of 15 mM HEPES, 1 mM EDTA, 15 mM dithiothreitol (DTT), 0.5 mM spermine, 80 mM KCl, 20 mM NaCl, 300 mM sucrose and 500 mM hexylene glycol at pH 7.0. Pass the protoplasts gently through a 27-gauge hypodermic needle 3 or 4 times until the cell membrane ruptures. Centrifuge at $200 \times g$ for 15 min to remove cellular debris. Then centrifuge at $2500 \times g$ for 10 min and collect chromosomes in the pellet.

For isolating chromosomes from cells in suspension culture, expose actively growing cells to 2 μg/ml fluorodeoxyuridine and 1 μg/ml uridine for one cell generation. Wash the cells in tissue culture medium supplemented with 2 μg/ml thymidine to terminate the reaction. Then expose the cells to the colchicine-enzyme mixture as described above.

To confirm the yield of chromosomes, stain with specific dyes like Schiff's reagent and DAPI. In the lysis buffer, EDTA removes the divalent Ca, Mn and Mg cations which act as nuclease cofactors and also prevents cation-induced chromosome condensation. Polyamine spermine is added to prevent condensation. KCl and NaCl maintain the correct ionic equilibrium. DTT preserves the protein structure. Hexylene glycol helps to maintain protein structure and to rupture the membrane and disperse the chromosomes. Sucrose keeps interphase nuclei intact, preventing contamination of chromosome preparations. HEPES maintains the pH near neutrality (after Griesbach *et al.*, 1982).

Alternative schedule

Sterilize the surface of the seeds. Place the seeds on moist filter paper and keep at 4°C for two days or up to one week. Germinate at 22°C in the dark until the root tips become visible. The subsequent synchronization steps are carried out at 22°C in the dark.

Transfer seedlings on filter paper soaked with 1.25 mM hydroxyurea and incubate for 18 h. Rinse two to three times in tap water. Place for 6 h on moist filter paper. Transfer to filter papers soaked with 4 μM APM and incubate for 4 h rinse the seedlings two to three times in tap water. Cut off the roots and store them in ice water overnight. Fix the roots in 3:1 fixative or 70% ethanol and store at −20°C for a minimum of one day before use. Alternatively, rinse the roots in tap water and keep them in water for at least 30 min. Cut off the tips and chop them. Avoid drying of the plant material. Digest 3 to 30 tips in the enzyme solution at 25°C for 50 to 60 min.

Filter the lysate through a nylon mesh into conical tubes and adjust to a final volume of 3 ml with 75 mM KCl. Spin down the suspension using a swing-out centrifuge at $80 \times g$ for 5 min. Discard the supernatant and resuspend the protoplasts carefully in 3 to 5 ml freshly prepared 3:1 ethanol/acetic acid fixative or 70% ethanol; spin down again under the same conditions.

Repeat the last step at least twice, pour off the supernatant, and resuspend the protoplasts in 3:1 fixative. Drop the protoplast suspension onto ice-cold, wet slides. The height must not exceed 20 to 40 cm. Check the protoplasts under a light microscope. When the fixative starts to evaporate, cells will burst, and the chromosomes are released. When only very little fixative is left, dip the slide shortly into ethanol and let it dry.

The specimen can be stored at room temperature for several weeks or at −20°C for longer periods. Rinse the roots in tap water and keep them in water for at least 30 min (after Busch *et al.*, 1996).

Mass isolation from plant protoplasts
Culture cell suspensions from cell lines maintained in suitable medium. Culture suspensions in continuous light (3000 lx) on a rotary shaker (120 rpm) at 25°C. Subculture at 2-day intervals, using 2 ml of settled cells in 100 ml culture medium. *For cell synchronization*, harvest 5 ml of settled cells from 1-day old suspensions and resuspend in 100 ml fresh medium containing 2.5 mM hydroxyurea (Sigma). Incubate for 24 h. Remove hydroxyurea by three successive washes with fresh media. Incubate cells in fresh medium supplemented with 0.05% colchicine (Serva, Heidelberg) for another 11 h on a shaker (180 rpm) in the dark. The step with hydroxyurea can be omitted if needed.

For isolating the protoplasts, harvest colchicine-treated cells by centrifuging at 100g for 5 min. Resuspend the pelleted cells in the supernatant medium in a ratio of 1:1. Mix cell suspension with an equal volume of enzyme solution [(5 mM MgCl$_2$, 2 mM CaCl$_2$, 3 mM 2-N-morpholinoethane sulphonic acid, 170 mM mannitol, 250 mM glucose, 6% cellulase (Onozuka R-10); 2% rhozyme (Rohm & Haas); 2% pectinase (Serva); 5% driselase (Fluka) at pH 5.6], containing 0.05 to 0.1% of colchicine. Incubate for 2–3 h on a shaker at 50 rpm in the dark. Collect protoplasts by centrifuging at 100g for 3 min. Wash protoplast pellets with a solution containing 100 mM glycine, 2.5 mM CaCl$_2$ and 5% glucose at pH 6.0. Pellet protoplasts again by centrifuging at 100g for 3 min.

For rupturing protoplasts, carry out the next steps on ice or at 0–4°C, unless otherwise noted and use siliconized glass tubes. Resuspend 2 ml of pelletted protoplasts in 100 ml of hypotonic glycine-hexylene glycol buffer (GHB) containing 100 mM glycine, 1% hexylene glycol at pH 8.4–8.6, adjusted by a saturated solution of calcium hydroxide. In some cases, it can be supplemented by 2.5% glucose. Incubate the protoplasts in GHB for 10 min at room temperature. Chill the suspension in ice water. Add Triton X-100 (Sigma) detergent from a 10% stock solution to a final concentration of 0.1%. Repeatedly pipette suspension to new tubes by plastic pasteur pipettes to disrupt protoplasts and liberate chromosomes into the suspension (after Hadlaczky, 1984).

Purification of chromosomes: Centrifuge suspension of ruptured proto- plasts at 1000g for 20 min. Resuspend the pellet containing chromosomes, nuclei and cellular debris in GH buffer supplemented with 0.1% Triton X-100 (GHT) and containing 1mM phenyl methyl sulphonyl fluoride (Sigma) and 1% isopropyl alcohol. Centrifuge at 200g for 10 min. If necessary, repeat this differential centrifugation until nuclei and cellular debris are totally elimin- ated from the chromosome suspension. Mix the supernatant, which contains only chromosomes and slowly sedimenting contaminants, with a 1 M sucrose solution in a ratio 1:1. Layer onto top of a 1 M sucrose solution made up in GHT buffer and centrifuge at 1000g for 20 min. Repeat this step 2 or 3 times to purify the chromosome suspension.

For observation under light microscope, place a drop of sample on a slide, immerse in 0.1 N HCl for 30 s, rinse in distilled water and stain with 2% Giemsa stain in Sörensen's phosphate buffer (pH 6.8) for 2–5 min at room temperature. Alternatively, place a drop of suspension on a slide, fix with acetic acid–ethanol (1:1), hydrolyse in 1 N HCl at 60°C for 10 min and stain with Schiff's reagent at room temperature for 1 h.

For electron microscopy of whole mount preparations, spread concentrated chromosome suspension (10–15 µl in GHT buffer) on a hypophase made from double glass distilled water. Pick up the samples on Formvar-coated 150-mesh copper grids and dehydrate through 30, 50, 70, 90 and 100% ethanol, a mixture of ethanol and amyl acetate (1:1), and 100% amyl acetate. Dry the grids by a critical point dryer, using carbon dioxide and then coat with carbon in a vacuum evaporator. Observe by transmission electron microscopy.

For scanning electron microscopy, fix preparations onto a coverglass with 2.5% glutaraldehyde, dehydrate with ethanol series and dry by critical point drying, as described above. Cut coverglass carrying dry samples into small pieces with a diamond, mount on the preparation holder with a conductive paint and coat with gold. Store isolated chromosomes in GHT buffer in cold upto over 48 h.

Synchronization of meristems and preparation of chromosome suspensions
Incubate seedlings of *Vicia faba*, with about 2 cm long main roots, in aerated Hoagland solution containing 1.25 mM hydroxyurea, which blocks the cell cycle in S-phase, for 18 h at 24°C. Rinse in distilled water. Incubate root tips for 6 h in fresh Hoagland solution and for 3 h in 0.05% colchicine to arrest cells at metaphase. Rinse in distilled water. Fix the root tips in 6% (v/v) formaldehyde in 15 mM Tris buffer, pH 7.5 for 30 min at 4°C. Wash twice for 20 min in Tris buffer, at 4°C. Chop up the meristems of about 30 root tips with a scalpel in a petri dish containing 1 ml LB01 lysis buffer (Dolezel *et al.*, 1989). The buffer contains 15 mM Tris-HCl, 80 mM KCl, 20 mM NaCl, 2 mM disodium EDTA, 0.5 mM spermine, 0.1% Triton X-100, 15 mM mercapto-ethanol, pH 7.5. Pass the resultant suspension of released chromosomes and nuclei through a 50 µm nylon filter to remove tissue and cellular fragments. Syringe twice through a needle of 0.7 mm diameter. In a glass tube, layer 0.7 ml of the suspension on the top of 0.7 ml 40% sucrose and centrifuge at 200 rpm for 15 min to remove nuclei and chromosome clumps. Carefully transfer the supernatant to Eppendorf tubes. The suspension contains upto 500 chromosomes per ml. Drop about 5 µl portions immediately on clean ice cold slides. Air dry.

For scanning electron microscopy: Place one drop of chromosome suspension on a glass slide and cover with a coverglass. Remove coverglass by freezing on solid CO_2 or liquid nitrogen. Immerse the slides in 2.5% glutaraldehyde fixative buffer (50 mM cacodylate, 2 mM $MgCl_2$, pH 7.2) and wash three times in buffer. Follow the usual schedules of osmium—TCH impregnation, dehydration, critical point drying and sputter coating (Wanner *et al.*, 1991).

In situ hybridization, Giemsa banding and restriction enzyme banding can be carried out on the air-dried slides. The slides can also be subjected to immunostaining of chromosomal antigens following the usual procedures described in Chapter II.8. This method gives a very good basis for production of chromosome-specific DNA libraries *via* microcloning and PCR amplification.

ISOLATION OF DNA

Deoxyribonucleic acid (DNA) forms the essential component of the chromosome. It is the component of which genes, the material basis of heredity, are composed. Being the genic material, it is endowed with the capacity of autocatalysis, heterocatalysis and mutation. In the plant system, it is present in the genome of the nucleus, chloroplastids and mitochondria. The present treatise is principally concerned with the chromosome only within the nuclei of the plant system.

Isolation and purification of nuclear DNA are essential for a variety of reasons. The differences in gene sequences form the basis of biodiversity. The study of biodiversity, therefore, requires a comparative analysis of DNA—its amount and sequence complexity in diverse organisms. Such genotypic diversity may be reflected to a certain extent at the chromosomal level, but *invariably* at the DNA level. The manifestation of biodiversity at the DNA level can be analysed, either through *in situ* DNA technique as discussed elsewhere, or after isolation and purification. Recent studies indicate that even for repeated sequences, not merely the amount, but the nature of repeats, with the quantity remaining constant, may reflect genetic diversity. The highly fast repeats often remain fixed, ensuring possibly stability. But the moderate and minor repeats may vary in ratio, contributing to diversity at the intraspecific level. Such results have been obtained through highly purified DNA and study of reassociation kinetics.

The study of reassociation kinetics requires denaturation of the double helix as a prerequisite, which is often achieved by heating, or treatment with formamide, or even with strong acid and alkali solution. The separation of the two strands is reflected in UV Spectrophotometer by rise in ultraviolet absorbance at 260 nm. The term T_m or melting temperature is used to denote the midpoint of the absorbance. As the T_m value is directly correlated with the GC ratio of DNA, which confers greater stability to DNA than its AT counterpart, T_m analysis has been utilized to detect biodiversity at the DNA level through its specific nucleotide content. However, the ionic concentrations, as well as DNA methylation, often affect the melting temperature to a certain extent.

The thermal stability is characterized by the mean thermal denaturation temperature (T_m), the temperature at which 50% of the hybrids dissociate. Comparative studies of *in situ* hybridization and solution or filter hybridization have shown that the same parameters control hybridization efficiency and that the *in situ* T_m of heterologous duplexes is 5°C below the solution or filter T_m. RNA:DNA hybrids are more stable than DNA:DNA hybrids and so dissociate at higher temperatures. Typically the difference in T_m is 11°C. Theoretically the T_m of any *in situ* hybridization reaction can be calculated using the following relationship: (These have been derived by combining several results for DNA:DNA and DNA:RNA hybrids).

$T_m = 87.5 + 16.6(\log M) + 0.41(\%G + C) - 0.61 \ (\%\text{formamide}) - 650/n$, where M is the molarity of the monovalent cation and ($\% G + C$) is the percentage of guanine and cytosine residues in the probe DNA. The reduction in T_m by formamide is not linear for RNA : DNA hybrids and is not valid above 50% formamide. The last term corrects for probe length, where n is the length of the probe in bases.

The reverse process whereby the two strands are reannealed by lowering temperature or using stronger salt concentration, is termed as *Reassociation*. The temperature required is much lower than the T_m value. The repeated sequences, because of their comparative homogeneity, anneal more quickly than the unique or low copy sequences. As such, the time required for reassociation or rate of reassociation is directly related to the sequence complexity. *Reassociation kinetics* is utilized to measure the repeat DNA content and detection of the fall in absorbance at 260 nm is a very convenient method for its study.

The value is expressed as *Cot*, where *Co* denotes the initial concentration and t implies the period of treatment for reassociation. It is expressed in terms of litre namely mol/s/l, and the midpoint, that is, *Cot 1/2* is the reassociation value.

The other way through which reassociation, *vis-à-vis* the amplified sequences, can be quantified in isolated DNA, is to pass the reassociated DNA through calcium phosphate based hydroxyapatite column, which holds back the duplex, allowing the single strands to pass through. The studies of reassociation kinetics in the purified and denatured DNA, provide a clear index of the amount of fast, moderate and minor repeats which is essential for the study of biodiversity.

In addition to the importance mentioned above, purified DNA is the basic feedstock for all experiments on genetic manipulation and engineering. It is a prerequisite for the preparation of probes, *in situ* hybridization, Southern Blot analysis, PCR amplification, restriction cuts, sequence analysis, and most important of all, gene mapping.

Schedules for DNA isolation

1. *Density gradient centrifugation*
The method is based on the principle of separation of cell fractions through density gradient centrifugation. Several media such as sucrose, cesium chloride, Ficoll (a sucrose polymer), dextran, potassium tartrate, sodium bromide as well as silica gel have so far been used. The change in osmotic pressure as well as the induced chemical toxicity are often common problems with these techniques. For tissues with heavy cytoplasmic content, cesium chloride method for DNA extraction is suitable for separating different nucleic acids, polysaccharides and proteins.

Homogenize tissues (2 ml/10 ml of CsCl) in 4 M CsCl (density 1.40 g/ml) at 4°C. Centrifuge the homogenate at 40,000 rpm for 24 h at 20°C to equilibrium;

polysaccharides and nucleic acids form a pellet. Separate the pellet and dissolve in $1 \times$ SSC (9 ml SSC, to pellet from 2 ml egg white). Digest successively in the following enzymes for 1 h each at 37°C: (a) 0.01 vol amylase (10 mg/ml); (b) 0.1 vol ribonuclease (0.01 mg/ml); (c) 0.01 vol pronase (50 mg/ml). Centrifuge initially at low speed to remove insoluble material and then at high speed (40,000 rpm) for 1 h at 4°C to pellet DNA. Dissolve DNA in $1 \times$ SSC and store at −30°C.

Due to the strong concentration of CsCl used, which inhibits the action of nuclease, the method does not allow enzymic degradation of DNA.

2. *Column with hydroxyapatite*

The methods are simpler than the previous one and involve preparation of the cell lysate and passing the lysate through a column for selective absorption, from which the sample can later be eluted.

This method utilizes absorption on hydroxyapatite (HAP) for separation of DNA directly from lysates. The technique requires, in addition to buffer, the use of (a) urea, for disrupting the cell, denaturing chromosome proteins, and inactivating enzymes, the last two properties being shared also by, (b) sodium lauryl sulphate (SLS) and (c) ethylene diamine tetraacetate (EDTA)—the chelating agent for binding bivalent and polyvalent metal ions. SLS is normally purified by recrystallization from hot ethanol, dried after washing in ether and stored as 25% at 4°C in a solid form. The only limitation of the technique is the limited capacity of HAP, which can recover only a small amount of DNA from a large mass of tissue. Suspend the tissue in a mixture of 8 M urea, 0.24 M phosphate buffer, 1% SLA and 0.01 M EDTA and homogenize in a blender. Pass the homogenate on HAP, and stir to check channelling. Wash HAP with urea buffer mixture (8 M urea, 0.02 M phosphate buffer). Wash with 0.14 M phosphate buffer to remove urea. Elute DNA with 0.04 M phosphate buffer.

Hydroxyapatite can distinguish between single and double stranded DNA at a wide range of temperatures. In view of this property, both thermal denaturation and reassociation kinetics analysis (T_m and *Cot*) are performed through the use of this column. The binding is not affected by the inclusion of solubilizing agents such as urea and detergents.

3. *Extraction with SDS–phenol*

SDS is used with phenol to denature and dissolve macerated hyphae leaving the DNA intact, which is precipitated from solution with isopropanol. It is mainly employed for plant materials and fungi.

Collect hyphae by filtration of fungal culture through a Buchner funnel. Rinse with 20 mM EDTA pH 8.0, remove liquid, freeze in liquid N_2 and lyophilize. Grind the dried material in a small mortar. Resuspend 50 mg in a microfuge tube by stirring in 0.5 ml extraction buffer (0.2 M Tris HCl pH 8.5; 0.25 M NaCl; 25 mM EDTA, 0.5% (w/v) SDS and 0.35 ml phenol) and mix by inverting the tube several times. Add 0.15 ml chloroform and again invert to mix. Centrifuge for 1 h at 15,000g. Immediately transfer the upper aqueous layer to a tube containing 25 μl RNase A (20 mg/ml) and incubate for 10 min at 37°C. Add an equal volume of chloroform–isoamyl alcohol, mix, and centrifuge for 10 min at 15,000g. Transfer the aqueous supernatant to a sterile microfuge tube and record its volume. Add a 0.54 vol of isopropanol to the sample and invert to mix. DNA precipitates and forms a clump in the tube. Centrifuge by pulsing for 5 s and remove the liquid using a pasteur pipette. Rinse the tube contents with 70% (v/v) ethanol, and recentrifuge to settle the pellet of DNA. Remove the liquid and dry the sample in vacuum. Dissolve the DNA in 100 μl TE, pH 8.0, by incubating at 4°C for several hours (see Brown, 1991).

4. *Isolation of DNA with CTAB buffer I*

It is a rapid and inexpensive method for extraction of total genomic DNA (nuclear, chloroplast and mitochondrial) from a wide variety of plant groups (after Doyle, 1991).

Preheat 5–7.5 ml of CTAB Isolation buffer (2% CTAB Sigma H-5882; 1.4 M NaCl; 0.2% 2-mercaptoethanol; 20 mM EDTA; 100 mM Tris-HCl, pH 8.0) in a 30 ml glass centrifuge tube to 60°C. Grind 0.5–1.0 g fresh leaf tissue in liquid nitrogen in a chilled mortar and pestle and transfer to preheated buffer. Alternatively, grind fresh tissue at 60°C in CTAB solution in preheated mortar. Incubate sample at 60°C for 30 min with occasional gentle swirling. Extract once with chloroform–isoamyl alcohol (24:1), mixing gently but thoroughly. Centrifuge at room temperature at 6000g for 10 min. Remove supernatant, transfer to clean glass centrifuge tube, add 2/3 vol cold isopropanol. Mix gently to precipitate nucleic acid. If possible, spool out nucleic acids with a glass hook and transfer to 10–20 ml of wash buffer (76% ethanol, 10 mM ammonium acetate). Alternatively, centrifuge at low speed for 1–2 min, remove supernatant, add wash buffer directly to pellet and resuspend by gently swirling. After 20 min of washing, spool out or centrifuge nucleic acids (6000 rpm, 10 min). Remove supernatant and dry residue briefly in air. Resuspend nucleic acid pellet in 1 ml/TE (10 mM Tris-HCl, 1 mM EDTA, pH 7.4). Add RNase A to a final concentration of 10 μg/ml and incubate at 37°C for 30 min. Dilute sample with 2 vol of distilled water or TE, add ammonium acetate (7.5 M stock, pH 7.7) of a final concentration of 2.5 M, mix, add 2.5 vol of cold ethanol and gently mix to precipitate DNA.

5. *Isolation of DNA with CTAB buffer II*

Prepare CTAB isolation buffer, according to the type of sample used (modified by Dr. Mark Chase at Kew). Preheat 10 or 20 ml of isolation buffer containing 40 or 80 µl of betamercaptoethanol in 50 ml Blue Cap tubes in a 65°C water bath. Grind 0.5 to 1.5 g of fresh leaf tissue in a mortar and pestle, preheated to 65°C, using a portion of the isolation buffer. Add remaining buffer and shake to suspend the slurry. Pour into the 50 ml Blue Cap tube, incubate at 60–65°C for 15–20 min. Alternatively, grind frozen leaf tissue in precooled mortar and pestle in liquid nitrogen and transfer to a tube containing isolation buffer. Shake to suspend and incubate at 60–65°C for 15–20 min. Extract once with equal volume of SEVAG (24:1 chloroform: isoamylalcohol), mixing gently but thoroughly. Extract for upto 30 min for mucilaginous samples. Centrifuge at setting 7 in IEC clinical centrifuge for 5–10 min. Transfer aqueous top phase containing DNA with a pasteur pipette to a Blue Cap tube. Add 2/3 vol isopropanol at −20°C and mix gently to precipitate DNA. Keep at −20°C for 30–60 min. Centrifuge at speed 7 for 10 min. Discard supernatant. Add 3 ml wash buffer (70% ethanol, 10 mM ammonium acetate). Shake to suspend pellet and wash for 5–60 min. Centrifuge DNA at speed 5 for 5 min, remove supernatant and allow pellet to dry by evaporation of ethanol. Resuspend DNA in 3 ml TE buffer (10 mM Tris-HCl, pH 8) and 0.25 mM EDTA). Heat, if needed, to 65°C for a few min to suspend DNA. In case of DNA not dissolving, add 3.4 g CsCl and continue to incubate at 65°C for upto a few hours. On the dissolution of the pellet, proceed with gradient centrifugation. Alternatively, for immediate restriction enzyme digest, resuspend DNA in 0.5–1.5 ml TE.

6. *Isolation of DNA with CTAB buffer III*

Limitations of traditional methods for DNA isolation include low yield, shearing of DNA, and are time-consuming and relatively expensive due to use of CsCl gradients. Hexadecyltrimethylammonium bromide (CTAB, Sigma), used in the DNA isolation buffer, significantly reduces carbohydrate contamination. The DNA isolated has a higher molecular weight, maximum purity, and a high yield. For a genomic library construction, PCR, or Southern blot analysis, etiolated seedlings grown in dark should be used to avoid chloroplast DNA contamination and to reduce polysaccharide content. A high level of aseptic conditions is required to prevent contamination. The DNA yield depends on the source of tissue and plant species. Generally, young tissue with meristematic cells has a higher yield of DNA (after Kaufman *et al.*, 1995).

** For most samples; 2X CTAB buffer = 100 mM Tris-HCl pH 8.0; 1.4 M NaCl; 20 mM EDTA, 2% CTAB (hexadecyltrimethyl ammonium bromide). For samples with very high water content, use 3X CTAB buffer. For samples with abundant mucilaginous polysaccharides, use Wendel's CTAB = 100 mM Tris-HCl pH 8.0; 1.4 M NaCl, 20 mM EDTA, 2% CTAB, 2% PVP 40 (polyvinylpyrrolidone); 50 mM ascorbic acid; 40 mM DIECA.

Reagents needed are:

DNA Isolation Buffer: 2% (w/v) CTAB (Sigma); 1.5 M NaCl; 25 mM EDTA; 0.2% (v/v) 2-mercaptoethanol; 100 mM Tris-HCl, pH 8.0.

TE Buffer: 10 mM Tris-HCl, pH 8.0; 1 mM EDTA.

Tris-Buffer-Saturated Phenol/Chloroform: Melt crystals of phenol in 65°C water bath and add 1 vol of Tris-HCl buffer, pH 8.0. Mix and keep for 30 min. Remove supernatant and add an equal volume of Tris buffer. Transfer the bottom phase into a fresh brown bottle and add an equal volume of chloroform. Mix and store at 4°C until use. The bottom phase is used for DNA extraction.

Immediately fix harvested tissue (2 to 5 g per extraction) in liquid nitrogen. If needed, store the frozen tissue in plastic tubes or bags at -20 or -80°C until use. Place frozen tissue in a mortar and add liquid N_2; vigorously grind the tissue. Repeat, adding liquid N_2 and grinding the tissue until a fine powder is obtained. Briefly warm the powder at room temperature for 2 min. Transfer to a conical bottom plastic chloroform-resistant centrifuge tube containing 5 ml/g tissue of DNA isolation buffer. Gently resuspend the powder by vortexing and then cap the tube loosely. *Alternatively*, grind frozen tissue in DNA isolation buffer preheated at 60°C. Incubate the samples in a 60°C water bath for at least 40 min with an occasional gentle swirling.

Remove the samples to room temperature and add 1 volume of chloroform: isoamyl alcohol (24 : 1, v/v) in the fume hood. Cap the tubes and gently invert four to five times. Loosen caps to release pressure and keep the tubes at room temperature for 10 min. Centrifuge samples at $7000 \times g$ or 5000 rpm for 10 min at 4°C, or $1600 \times g$ for 15 min at room temperature. Slowly and gently decant supernatant to a fresh tube with a wide bore pipette. Slowly pour the supernatant into a fresh tube or beaker containing 2 to 2.5 volumes (of the supernatant volume) of 100% chilled ethanol.

Do not vortex or invert the tube until the DNA precipitate floats up to the surface of the ethanol (15 to 30 min at room temperature). Transfer the DNA into a fresh tube with a glass hook and briefly dry under vacuum for 5 min. Gently resuspend the DNA in 4 ml/g original tissue of TE buffer at 37 to 40°C. *The DNA is then purified as follows*:

Add RNase A (DNase-free) to a final concentration of 20 µg/ml DNA sample and incubate at 37°C for 30 min to hydrolyse the RNAs. Extract RNase A, proteins, or polysaccharides once with 1 volume of tris-buffer-saturated phenol/chloroform and gently mix. Centrifuge at $10,000 \times g$ for 10 min at 4°C and slowly transfer the supernatant into a fresh tube. Add 1 volume of chloroform : isoamyl alcohol (24 : 1, v/v) to the supernatant and gently mix to avoid DNA shearing. Centrifuge. Add 0.5 volume of 7.5 M ammonium acetate or 0.15 vol of sodium acetate (pH 5.2) to the supernatant and gently mix. **Slowly and gently** pour the mixture into 2.5 vol 100% ethanol.

Transfer the DNA into a fresh tube with a glass hook or centrifuge the DNA at $5000 \times g$ for 5 min at 4°C. Briefly and gently rinse the DNA with 4 ml of 70% ethanol and dry the sample under vacuum for 15 to 30 min. This process removes traces of ethanol and dehydrates the carbohydrates.

Resuspend the DNA in TE buffer (100 µl/g tissue). Because of its high molecular weight, the DNA is usually not easily resuspended. In that case, the DNA sample can be placed at 45 to 50°C for 15 to 30 min with gentle and occasional shaking until it is dissolved. Keep the tube open to let remaining ethanol evaporate. The DNA is now suitable for restriction enzyme digestion and genomic library construction. Measure the quantity and quality of the DNA with spectrophotometer at 260 and 280 nm. A pure DNA preparation should have a ratio of 1.85 to 2.0 of A_{260}/A_{280} reading numbers. The sample can be stored at $-20°C$ until use. The quality of the preparation may be checked by 0.6 to 0.8% of agarose gel electrophoresis.

7. DNA from higher plants: minipreparation

Rapid microscale method has been developed for isolation of plant DNA from higher plants without ultracentrifugation with CsCl (Dellaporta et al., 1983) from the procedure commonly used for yeast DNA preparation (Davis et al., 1980).

Weigh 0.5 to 0.75 g of leaf tissue, quick freeze in liquid nitrogen and grind to a fine powder in a mortar and pestle. Transfer powder and liquid nitrogen to a 30 ml Oak Ridge tube. Add 15 ml extraction buffer (EB = 100 mM Tris pH 8.0; 50 mM EDTA pH 8.0; 500 mM NaCl; 10 mM mercaptoethanol). Add 1.0 ml of 20% SDS, mix thoroughly and incubate at 65°C for 10 min. Add 5.0 ml potassium acetate. Shake tube vigorously and incubate at 0°C for 20 min. Centrifuge at 25,000g for 20 min. Transfer supernatant through a Miracloth filter (Calbiochem) into a clean 30 ml tube containing 10 ml isopropanol. Mix and incubate at $-20°C$ for 30 min. Centrifuge at 20,000g for 15 min. Remove supernatant and dry pellets by inverting the tubes on paper towels for 10 min. Dissolve pellets with 0.7 ml 50 mM Tris, 10 mM EDTA, pH 8.0. Centrifuge in a microfuge for 10 min to remove insoluble debris. Transfer supernatant to new tube, add 75 µl 3M sodium acetate and 500 µl isopropanol. Mix. Centrifuge in a microfuge for 30 s. Wash pellet with 80% ethanol, dry and dissolve in 100 µl 10 mM Tris, 1 mM EDTA, pH 8.0.

For difficult materials, like soybean, instead of the last step with sodium acetate and isopropanol, add 50 µl of 3M NaOAc and 100 µl of 1% CTAB which will precipitate the nucleic acids. Pellet the precipitate for 30 s in microfuge and wash with 70% ethanol. Redissolve pellet in 400 µl TE. Precipitate the DNA with 50 µl 3M NaOAc and 1 ml ethanol. Repeat this step to remove residual CTAB, leaving DNA in sodium form. Redissolve dry

DNA pellet in 10 mM Tris, 1 mM EDTA, pH 8.0, per g starting material. Such miniprep DNA can be stored for months and cut with a variety of restriction enzymes. However heat-treated RNase is needed to digest contaminating RNA. A typical reaction would contain Miniprep DNA 10.0 μl; 10X restriction buffer 3.0 μl; 0.5 mg/ml RNAase 2.0 μl; *Eco*RI 8.0 units and distilled water to make upto 30 μl. Digest for 3 h at 37°C.

8. *DNA from callus and callus buds*

Isolation of DNA from callus or developing buds from callus base has been carried out. Such callus, grown in sterile cultures, both provides profuse materials and is free from contaminating bacteria and fungi.

Prepare sterile cultures from the seeds in a suitable basal medium (*see* chapter on tissue culture) supplemented with 10 ml/l myoinositol and 50 ml/l coconut milk. Expose the cultures (masses of buds at the base of the callus) to short photoperiods (10 h light/14 h dark) followed by long days (19 h light/8 h dark) which results in the profuse growth of buds, as compact masses of cells.

Collect the buds and follow the DNA isolation procedure as outlined below:

Grind the tissues in cooled ethanol (− 20°C) in a ground glass homogenizer till the homogenate becomes a mixture of cell clumps, single and broken cells. Wash twice with 70% ethanol. Disrupt the cells by stirring in a mixture of 2 vol of 1% sodium lauryl sulphate (SLS), 0.01 M ethylenediamine tetra-acetic acid (EDTA) and 0.05 M Tris buffer, pH 8.5. Keep the mixture at 65°C for 20 min. Centrifuge and re-extract the pellet in the original extraction fluid. Add 2 vol of 95% ethanol, centrifuge and collect the precipitate by centrifugation. Wash twice with 70% ethanol. Redissolve the precipitate by slow stirring at 21–23°C, in a mixture of 0.01 M Tris, pH 7.5, 0.01 M EDTA, 1M NaCl, and 1 mg/ml pronase. Pronase should be pre-treated earlier at 65°C for 10 min to remove any contaminating deoxyribonuclease.

Incubate for at least 4 h. Deproteinise the solution twice in a mixture of chloroform and isoamyl alcohol (24 : 1). Spool out DNA by slowly adding 0.55 vol of isopropanol. Follow further purification through two ribonuclease treatments (50 μg/ml, 15 min, 37°C) followed by pronase treatment (200 μg/ml, 2 h, 37°C).

Extract with chloroform and isoamyl alcohol till no precipitate is formed at the interface. Finally precipitate with isopropanol. Store the extracted DNA at − 30°C in 95% ethanol.

9. *DNA from other plant tissues*

For purification of DNA from frozen aerial tissues of wheat according to the schedule of Smith and Flavell (1974), after treatment with amylase, ribonuclease and pronase, continuously shake the DNA with phenol saturated with $0.2 \times$ SSC in the presence of 2% sodium lauryl sulphate till no further precipitate appears at the interface on centrifugation. Carry out dialysis of the final aqueous layer overnight against $0.1 \times$ SSC at 3°C. Recover the DNA by ethanol precipitation or centrifugation at $120,000 \times g$ for 8 h. Further purification can be carried out as outlined later.

Extraction from subcellular samples   In order to extract purified DNA or RNA from the subcellular samples, centrifuge fractions stored in 70% ethanol at -20°C at $8000 \times g$ for 10 min at 2–4°C. Suspend the freed pellet through mixing with equal volume of $1 \times$ SSC in cold. Resuspend pellet in 2.5% sarkosyl (sodium-*n*-lauryl sarcosinate Geigy NL 97) and disperse thoroughly. Heat at 70°C for 20 min. Centrifuge at $8000 \times g$ for 10 min at 2 -4°C. Decant supernatant and dilute it with 4 vol of $1 \times$ SSC so that the final sarkosyl concentration is 0.5%. Add T_1 ribonuclease (20 µg/ml) and DNase free ribonuclease A (25 µg/ml).* Keep at 37°C for 30 min.

Gradually add through mixing 0.2 vol of 5 M NaCl and then predigested Pronase† (500 µg/ml) and heat at 60°C for 1 h. This digested lysate can be used for density analysis or hydroxyapatite column filtration.

In order to remove protein and precipitate DNA, add equal part of water saturated freshly distilled phenol to the lysate. Close tightly the tube with rubber stopper and mix it thoroughly by repeated inversions several times, per min. Add equal part for chloroform after 30 min. Shake thoroughly and centrifuge the mixture at $8000 \times g$ for 10 min. Remove the top layer with pipette and repeat the above phenol–chloroform extraction method till no precipitate is found in the interface.

Take the upper layer from the finally separated phase into a centrifuge tube through pipetting and add slowly an equal volume of 95% ethanol in the DNA solution. Winding out the fibres at the interface may be necessary for high concentration of DNA, followed by washing with 70% ethanol. With low concentration of DNA, it would be necessary to mix the phases, chilling and centrifuging the solution at $10,000 \times g$ for 10 min at 2–4°C followed by washing the pellet with 70% ethanol and then discarding the excess ethanol.

Take 2–4 ml of DNA and dissolve in $0.1 \times$ SSC, keeping in a shaker for overnight at 2–4°C. Test the purity of the preparation in a spectrophotometer where for pure DNA the ratio will be two in 260–280 nm. Protein contamination is indicated in lower ratios and peak at 270 nm is observed if phenol is not completely removed.

---

* Recrystallise ribonuclease A five times at 2 mg/ml in $1 \times$ SSC to make it DNase free. Heat the enzyme solution to 90°C for 10 min and freeze for storing.
† Incubate B grade pronase at 2.5 mg/ml in $1 \times$ SSC at 65°C for 20 min to prepare predigested pronase and freeze for storing.

10. *Localization of specific repetitive DNA sequences*

Isolate total DNA from *Oryza sativa* var. IR36 and construct a lambda EMBL4 rice genomic library as described by Xie and Wu (1989). Isolate genomic clones from the library by using the repetitive sequence pOs48 (Os stands for *O. sativa*) as the probe (Wu and Wu, 1987). For gel blots, digest DNA with *Eco*RI and *Sal*I, fractionate by agarose gel electrophoresis and blot onto a nitrocellulose filter. Hybridize the filter to a ^{32}P-nick-translated pOs48 DNA fragment. Perform DNA sequencing by using the di-deoxy-nucleotide chain termination procedure adapted to single-stranded M13 phase DNA (after Wu *et al.*, 1991).

Determine the copy number of repetitive DNA sequences by using slotblot hybridization with defined amounts of total rice DNA and recombinant plasmid DNA (Zhao *et al.*, 1989).

11. *Determination of purity of DNA*

Dilute the DNA sample in 1× TE (10 ml Tris HCl, 1mM EDTA, pH 8.0) or distilled water 1:100; take the aliquot in a cuvette. Determine the optical density OD at 260, 280 and 320 nm against a blank (1× TE).

Calculate the concentration of DNA using the formula—1.0 OD260 = 50 µg/ml under standard conditions. Pure DNA preparation should give a ratio of 1.8–2.0µ OD260/OD280. Lower value may indicate protein contamination. OD320 may yield a zero value.

12. *Isolation of chromosomal DNA from yeast and*
pulse field gel electrophoresis (PFGE)

In lower unicellular organisms including yeast, chromosomes often fail to condense. Prior to the advent of PFGE, estimates of genome size and chromosome number were based on genetic linkage analysis DNA, reassociation kinetics, and in some cases, electron microscopy. The first full electrophoretic karyotype worked out was of the common yeast *Saccharomyces cerevisiae* by Carle and Olson in which the existence of 16 chromosomes ranging in size from 200 kb to over 2 Mb was recorded. Electrophoretic karyotypes are now available for many yeasts and other fungi including *Schizosaccharomyces pombe* and *Candida albicans*.

The accurate sizing of chromosomes in yeast, based on rare cutter restriction enzyme mapping, means that chromosomes from these yeasts can be used as size markers on pulsed field agarose gels.

The success of PFGE in separating DNA molecules involves also embedding intact DNA in agarose plugs and loading directly onto a gel. Agarose protects the DNA from shearing forces, which normally break the molecules into fragments of about 200 kb in size. The digestion of DNA, in agarose plugs, by restriction endonucleases, is also a crucial step. Accurate measurement and sizing of long chromosomes depend on digestion by restriction enzymes to produce fragments, which are separated by PFGE. Several cutting restriction enzymes are now available.

PFGE has been used extensively as a technique in the construction of long-range restriction maps in eukaryotes. Digestion of DNA in single and double digests, can be used to map mutations and chromosome rearrangements. The technique differs from conventional agarose gel electrophoresis in that the direction of the electric field periodically changes relative to the gel. Smaller DNA molecules respond quicker to this perturbation than larger ones and therefore migrate faster.

The most successful instrument is the CHEF (Contour-clamped Homogeneous Electric Fields) in which the field switches in direction through 120°. The CHEF can provide a versatile, reliable, high-sample-capacity apparatus, ideal for both analytical and preparative gels.

The technique followed for isolation of chromosomal DNA from yeast and PFGE analysis is outlined below (Maule 1994) :

Pick a single colony from a freshly grown culture, streaked on a YPD plate, and inoculate 100 ml of YPD medium in a 500 ml flask. Shake for 24 h at 30–33°C, at approx 200 rpm. Dilute an aliquot 1/10 in YPD, and either count the number of cells using a hemocytometer or measure the OD_{600}. Cell count should be ~1×10^8 cell/ml and $OD_{600} \cong 0.45$.

Chill the culture on ice for 15 min, and then harvest the cells by spinning at $2000g$ for 10 min at 4°C. Discard the supernatant, and gently disrupt the pellet with a sterile loop before adding 50 ml of chilled 50 mM EDTA, pH 7.5. Make sure the cells are thoroughly dispersed. Spin at $2000g$ for 5 min at 4°C. Repeat from discarding supernatant to spinning.

Finally, discard the supernatant and take up the pellet in 3 ml of ice-cold 50 mM EDTA, pH 7.5 (gives a final vol of approx 3.5 ml). Transfer the cells to a 20 ml universal container, with a fine-tip sterile Pastet (to disrupt any cells that are clumped), and heat to 37°C. Add 6 ml of 1% low melting temperature agarose (in 0.125 M EDTA, pH 7.5), which has been cooled to 50°C. Finally, add 1.2 ml of cell wall digestion solution, which has been freshly prepared and stored on ice until required. Immediately, mix thoroughly and dispense into plug molds.

Eject the plugs into a 50 ml Falcon tube containing 25 ml of ETM solution, and incubate overnight at 37°C in a water bath. Replace ETM with 20 ml 1% NDS containing 1 mg/ml proteinase K. Finally store plugs in 20 ml ETM at 4°C. Plugs should be equilibrated for at least 1 h in gel running buffer before use.

Restriction endonuclease digestion of DNA embedded in agarose plugs   Preparation of agarose plugs: Agarose plugs can be conveniently formed, using a mold made from a 96-well microtiter plate.

Spray the mold with 70% ethanol and air-dry. Cover the base of the plate with a sealing strip and place on ice. Pour the molten mixture of cells and agarose into a trough, and using an eight-channel micropipette, dispense 100 μl aliquots into the mold, one row after another. Tilt the trough on end to remove the remaining liquid using a standard micropipette.

Allow the plugs to set, on ice, for 20 min. Remove the sealing strip and, using a sterile yellow plastic micropipette tip, eject the plugs from the mold into a Falcon tube. Clean the mold can by soaking in 0.1 N HCl and then in distilled water. Incubate the plugs in a detergent + proteinase K solution. Up to 96 plugs are incubated in 20 ml of 1% NDS + 1 mg/ml proteinase K for 48 h at 50°C in a water bath; change the solution after 24 h. Finally, rinse the plugs in the storage solution and store at 4°C.

Restriction endonuclease digestion Soak the plugs for 10 min in a large excess of sterile TE, at room temperature. Typically, up to ten plugs are immersed in 20 ml of TE. Invert the tube frequently to achieve efficient mixing. Immerse the plugs in 5 ml TE containing phenylmethylsulfonylfluoride (PMSF) at 40 μg/ml. Incubate at 50°C for 30 min. Repeat the two earlier steps.

Soak the plugs for 2 h in 10 vol of 1X restriction enzyme buffer at room temperature, inverting the tube frequently. This step can be reduced to 1 h, if the PMSF treatment is performed in 1X restriction enzyme buffer. The plugs are now ready for digestion, which is performed in 1.5 ml microcentrifuge tubes at 1 plug/tube. Mix in each tube, on ice, 1X restriction enzyme buffer containing 0.1% Triton X-100 and 200 μg/ml BSA. Add 20 U of restriction enzyme to make upto 100 μl.

Transfer the plug to the tube, checking that it is completely immersed in liquid and that no air bubbles are trapped around the plug. Incubate overnight, in a water bath, at the recommended temperature; cool the tube on ice for 15 min and then remove the liquid with a fine-tip Pastet. Add 1 ml of ice-cold TE/tube, invert once, and remove TE. Add 200 μl of stop buffer per tube, and maintain on ice for 20 min. The plug is now ready for loading onto the gel.

Use CHEF for separation of large DNA molecules.

Southern transfer from pulsed field gels The large DNA molecules separated by PFGE need to be fragmented to allow efficient transfer to the hybridization membrane. Immerse the gel in 2.5 times its vol of 0.25 M HCl, and agitate gently for 20 min at room temperature. This process leads to the partial depurination of the DNA. Rinse the gel in distilled water. Immerse in 2.5 times its volume of denaturant and agitate gently for 20 min. Repeat with fresh denaturant. Rinse in distilled water. Immerse the gel in 2.5 times its volume of neutralizer and agitate gently for 40 min. Set up a device for southern transfer by filling a plastic tray to the brim with 20× SSC. A sheet of 5 mm thick plastic rests across the top of the tray lengthwise, with a gap on each side to allow two sheets of Whatman 17Chr paper, placed on top of the plastic sheet, to dip into the buffer on both sides. This forms a wick that draws buffer up from the tray. Allow the paper to become saturated before the next stage.

Invert the gel, so that the base is now the top, and lay on the paper wick, checking that there are no air bubbles trapped between the gel and the paper. Cover the surrounding paper with cling wrap to prevent evaporation of the buffer. Lay the hybridization membrane onto the gel, following manufacturer's instructions.

Mark the position of the slots, on the membrane, with a ballpoint pen, numbering the first and last tracks so that the membrane can be oriented relative to the autoradiogram. Lay on two sheets of Whatman 17Chr paper, slightly larger than the gel. Lay on paper towels to cover the sheets of Whatman paper, stacked to a height of 5 cm. Finally, lightly compress the paper towels with a 1kg weight, resting on a sheet of plastic or glass to distribute the weight over the whole area of the gel. After 24 h, replace wet towels with dry ones.

After further 24 h, remove the membrane, and wash in 2× SSC for 5 min. The gel may be stained in ethidium bromide to check the efficiency of transfer. Crosslink the DNA to the membrane by treating with UV light or baking or both. Follow the manufacturer's instructions.

13. Study of genomic DNA after extraction

The isolation of DNA is necessary for identifying individual genes. However, it is also necessary to understand the overall genomic organization. Considerable diversity exists in the amount and type of repetitive sequences, along with single copy DNA. These can be analysed through reassociation kinetics. The method for analysing genomic DNA, specially repetitive sequences, is outlined below (Pasternak, 1993).

Determination of genomic GC content and melting temperature   Load native, sheared DNA (25 to 50 µg/ml) in 0.12 M phosphate buffer, pH 6.8 into quartz cuvettes and overlay with mineral oil. Use a UV spectrophotometer with a water-jacketed cuvette holder and fitted with a temperature probe for thermal denaturation. For each experiment keep a reference cuvette containing 0.12 M phosphate buffer, pH 6.8. With all cuvettes in place, raise the temperature to 60°C. Allow the temperature at 60°C to equilibrate and raise it thereafter 0.3 to 0.5 degrees/min up to 100°C. The absorbance at 260 nm is monitored in each cuvette at each temperature point. The measure of hyperchromicity (H) is

$$\frac{A_{260}[100°C] - A_{260}[60°C]}{A_{260}[100°C]}$$

The median dissociation temperature (T_m) is the temperature at which 50% hyperchromicity occurs. Using the equation

$$\%GC = 2.44(T_m - 81.5 - 16.6\log M)$$

where M is the concentration of monovalent cations in the solution, the %GC content of the DNA can be determined.

DNA Reassociation Seal sonicated DNA samples (approx. 2 ml) with an average size of $\cong 500$ np at concentrations from 0.25 to 1750 µg/ml in 0.12 M sodium phosphate, pH 6.8 in Pasteur pipettes; denature by boiling for 5 min and then maintain at 60°C until a specified *Cot* value is reached. Roughly, a particular *Cot* value is equivalent to $0.5A_{260}$ at zero time of the DNA sample, multiplied by the time of the reaction in hours. Although 60°C is the standard temperature for reassociation kinetics, optimal reassociation occurs at about 25°C below the median melting temperature (T_m) of the genomic DNA. After reassociation to a particular *Cot* value, each sample is quickly frozen by immersion in a solid CO_2 ethanol mixture. To determine the fraction of nonreassociated (i.e., single-stranded) and reassociated DNA (i.e., duplex fraction), thaw and slowly apply each frozen DNA sample (100 µg of DNA per 2.5 ml) to reagent grade hydroxyapatite (0.5 g) in a 5 ml water-jacketed column (equilibrated with 0.12 M phosphate buffer, pH 6.8, and maintained at 60°C). Wash the column three times with 2.5 ml 0.12 M phosphate buffer, pH 6.8, to elute single-stranded DNA. To elute double-stranded (renatured) DNA, raise the column temperature to 97°C and wash four times with 2.5 ml aliquots of 0.12 M sodium phosphate, pH 6.8. Alternatively, double-stranded DNA can be eluted from the hydroxyapatite column at 60°C with four 2.5 ml aliquots of 0.5 M sodium phosphate buffer. Recoveries from hydroxyapatite columns usually are greater than 95%. To reduce nonspecific binding and light scattering, boil and wash the hydroxyapatite in 0.12 M sodium phosphate buffer before packing the column.

The absorbance of each eluted sample is read at both 260 and 320 nm. Assume that one absorbance unit at 260 nm (A_{260}) is equivalent to approximately 50 µg DNA. The reading at 320 nm is used to correct for light scattering effects. The fraction of DNA that is single-stranded at each *Cot* value is determined from the relationship

$$\frac{A_{260}\text{ss-DNA}}{A_{260}\text{ss-DNA} + A_{260}\text{ds-DNA}}$$

where ss-DNA is the single-stranded DNA fraction and ds-DNA is the double-stranded DNA fraction. For plant genomic DNA, the x-axis, i.e., log *Cot*, of a *Cot* curve can range over seven log intervals: therefore, about 30 to 50 or more *Cot* points are often required to generate a reliable *Cot* curve.

Determination of a Cot curve in spectrophotometer   Denature sonicated DNA ($\cong 300$ np) in 0.12 M phosphate buffer, pH 6.8, by boiling and then reanneal at 60°C in a water-jacketed cuvette holder of a UV spectrophotometer. Take absorbance readings at 260 and 320 nm during the reassociation period. To obtain a complete set of *Cot* points in a reasonable time, a range of initial DNA concentrations (e.g., 20 to 1500 µg/ml) is needed. With corrected absorbance values, the decrease in hyperchromicity is calculated at each *Cot* value and converted to the fraction of the sample that is single stranded DNA. The results are plotted as the fraction of single stranded DNA as a function of the log equivalent *Cot*. With one of the available computer programmes, the best-fit curves for the various kinetic components can be generated.

Determination of repetitive DNA sequences   Renature a genomic DNA sample to a *Cot* value at which only repeated sequences are fully renatured, e.g., *Cot* 10. Size-fragmented DNA samples with either short or long lengths can be used for these experiments. After reassociation, digest the sample with Sl nuclease to remove single-stranded regions and then isolate Sl-resistant duplex DNA by hydroxyapatite chromatography. The length of DNA can be determined by agarose gel electrophoresis.

Sl nuclease digestion entails treating the *Cot* 10 DNA fraction in 0.15 M NaCl, 0.05 PIPES, 0.025 M Na acetate, and 0.1 mM $ZnSO_4$, pH 4.55, with 100 units Sl nuclease per microgram DNA for 45 min at 37°C. The reaction is terminated by chilling and bringing the reaction mixture to 0.1 M phosphate buffer, pH 6.8.

APPENDIX

Rare cutter restriction enzymes	
Enzyme	Recognition sequence[a]
Not 1	GC/GGCCGC
Asc 1	GG/CGCGCC
BssH 11	G/CGCGC
Ssc 83871	CCTGCA/GG
Sfi 1	GGCCNNNN/NGGCC
Rsr 11	CG/GwCCG
Sgr A 1	Cr/CCGGyG
Xma 111, Eag 1, Ecl X1	C/GGCCG
Sst 11, Sac 11, Ksp 1	CCGC/GG
Mlu 1	A/CGCGT
Pvu 1	CGAT/CG
Nru 1	TCG/CGA
Aat 11	GACGT/C
Sal 1	G/TCGAC
Nae 1	GCC/GGC
Nar 1	GG/CGCC
Sna B 1	TAC/GTA
BsiW 1, Spl 1	C/GTACG
Sma I	CCC/GGG
Xho 1	C/TCGAG
Bse A1, Acc, Mro 1	T/CCGGA
Cla 1	AT/CGAT
Sfu 1, Asu 11	TT/CGAA
Swa 1	ATTT/AAAT
Pac 1	TTAAT/TAA
Pme 1	GTTT/AAAC

[a]w = A or T, r = A or G, y = C or T, N = A, C, G or T.

CHAPTER II.7

MOLECULAR ANALYSIS THROUGH *IN SITU* HYBRIDIZATION OF CHROMOSOME SEQUENCES AND IMMUNOLOGICAL DETECTION

PRINCIPLES

The localization and mapping of functionally differentiated segments of chromosomes and gene loci have been greatly facilitated through the application of molecular hybridization technique at the chromosome level. Such knowledge of the operational mechanism of chromosomes at the molecular level has been obtained basically due to important advances in methodology, one involving the annealing technique for analysing the sequence complexity of DNA and the other the hybridization of RNA and DNA molecules.

The complementarity of the DNA strand is the basis of *in situ* molecular hybridization. The method principally involves the application of probe sequences which have been tagged with radioisotopes or a chemical reporter. This step is preceded by the denaturation of the target chromosomal DNA mostly in metaphase, to facilitate access of the probe to the target. Complementary sequences undergo pairing, followed by hybridization. The hybridized sites are localized either through autoradiography or immunofluorescence, depending on the type of probe used (see Levi and Mattei, 1995).

Types of hybridization

The localization of a gene at the chromosomal level is necessary to identify its function, its relationship with other genes as well as for isolation and its dissection. Of the two types of hybridization used, the advantage of *isotopic in situ* hybridization principally lies in its ability to detect small length probes. It is mainly because longer exposure ensures the formation of more silver halide crystals. The availability of cloned probes and improvement in the rate of hybridization with the aid of dextran sulphate in the buffer have made *isotopic in situ hybridization* a convenient method for localizing single copy sequences.

The non-isotopic method is used widely (Gosden and Lawson, 1994). Its underlying principle is to make the specific loci of DNA antigenic.

Discovery of various chemical reporter molecules which can be tagged with the probes and final detection through appropriate immunofluorescing compounds have made non-isotopic hybridization a very convenient and fascinating technique, widely pursued in various centers.

Nature of probes

Length of the probe is an important factor in the detection of hybridization. The efficiency of detection is not high if the probe is short. For complex and repeat DNA probes, it is preferable to apply non-isotopic methods of hybridization (Levi and Mattei, 1995).

For shorter probes upto 1000 bp isotopic hybridization is recommended. A network of probe fragments incorporated in plasmid DNA yields very intense signals as well. In view of the fact that nick translation involves both tritiated 11 dCTP and 16 dTTP, the probe-specific activity is very high and labelling is optimum. The size of the insert however determines, to a great extent, the concentration of the probe and exposure time required. For convenience, liquid emulsion technique for detection of signals is suggested.

For complex larger probes, non-isotopic labels are preferred, such as biotin, in view of the convenience of immunocytochemical detection of this compound. The objective underlying labelling is to incorporate biotinylated analogue of dUTP into probe by nick translation. Biotin is linked by a linker to C^5 position of the pyrimidine. With short linkers, the linking is strong whereas with long linkers, their detection through detector is more efficient (Levi and Mattei, 1995).

The detection of hybridization sites is done by application of avidin, conjugated with fluorochrome or antibiotin antibodies.

For single copy probes, amplification becomes necessary through biotinylated anti-avidin antibodies, followed by avidin fluorophore such as fluorescin isothiocyanate (FITC), rhodamine isothiocyanate (TRITC). Counterstaining is done with red propidium iodide for FITC-stained probes or DAPI for TRITC-stained probes. The dilution of the counterstain with antifade solution at an alkaline pH (8.0) is necessary at the end. A fine layer of antifade reveals a crisp image. Various banding techniques can be adopted after detection of probe signal. Observations are carried out using fluorescent microscope with different combinations of emission and excitation filters suitable for different fluorophores.

The identification of complex DNA sequences is often hampered by the repetitive sequences which, because of their non-specificity, often hybridize with their corresponding sequences present in the genome. In order to eliminate this error, chromosomal *in situ* suppression (CISS) or competitive *in situ* hybridization (CISH) is adopted. This technique involves preannealing the probe with appropriate competitor DNA to render the repetitive sequences unavailable for hybridization. Normally, the competitive DNA is the total fragmented unlabelled genomic DNA, enriched with repetitive sequences as in human genome (Levi and Mattei, 1995). This competitive DNA is fragmented through sonication to bring it to desired size (200–400 bp). The desired length of the biotinylated probe is achieved by nick translation and by varying the concentration of DNA. The probe can then be amplified through PCR method.

For detection of telomeric sequences through *in situ* hybridization, tandem arrays of hexanucleotide sequence (TTAGCG), as found in trypanosomes, have been observed to be useful in higher eukaryotes. The telomeric sequences may prove to be of value because of their involvement in chromosome rearrangements and use in locating probe sequence situated at chromosome ends.

Polymerase chain reaction and *in situ* hybridization

Polymerase chain reaction, associated with *in situ* hybridization in plant chromosomes (Mukai and Appels, 1996), has been a new approach for physically localizing gene sequences in chromosomes. It is specially useful for low single copy genes. The technique involves preparation of cells in such a way that their morphology is preserved, while permitting diffusion of primers and enzymes into intracellular sequences and the amplification of the latter. The largest sequences remaining fixed on the slide are allowed to undergo amplification overlaid with the PCR reaction mixture. The final visualization of the PCR product on the slide is carried out by directly using fluorescence-labelled nucleotide in the amplification mixture, or by using biotinylated nucleotides in the mixture to be visualized later through antibody-labelling. *In situ* PCR studies have proved to be of much value in the detection of low copy DNA in single cell preparations.

In another method, the oligonucleotide primed *in situ* synthesis (PRINS) is carried out to detect tandem repeated sequences, in addition to its application for low copy sequences (Koch, 1995). In this method, the oligonucleotide primers are annealed to the target denatured DNA and extension is carried out in presence of labelled nucleotides and DNA polymerase. In cases where more extension does not give adequate signals, further cycling is carried out through PRINS *in situ* PCR method. The amplified copies give strong signal. Immunocytochemical detection of the site of reporter molecules indicates the site of low copy repeats.

Multicolour fluorescent *in situ* hybridization

The study of different gene loci and segments of complex genomes in the chromosomes has been facilitated by the development of probes with differential fluorescence. Such multicolour techniques have been applied to localise repeated DNA and other sequences on chromosomes of various crop plants, such as wheat, tobacco, Aegilops, barley, sugarbeet and *Triticale* (Kenton *et al.*, 1993; Mukai *et al.*, 1993).

There are two sets of methods for such multicolour preparations also termed as chromosome painting. The *indirect* methods use biotin, digoxigenin and dinitrophenol (DNP) as reporter molecules, which are detected by fluorochrome-conjugated avidin or antibodies. The common fluorochromes are FITC (fluorescin-isothiocyanate), rhodamine and AMCA (amino-4-methyl coumarin-3-acetic acid).

The D-isomer of biotin (mol. wt. 244.31) is a natural vitamin of the cell, specially of liver, kidney and pancreas. It has a few chromophoric groups (vide Meier and Fahrenholz, 1996). Avidin (mol. wt. 68,000) is a tetrameric glycoprotein, originally obtained from chicken egg white. Each subunit is identical and the oligosaccharide present in each subunit is characteristic of avidin. All four tetramers have four biotin binding sites.

The avidin–biotin interaction is the strongest noncovalent biological recognition so far known between ligand and protein. The reaction is rapid; not affected by pH, solvents or denaturing agents and can withstand short exposure to high temperature more than 100°C. The common detergents such as SDS, Tween 20 or Triton X do not affect its binding efficacy. However, avidin denaturation may affect the binding capacity (Meier and Fahrenholz, 1996).

The direct method involves probe labelling by fluorochrome-labelled antibodies. Such direct coupling of fluorochromes to probes does not need detection through immunocytochemical methods. In the indirect procedure, the long immunological detection method is avoided by using fluorescin—12 dUTP instead of dinitrophenol. However, direct labelling of fluorochrome into the probe is a rapid method ensuring good resolution.

It has further been shown that by using combinations of three primary colours (red, green and blue) seven different colour combinations can be obtained. This method has been advantageously employed by Mukai (1996) to locate seven different sequences in wheat-rye translocations. In polyploid species of *Triticum* and *Aegilops*, several translocations and insertions have been revealed using multicolour probes in chromosome painting. With the use of cloned repetitive DNAs it has been possible to locate clusters of homogeneous major repeats at certain loci in chromosomes of the complement in *Triticum*. In fact, multicolour FISH technique has become a powerful tool for gene mapping. Moreover, it has been possible to work out abnormalities, translocations, insertions and breakage points by using genome-specific dispersed probes. The probes generated from aberrant chromosome and their hybridization on metaphase may indicate branchpoints, deletion, translocation as well as confirm the normal sequences in the chromosome. However, aberrant chromosomes need to be sorted through flow sorting followed by labelling and amplification through PCR, utilizing oligonucleotide primer and hybridization as well. The method is termed as *Reverse Chromosome Painting*.

Certain factors (Mukai, 1996) need to be considered while using multicolour technique. Because of so many colours, propidium iodide counterstaining is not possible, and may not be necessary. Moreover, such chromosome painting for detecting several targets is most effective with repeated sequences and multigene families. The technique is effective in polyploids, provided at least the origin of a certain genome is known and constituent genomes are rather diverse from each other.

In situ extended DNA fibre

Further refinements of *in situ* technique lie in the hybridization of probes to extended DNA fibres. This method (Fransz *et al.*, 1996) has enabled physical mapping at high resolution in *Arabidopsis thaliana* through fluorescence hybridization. In this species, a contig of 3 cosmids, covering genomic segment of 89 kb length, has been analysed. The method has been combined with multicolour FISH as well and the lower limit of resolution goes upto 0.7 kb.

The method involves, in principle, nuclear lysis, release of DNA fibres from lysed nuclei, spreading the DNA on the surface of the slide and hybridization of probes following the standard schedule for fluorescence detection. This method permits physical fine mapping of sequences in plasmids, cosmids and YACs, ranging from a few to several hundred kilobases. Repetitive and single copy DNA fragments of different sizes can be mapped with high resolution at the molecular level.

Prerequisites, steps and techniques of hybridization

Hybridization at the chromosome level *in situ* involves a number of steps, all of which have to be critically followed. The essential prerequisites are *preparation of proper DNA probes* as well as *well spread metaphase chromosomes.*

Probe sequences

The probe sequences may be one or a few copies or even repetitive. The entire genomic DNA can also serve as a probe. For single copy sequences, special methods, involving amplification and *in situ* detection for increasing sensitivity, are needed, over and above critical control of the denaturation and hybridization set up. A 10^6 increase is needed for single detection sensitivity on Southern Blot (Schwarzacher *et al.*, 1994). The specificity of the repeat sequences may be restricted to a single chromosome, to a single species or to many allied species. pTa71—a ribosomal DNA probe, isolated from wheat, contains highly repeated sequences. The dispersed repeats within a genome may exist in addition to restriction to a single locus. Their localization at the chromosomal level through proper probes is possible.

Further refinements have led to the use of entire genomic DNA as probe. The use of genomic probe serves to identify the parental genome in hybrids, alien chromosomes or chromosome segmental translocations, as well recombinant lines. With the genome *in situ* hybridization a number of alien segments can be identified at interphase as well.

With the gradual refinements in flow cytometers and flow sorting technique, the possibility of isolation of a few chromosomes and their amplification by PCR, followed by use as probe in hybridization has been opened out. Similarly, single chromosome isolation and preparation of DNA library of individual chromosomes have been possible. The sequences from such libraries can serve as probes for application in individual chromosomes through multiple probe method. This strategy, utilizing differential fluorescence of different probes, is of considerable advantage in mapping of gene sequences at the chromosome level.

Possibly the most important advances in *in situ* molecular hybridization are to be achieved with the use of synthetic oligomers. Such oligomers, serving as primers as in primed *in situ* hybridization, permit the use of short consensus sequences as probes. The development of oligonucleotide synthesis in different laboratories has increased the scope of the application of oligomers in the detection of short and long sequences, both coding and noncoding, of gene families as well as introns and promoters. It has been visualized that in the near future, any site with a sequence from 20 bp onwards, would be detectable through oligomers, alleviating the use of cloned probes (Schwarzacher *et al.*, 1994).

Metaphase chromosome

The other partner in hybridization is the spread metaphase plate serving as the target. Single layered single celled preparations, with a clear background and well spread metaphase chromosomes, are essential. Large numbers of metaphase plates with non-overlapping chromosomes and well-clarified constrictions are necessary.

In order to achieve these objectives, pretreatment of chromosomes, before fixation in ethanol or methanol–acetic acid mixture, is required. The pretreatment agents commonly used are ice-cold water; colchicine upto 1%; 8-hydroxyquinoline—0.2 M, and saturated solutions of paradichlorobenzene and aesculine. Treatment for 24 h at 0°C in ice water; 10–20 h at 4°C in colchicine and 30 min to 2 h at room temperature followed by 20 min to 2 h at 4°C in 8-hydroxyquinoline has been recommended (Schwarzacher *et al.*, 1994).

Following fixation in acetic–methanol or ethanol (1 : 3), enzyme treatment for cell separation and cleaning is recommended. It involves washing twice in enzyme buffer (0.1 M citric acid and 0.1 M trisodium citrate 40 : 60—diluted 10 times); treatment in enzyme mixture (2% cellulase and 20% pectinase in enzyme buffer) and incubation at 37°C for 30 min to 2 h. Either squashing or dropping method can be applied to secure well spread plates. During squashing in 45% acetic acid, short digestion is required, followed by washing once more in enzyme buffer and removing the root cap under stereo microscope. The material is squashed under a coverglass. After observation, treatment with dry ice or liquid nitrogen is recommended for separation of coverglass and air drying. Slides can be checked under phase contrast, stained by DAPI stain and analyzed under fluorescence microscope.

When preparing slides following the dropping method, digestion is carried out in microcentrifuge tubes and the material dispersed through pipette or with a glass rod, at 15–20°C. Subsequent steps include centrifugation at 200 rpm for 3 min; discarding the supernatant; centrifugation of the pellet three times in enzyme buffer and three times in the fixative, keeping for 3 min in each, and finally resuspension in the fixative to a volume of 50–100 μl. A few drops of suspension can be spread on a clean slide and air dried. The extent of spreading and cell density can be monitored under phase contrast microscope.

High resolution studies can also be carried out on molecular hybridization *in situ*, both in sections and whole cells. For such studies the steps involve: crosslinking glutaraldehyde fixative in phosphate buffer (0.8 ml 25% glutaraldehyde in phosphate buffer, pH 9.2) with picric acid addition (1:2) for 2 h at 15–20°C; washing in phosphate buffer; dehydration in ethanol series, and embedding in L:R white resin by passing through ethanol–resin mixture with gradual replacement of ethanol; polymerization at 65°C for 15 h; cutting sections using glass or diamond knife, floating in benzyl alcohol for expansion and finally drying section on poly-lysine coated slides.

Labelling of probes
For probe labelling, it is possible to use direct fluorochrome labelled probes available commercially as conjugated nucleotides at different loci. Labelling of probes can be done following different methods, both for isotopic and non-isotopic ones. Standard kits are also available for the purpose. For *in situ* hybridization radioisotope labelling and detection are less practised because of the exposure and complex protocol. The different methods include random primer, nick translation, PCR-amplified sequence incorporation and tailing at the ends for short oligomers. Of these, nick translation is useful for genomic DNA or long inserts whereas random primer labelling is convenient for short sequences. While screening for the extent of labelling, fluorochrome-labelled probes are detected through fluorescence whereas biotin or digoxigenin incorporation is tested through test blots after Southern hybridization. It is possible to combine direct fluorochrome-labelled probes with biotinylated ones to increase the capacity for detection. The principal steps involved in Random Priming and Nick Translation are outlined later (Schwarzacher *et al.*, 1994).

Hybridization
Prior to hybridization the materials require treatment to avoid non-specific probe binding as well as increasing access of the probe and the detection

reagents to the target. In order to achieve these objectives, slides are treated with ribonuclease to remove the endogenous RNA; pepsin to digest surface proteins and paraformaldehyde to reduce loss of DNA in chromosomes during denaturation. Non-specific binding is checked through the usc of blocking DNA in the hybridization mixture and repeated washing. The method adopted for the preparation of slides is outlined at the end of the chapter.

The preparation of the hybridization mixture is another crucial step. Ideally the process of hybridization should be stringent. It depends on the size and composition of probe DNA, temperature, salt concentration and destabilization of two strands of DNA which is achieved through formamide. These factors influence the stringence of hybridization. Moreover, the blocking DNA to be used should be clear and cut into smaller fragments 100–200 bp in length.

The hybridization mixture therefore may have a composition of formamide— 20 µl; 50% aqueous dextran sulphate, 8 µl; 20× SSC, 4 µl; 10% SDS (sodium dodecyl sulphate); probe DNA—1–5 µl (0.5–2 µl/ml final conc.) and blocking DNA (Salmon sperm)—1–5 µl (5–50 µg/ml final conc.), 10–200× of probe concentration. The final volume is made upto 40 µl with water. The mixture is then vortexed and kept at 70°C for 10 min for denaturation before transferring to ice and keeping for 5 min. This denatured hybridization mixture in excess—40 µl, is spread over each air-dried slide and covered with a plastic coverglass. The slide with material is transferred to preheated humid chamber kept at 85–90°C.

The process of hybridization in principle involves the following steps:

Transfer of the slide with hybridization mix to prewarmed humid chamber kept at 85–90°C; incubation for 5–10 min; followed by transfer to 37°C water bath for overnight. The coverglasses are then floated in 2× SSC at 42°C, washed in 20% formamide in 0.1 SSC twice for 5 min each; cooled for 5 min and then kept for 3 times in 2× SSC for 3 min each at room temperature. Denaturation and cooling can be carricd out in an automatic programme controller as well (Heslop-Harrison *et al.*, 1991).

Detection

The next step following hybridization and washing is the detection of hybridized sites. When isotopic probes are used, autoradiography along with chromosome staining for visualization of the entire chromosome is adequate. But for non-isotopic probes, the detection is either through the use of fluorochromes which have specific wavelengths for excitation and longer wavelengths for emission. They can be used as conjugated to nucleotides, antibodies as well as serving as stain for chromosomes. The direct nucleotide incorporation can be done with Coumarin (AMCA-amino-4-methyl-coumarin 3-acetic acid), yielding blue fluorescence, requiring 350 nm for excitation and 450 nm for emission. Similarly, green fluorescence of

Fluorescin (FITC—fluorescin-isothiocyanate) and red fluorescence of Rhodamine 600 (TRITC—tetramethyl rhodamine isothiocyanate) require 495 and 575 nm for excitation and 515 and 600 nm for emission respectively.

The commonly used fluorochromes attached to reporter and antibody molecules in addition to the three mentioned above, are Rhodamine with excitation and emission at 550 and 575 nm respectively for orange red fluorescence; R-phycoerythrin with excitation and emission at 525 and 575 nm respectively for orange red fluorescence; Texas Red with excitation and emission at 595 and 615 nm respectively for red fluorescence and Ultralite 680 with emission at 680 nm for infrared fluorescence. For staining the chromosome, *DAPI* (4,6-diamidino-2-phenylindole) at 355 and 450 nm and *Hoechst 33258* at 356 and 465 nm for blue; *Chromomycin A*, at 430 and 570 nm for yellow, *Quinacrine* at 455 and 495 nm for yellow green and *Propidium iodide* at 530 and 615 nm for excitation and emission respectively for red fluorescence are usually employed. Of these, DAPI and Propidium iodide have a wide application as chromosome stain with *in situ* hybrid detection.

The method in principle involves incubation of the slides in detection buffer (4× SSC and 0.2% Tween 20, 1:1) at 37°C; addition of 5% BSA block, covering with a coverglass; reincubation for 5 min at room temperature; removal of coverglass; addition of conjugated antibody (10–20 µg) anti-digoxigenin or avidin for digoxigenin and biotin respectively; covering the slide again and incubation in a humid chamber at 37°C. Finally slides are washed in detection buffer for 8 min at the same temperature.

Chromosome counterstaining is done with DAPI (2 µg/ml) in McIlvaine's buffer (18 ml Na_2HPO_4 + 82 ml citric acid—pH 2.0) on incubating the slide with coverglass for 10 min. This step is followed by washing in PBS, and staining in Propidium iodide (100 µg/ml in water stored at −20°C in microcentrifuge tubes). The next steps of incubation and washing are similar to DAPI. The addition of Antifade solution (AFI Vector Lab) and filter pressing and squeezing off the excess solution are necessary before viewing under the microscope.

METHODS FOR HYBRIDIZATION AND DETECTION

Labelling DNA by random priming or nick translation

Labelled nucleotides are incorporated into the DNA so that the site of probe hybridization can be visualized following *in situ* hybridization.

Reagents needed include:

Labelled nucleotides, one of the following is used:

(a) digoxigenin-11-dUTP (1 mM solution, Boehringer Mannheim)
(b) biotin-11-dUTP (powder, Sigma) or biotin-14-dATP (0.4 mM solution)

(c) fluorescein-12-dUTP (1 mM solution, Boehringer Mannheim) or fluo-
    rescein-11-dUTP (FluoroGreen, 1 mM solution, Amersham)
(d) rhodamine-4-dUTP (FluoroRed, 1 mM solution, Amersham)
(e) coumarin-4-dUTP (FluoroBlue, 1 mM solution, Amersham)

Unlabelled nucleotides: dATP, dCTP, dGTP, TTP (lithium salt, 100 mM
solution in Tris HCl pH 7.5, Boehringer Mannheim).

Modifying enzyme   (a) for random priming and (b) for nick translation:

(a) Klenow enzyme (DNA polymerase 1, large fragment, labelling grade,
    6 units/μl, Boehringer Mannheim)
(b) DNA polymerase I/DNase I mixture (0.4 units/μl, GIBCO BRL)

For random priming only: hexanucleotide reaction mixture (in 10× buffer
Boehringer Mannheim) or "random" hexanucleotides or primers.
3 M sodium acetate or 4 M lithium chloride
0.3 M EDTA pH 8; ice cold 100% and 70% ethanol, dithiothreitol (DTT).

Method (after Schwarzacher *et al.*, 1994)
Use clean total genomic DNA, sheared or sonicated to 50 kb fragments.
Cloned DNA should be either linearized whole plasmid or cut out and
cleaned insert. Labelled nucleotide mixture should contain:

(a) for digoxigenin: 18.5 μl 1 mM solution of TTP in Tris HCl pH 7.5 and
    mixed with 11.5 μl digoxigenin-11-dUTP.
(b) for biotin: 0.4 mM solution in Tris HCl pH 7.5.
(c) for direct labelled nucleotides: 10 μl 0.5 mM solution of TTP, mix with
    25 μl of fluorochrome conjugated dUTP.

Unlabelled nucleotide mixture contains the three nucleotides not present as
labelled nucleotide. 0.5 mM solution of each nucleotide in 100 mM Tris HCl,
pH 7.5 is mixed together at 1:1:1 and stored at −20°C.

Random priming
Prepare 50 μl hexanucleotide reaction buffer by mixing 0.5 M Tris HCl,
pH 7.2; 0.1 M MgCl$_2$; 1 mM dithiothreitol; 2 mg/ml BSA, 62.5A$_{260}$ units/ml
"random" hexanucleotide, stored at −20°C.
    Take 50–200 ng of DNA for labelling and mix with sterile distilled water
to make 12.5 μl in an 1.5 ml microcentrifuge tube.
    Denature the DNA in boiling water for 10 min, put on ice for 5 min and
centrifuge briefly.
    Add: 3 μl unlabelled nucleotide mixture, 1.5 μl labelled nucleotide mixture
and 2 μl of hexanucleotide of reaction buffer. Vortex briefly.
    Add 1 μl of Klenow enzyme and mix gently. Incubate at 37°C for 6–8 h or
overnight. Add 2 μl of 0.3 M EDTA, pH 8, to stop the reaction.

Nick translation

Prepare 150 µl 10× nick translation buffer (NTB) by mixing 0.5 M Tris HCl (pH 7.8), 0.05 M MgCl$_2$ and 0.5 mg/ml bovine serum albumin (BSA) Sigma, store at −20°C.

Start the reaction in a 1.5 ml microcentrifuge tube by adding 5 µl of 10 nick translation buffer; 5 µl unlabelled nucleotide mixture; 3 µl of labelled nucleotide mixture; 1 µl dithiothreitol: 1 µg of DNA, adjust to the final volume of 45 µl with distilled water and briefly vortex.

Add 5 µl of DNA Polymerase I/DNase I mixture and mix gently. Incubate at 15°C for 1–2 h. Add 5 µl 0.3 M EDTA to stop the reaction.

Removal of unincorporated nucleotides after random priming or nick translation

Add 1/10 volume sodium acetate or lithium chloride; add 2/3–3/4 volume cold 100% ethanol and mix gently; incubate at −20°C overnight or −80°C for 1–2 h. Prepare 10 ml 100× TE buffer by mixing 1 M Tris HCl (pH 8) and 0.1 M EDTA. For use, dilute 100 µl with 9.9 ml distilled water and filter sterilize. Spin the tubes containing precipitated DNA at −10°C for 30 min ×12,000g. Discard the supernatant and add 0.5 ml of cold 70% ethanol and spin at −10°C for 5 min at 12,000g. Discard supernatant and leave pellet to dry. Resuspend the DNA in 10–30 µl 1× TE.

Chromosome preparation, denaturation and hybridization
(after Schwarzacher *et al.*, 1994)

Pretreatment

Chromosome preparations are pretreated to reduce non-specific probe and detection binding, increase permeability to probe and detection reagent and to stabilize the target DNA sequences.

For spread preparations

Dry chromosome spreads at 37°C overnight. Make up 1 litre of 2× SSC stock 1:10. Make up 10 ml RNase A stock by dissolving 10 mg/ml of RNase in 10 mM Tris HCl, pH 7.5 and 15 mM NaCl. Boil for 15 min and allow to cool. Store at −20°C in aliquots. Dilute RNase stock 1:100 with 2× SSC for a final concentration of 0.1 mg/ml. Add 200 µl of RNase A to each preparation, cover with a plastic cover glass and incubate for 1 h at 37°C in a humid chamber. Start preparing paraformaldehyde. In the fume hood, add 4 g of paraformaldehyde to 80 ml water, heat to 60°C for about 10 min, clear the solution with about 20 ml 0.1 M NaOH, let it cool down and adjust the final volume to 100 ml with water. Wash slides in 2× SSC for 3× 5 min. This step and all further washing steps are carried out in Coplin jars using enough liquid to fully immerse the whole slide.

Make up 5 µg/ml pepsin in 0.01 M HCl. (A stock of 500 µg/ml in 0.01 M HCl can be stored at −20°C in aliquots and diluted prior to use.) Place slides in 0.01 M HCl for 2 min. Add 200 µl pepsin onto each preparation, cover with a plastic coverglass and incubate for 10 min at 37°C. Stop pepsin reaction by placing the slides in water for 2 min and then wash in 2× SSC for 2× 5 min. Place slides into paraformaldehyde in a Coplin jar and incubate for 10 min at room temperature. Wash slides in 2× SSC for 3× 5 min. Incubate slides 3 min each in 70%, 90% and 100% ethanol and then air dry. Slides can be kept for up to 24 h before denaturation and hybridization.

For sections

The earlier steps are similar to spread preparations upto washing the slides in 2× SSC for 3× 5 min.

Make up 200 ml proteinase K buffer consisting of 20 mM Tris HCl, pH 8 and 2 mM CaCl₂ and 300 ml proteinase K buffer containing 50 mM MgCl₂. Make up 1 µg/ml proteinase K in proteinase K buffer. Incubate slides in proteinase K buffer for 2× 5 min. Add 200 µl proteinase K, cover with a plastic coverglass and incubate for 10 min at 37°C. Stop reaction by placing slides in proteinase K buffer containing MgCl₂ and wash 3× 5 min. Place slides into paraformaldehyde in a Coplin jar and incubate for 10 min at room temperature. Wash slides in 2× SSC for 3× 5 min. Slides can be kept for a few hours before continuing with denaturation and hybridization.

Hybridization

For preparing the hybridization mixture (see Table 1)
Make 50% (w/v) dextran sulphate in water and filter (0.22 µm) sterilize. Store at −20°C.

Autoclave salmon sperm DNA to fragments of 100–300 bp (typically 5 min at 100 kgm⁻²). Store at −20°C.

Prepare the hybridization mixture in a 1.5 ml microcentrifuge tube by adding: formamide, 20 µl; 50% dextran sulphate, 8 µl; 20× SSC, 4 µl; 10% SDS, 0.5–1 µl; probe DNA 1–5 µl; (final concentration 0.5–2 µg/ml) and, salmon sperm DNA, 1–5 µl (final concentration 5–50 µg/ml, 10–200× of probe concentration)

Adjust to the final volume of 40 µl with distilled water and briefly vortex the mixture. Denature at 70°C for 10 min and then transfer to ice for 5 min. Take slides with sections out of 2× SSC. Rapidly remove excess liquid and proceed before slides are desiccated. Add 40 µl denatured hybridization mix to each slide and cover with a plastic coverglass. Ensure no bubbles are trapped.

Table 1 Hybridization mixture

	Amount per slide (total 40 µl)	Final concentration
100% formamide high grade	20 µl	50%
50%(w/v) dextran sulphate in water	8 µl	10%
20 × SSC	4 µl	2 ×
Probe	4 µl	25 to 100 ng per slide
Blocking DNA (autoclave from salmon sperm)		2 to 100 × probe concentration
10% (w/v) sodium dodecyl sulphate (SDS) in water	0.2 to 4 µl	0.05 to 1%
Water	in requisite amount to make final volume per slide	40 µl

Methods for denaturation and hybridization

Modified thermal cycler method
Programme the thermal cycler for one cycle at 75–80°C for 5–10 min using the mode which starts the timing after the target temperature is reached. Then programme to cool down slowly to 37°C. Maintain the temperature at 37°C. Place some filter paper in the chamber and moisten it with distilled water. Transfer the slides with hybridization mixture to the moist chamber and start the machine. After the programme has reached 37°C, slides can be left in the machine or can be transferred to an incubator or water bath at 37°C and kept overnight.

Floating dish method
Prepare a humid chamber by placing filter paper in the bottom of a metal tin with a lid and soak with distilled water. Place in a water bath to equilibrate the internal temperature to 85–90°C. Transfer slides with hybridization mixture to the preheated humid chamber, and incubate for 5–10 min. Monitor temperature carefully. At least 2/3 of incubation time should be the target temperature (e.g. 85°C). Transfer humid chamber to 37°C incubator or water bath and leave slides to hybridize overnight.

Washing after hybridization

After hybridization overnight, the slides are taken through several washing steps to remove non-specifically and weakly bound probe. They include a stringent formamide wash, followed by treatment with 2× SSC to remove the formamide.

Prepare 500 ml 2× SSC, 200 ml 0.1× SSC and 200 ml 20% formamide in 0.1× SSC (stringent formamide wash) and heat to 42°C. Remove slides from the moist chamber carefully, remove coverglass and place in a Coplin jar containing $2\times$ SSC at 42°C. Decant $2\times$ SSC and replace with stringent formamide wash. Incubate for $2\times$ 5 min at 42°C shaking gently. Wash slides in $0.1\times$ SSC for $2\times$ 5 min at 42°C and then in $2\times$ SSC for $2\times$ 5 min at 42°C. Allow to cool down to room temperature.

Detection of probe hybridization sites by fluorescence

Several fluorochromes are used as conjugates to nucleotides or antibodies and as DNA stains for chromosomes (see Table 2 from Heslop–Harrison et al., 1994).

Epi-fluorescence microscopes use lamps and filter blocks to select light of the correct wavelength for excitation and emission. Where several fluorochromes are used simultaneously, each requires a different pair of excitation and emission maxima. Digoxigenin or biotin-labelled probes are detected by

Table 2 Properties of fluorochromes used in signal generating systems for in situ hybridization and as stains for chromosomal DNA

Fluorochrome	Excitation (max. nm)	Emission (max. nm)	Fluorescence colour
(a) conjugated to detecting and antibody molecules			
Coumarin (AMCA)[1]	350	450	Blue
Fluorescein (FITC)[2]	495	515	Green
R-phycoerythrin	525	575	Orange-red
Rhodamine	550	575	Orange-red
Rhodamine$_{600}$ (TRITC)[3]	575	600	Red
Texas Red	595	615	Red
Ultralite 680		680	Infrared
(b) conjugated directly to nucleotides			
Coumarin (AMCA)[1]	350	450	Blue
Fluorescein (FITC)[2]	495	515	Green
Rhodamine$_{600}$ (TRITC)[3]	575	600	Red
(c) DNA stains			
DAPI[4]	355	450	Blue
Hoechst 33258	356	465	Blue
Chromomycin A3	430	570	Yellow
Quinacrine	455	495	Yellow-green
Propidium iodide	340,530	615	Red

1 = 7-amino-4-methyl-coumarin-3-acetic acid
2 = fluorescein isothiocyanate
3 = tetramethyl rhodamine isothiocyanate
4 = 4',6-diamidino-2-phenylindole

immunocytochemistry to attach marker molecules for visualization. For direct fluorochrome-labelled probes, this step is not needed and preparations are counter-stained immediately.

The *materials* needed are

20× SSC stock: 3 M NaCl, 0.3 M Na citrate adjusted to pH 7, Tween 20, Bovine serum albumin (BSA; Sigma Fraction V, globulin-free).

For *digoxigenin labelled probes*: Anti-digoxigenin conjugated to fluorescein or rhodamine raised in sheep (Boehringer Mannheim).

For *biotin labelled probes*: Avidin conjugated to fluorescein, Texas Red, rhodamine or coumarin (e.g. Vector Laboratories); plastic coverglass: pieces of appropriate size cut from autoclavable waste disposal bags.

The *steps* include:

Make up 1 litre detection buffer by diluting 20× SSC to 4× SSC and adding 0.2% (v/v) Tween 20 and heat to 37°C. Incubate hybridized slides in detection buffer for 5 min. Make up 5% BSA block in detection buffer. Add 200 μl of BSA block to each slide, cover with a plastic coverglass and incubate for 5 min at room temperature. Make up 10–20 μg/ml anti-digoxigenin and/or avidin in BSA block. Remove coverglass, drain slide and add 30 μl of the anti-digoxigenin and/or avidin solution. Cover and incubate slides in a humid chamber for 1 h at 37°C. Wash slides in detection buffer for 3× 8 min at 37°C.

Counterstaining of chromosomal DNA

The *materials* required are: 0.2 M Na_2HPO_4; 0.1 M citric acid; PBS (phosphate buffered saline): 0.13 M NaCl, 0.007 M Na_2HPO_4, 0.003 M NaH_2PO_4, adjust to pH 7.4; DAPI (4′,6-diamidino-2-phenylindole); PI (propidium iodide); Antifade solution: AF1 (Citifluor, City University, Northampton Square, London) or Vectashield mounting medium (Vector Laboratories).

The *steps* are as follows:

Make McIlvaine's buffer (pH 7.0, 18 ml Na_2HPO_4 and 82 ml citric acid), prepare DAPI stock solution of 100 μg/ml in water. Aliquot and store at −20°C. Prepare a working solution of 2 μg/ml by dilution in McIlvaine's buffer, aliquot and store at −20°C. Add 100 μl of DAPI per slide, cover with a plastic coverglass and incubate for 10 min. Wash briefly in PBS, and drain.

Prepare PI stock solution of 100 μg/ml in water. Aliquot 50 μl or 100 μl in 1.5 ml microcentrifuge tubes and store at −20°C. Dilute freshly with PBS to 2–5 μg/ml prior to use. Add 100 μl of PI, cover with a plastic coverglass and incubate for 10 min. Wash briefly in PBS, and drain the slide.

Apply 1–2 drops of antifade on the wet slide, cover with a coverglass. Squeeze out excess antifade from the slide with filter paper. Slides can be viewed immediately, but the signal stabilizes after storing for a few days in the dark at 4°C. Slides can be kept at 4°C for up to a year.

Calculation of the specific activity of the probe:

(a) During nick-translation, nucleotides are effectively excised and replaced, and there is usually no net synthesis of DNA. Thus the probe specific activity is calculated simply as:

$$\text{specific activity (d.p.m. } \mu g^{-1}) = \frac{\text{total activity incorporated (d.p.m)}}{\text{amount of substrate added } (\mu g)}$$

(b) 50% incorporation in the above reaction would give a specific activity of 2×10^8 d.p.m. μg^{-1} with 500 ng input DNA and 2×10^9 d.p.m. μg^{-1} with 50 ng ($1 \mu Ci = 2.2 \times 10^6$ d.p.m.).

Calculation of the probe yield and specific activity:

During random-primer labelling there is net synthesis of DNA, while the initial substrate remains unlabelled. Both participate in the subsequent hybridization so:

• Probe yield = ng initial substrate DNA + ng DNA synthesized. As the average mol. wt. of a nucleoside monophosphate in DNA is 350, for labelled nucleotide at $X \times 10^3$ Ci mmol^{-1}:

• $\text{ng DNA synthesized} = \dfrac{\mu Ci \text{ incorporated} \times 0.35 \times 4}{X}$

A multiplication factor of 4 is included as there are four nucleotides, only one of which is labelled.

Once the probe yield has been calculated, the specific activity can be determined:

$$\text{specific activity (d.p.m. } \mu g^{-1}) = \frac{\text{total activity incorporated (d.p.m.)}}{\text{probe yield } (\mu g)}$$

Complementary RNA

Prepare DNA solution 0.04 M Tris, pH 7.9, to have an optical density at 260 nm of between 1.0 and 2.0 (50–100 µg/ml). Place DNA in a boiling water bath for 5 min and cool on ice if transcription from single-stranded DNA is required. Prepare dried "XTPs" containing 100 µCi each of GTP, CTP, ATP and UTP, all tritium labelled with specific activities of 10–30 Ci/mM. Keep in small glass tube and vacuum dry. Add to the dried "XTPs", DNA 100 µl; β-mercaptoethanol 50 µl (Sigma cat. no. M-6250, about 14.3 M, diluted 20 µl in 10 ml water); salt solution 50 µl (Tris 1 M, pH 7.9, 4.0 ml; 2 M KCl 9.4 ml;

1 M MgCl₂ 0.58 ml; 0.01 M EDTA 0.88 ml, made upto a final volume of 25 ml with water, 80 µl of 0.125 MnCl₂ added per ml of salt solution before use); water 40 µl; RNA polymerase (*E. coli* Sigma cat. no. R5376) 10 µl. The total volume should be 250 µl.

Cover tube with parafilm and incubate at 37°C for 60–90 min, add 0.75 ml of 0.04 M Tris, pH 7.9 and 20 µl of DNASe I (Sigma cat. no. D-1126, 1 mg/ml in water), equivalent to 20 µg. Keep at 20°C for 15 min.

Remove two 10 µl samples and spot on nitrocellulose filter. Dry. Place one filter in cold 5% TCA for 5 min, wash in 70% ethanol and dry. Count both filters in a liquid scintillation system to determine the percentage of counts which are TCA insoluble (usually between 15 to 40%). Add 200 µl of 5 Sarkosyl or sodium dodecyl sulphate (SDS) 5% in water. Add 50 µg of *E. coli* RNA in 1 ml of 0.04 M Tris, pH 7.9, to act as carrier in cases where cRNA is being transcribed from eukaryotic DNA. Add 2 ml of redistilled water-saturated phenol.

Alternatively, Use a mixture of phenol 100g, 8-hydroxyquinoline 0.1 g, chloroform 100 ml and isoamyl alcohol 4 ml. Centrifuge for 10 min at 10,000 rpm. Remove supernatant, extract phenol with 200 µl 0.04 M Tris. Gently add supernatant to Sephadex column (G-50, 15× 1 cm) and collect 25 consecutive 30 drop fractions. Spot 10 µl from each fraction on nitrocellulose filter, dry and count. Pool peak fractions, keep in 85°C water bath for 3 min. Filter through 0.45 µm pore nitrocellulose filter. Store at −20°C.

In situ hybridization using radioactive probe in pachytene

Freeze slides containing chromosome spreads in liquid nitrogen and remove coverglass. Rinse with a freshly prepared mixture of acetic acid: 95% ethanol (1 : 3). Treat with 100 µg/ml RNAse in 2 × SSCP (1× SSCP=0.15 M NaCl, 0.015 M sodium citrate, 0.02 M sodium phosphate, pH 7.0) at 37°C for 1 h under a siliconized coverglass in a moist chamber. Wash slide thoroughly in 2× SSCP, dehydrate sequentially in 70, 80 and 95% ethanol and air dry.

For *in situ* hybridization, denature chromosome preparations in 70% formamide in 2× SSCP at 70°C for 4 min. Transfer quickly to chilled 70% ethanol and dehydrate in 80, 95 and 100% ethanol. Denature probe mixture, containing 50% formamide, 10% dextran sulphate, 100 ng/100 µl probe DNA, 10 µg/100 µl carrier DNA (sheared *E. coli* DNA) at 70°C for 6 min and quickly cool in an ice bath. Add 20 µl probe mixture to each slide and cover with a coverglass. Incubate in a moist chamber at 37°C for 20 h. Rinse in two changes of 50% formamide in 2× SSCP at 39°C, keeping 15 min in each. Wash in 2× SSCP at 39°C for at least 7 changes, keeping 15 min in each.

For autoradiography, dehydrate slides and coat with photographic emulsion (diluted L4 photographic emulsion, Polyscience, L4 : water=1 : 1) at 45°C. Airdry slides in dark for 1 h, and keep in light-tight box at −20°C for

2–10 weeks. Develop in Kodak D-19 for 4 min, treat with 1% acetic acid, fix with Kodak fixer, rinse thoroughly in distilled water and stain with Giemsa for light microscopy (after Shen *et al.*, 1987).

Fluorescence *in situ* hybridization (FISH) with centromeric satellite DNA

Denature metaphase chromosome preparations for 4 min at 70°C in 70% formamide (IBI, New Haven, Conn), 2× SSC pH 7.0. Prepare hybridization mixture (55% formamide, 10% dextran sulphate, 1 µg/µl sonicated herring sperm DNA, 2× SSC, pH 7.0). Add biotinylated probe (1 µl, 20 ng/µl) to 9 µl of the hybridization mixture. Denature DNA in hybridization mixture at 72°C for 5 min. Chill on ice. Add chilled mix to metaphase spread, cover with coverglass and hybridize overnight at 37°C in a moist chamber. Wash slide in 50% formamide, 2× SSC pH 7.6 at 42°C for 10 min each, followed by two washes of 20 min each in PN buffer (0.1 M sodium phosphate, pH 8.0, 0.1% NP40) at 37°C. Detect biotinylated probe by incubating for 20 min in 5 µg/ml avidin—FITC (fluorescein isothiocyanate, Vector Lab, Burlingame, Calif) in PN buffer plus 5% nonfat dry milk and 0.01% sodium azide at room temperature. Wash twice in PN buffer at room temperature. Counterstain DNA with 0.5 µg/ml DAPI (Calbiochem, La Jolla, Calif) in antifade solution. Observe under Zeiss Universal fluorescence microscope with a plan-Neofluar 63 × 1.20 oil objective using appropriate filters to simultaneously observe FITC and PI fluorescence and separate filters for DAPI. Wash the slide briefly in PN buffer and add PI (2 µg/ml in antifade solution, Sigma).

MICRO-FISH—rapid regeneration of probes by chromosome microdissection

Chromosome microdissection: Harvest metaphase cells following conventional cytogenetic techniques and prepare spreads on 24× 60 mm coverglasses. Stain by trypsin-Giemsa banding.

For microdissection, use glass microneedles controlled by a Narashige micromanipulator (Model MO 302) attached to an inverted microscope. Before use treat microneedles with UV light for 5 min. Dissect out chromosome fragment and transfer to a 20 µl collecting drop (containing proteinase K 50 µg/ml^{-1}) in a 0.5 ml microcentrifuge tube. Use fresh microneedle for dissecting out each fragment.

PCR amplification: Incubate collecting drop with chromosome fragment at 37°C for 1 h and then at 90°C for 10 min. Add the components of the PCR reaction to a final volume of 50 µl in the same tube (1.5 µM universal primer 200 µM each dNTP, 2 mM MgCl$_2$, 50 mM KCl, 10 mM Tris-HCl, pH 8.4; 0.1 mg/ml^{-1} gelatin and 2.5 U Taq DNA polymerase, Perkin Elmer Cetus).

Overlay the reaction with oil and heat to 93°C for 4 min and cycle for 8 cycles at 94°C for 1 min; at 30°C for 1 min; at 72°C for 3 min; followed by 28 cycles at 94°C for 1 min, at 56°C for 1 min and at 72°C for 3 min with a final extension at 72°C for 10 min.

To prepare probes: Label 2 µl of the microdissection PCR in a secondary PCR reaction identical to the first except for reduction of the TTP concentration to 100 µM and the inclusion of 100 µM biotin-11-dUTP cycled for 8 cycles at 94°C for 1 min, at 56°C for 1 min and at 72°C for 3 min with a final extension at 72°C for 10 min. Purify the products of this reaction with a Centricon 30 filter and use for *in situ* hybridization.

Alternatively, Probe label through biotinylation with biotin-11-dUTP in a nick translation reaction, as described earlier.

In situ hybridization of micro-FISH probes: Micro-FISH probes are based on standard Imagenetics (Naperville, Illinois) protocol after Pinkel *et al.*, 1986. Denature slides with metaphase plates at 72°C for 2 min in a bath of 70% formamide, 2× SSC, dehydrate through ethanol series and airdry. Place in a slide warmer at 37°C, with 20 µl of hybridization mixture (10 ng µl^{-1} probe, 50% formamide, 10% dextran sulphate, 1× SSC, 3 µg Cot1 DNA, BRL), place a coverglass, seal with rubber cement and incubate overnight at 37°C in a moist chamber. Wash slides three times in 2× SSC/50% formamide at 42°C for 3 min, then rinse for 3 min in 2× SSC at ambient temperature.

To detect probe, wash slide in 0.1 M sodium phosphate, 0.1% NP- 40, pH 8.0 (PN) at 45°C for 15 min, at ambient temperature for 2 min and then incubate in a bath containing 5 µg/ml^{-1} fluorescein-conjugated avidin (Vector Laboratories) in PN; with 5% nonfat dry milk and 0.02% sodium azide for 20 min at ambient temperature. Wash slide twice in PN for 2 min each. Incubate in a bath containing 5 µg/ml^{-1} biotinylated anti-avidin (Vector Laboratories) in PN with 5% nonfat dry milk and 0.02% sodium azide for 20 min at ambient temperature. Wash twice in PN for 2 min each and repeat incubation in fluorescein-conjugated avidin. Finally apply 10 µl of fluorescence antifade solution (10 µg/ml^{-1} p-phenylamine dihydrochloride in 90% glycerol, pH 8.0) containing 0.2 µg/ml^{-1} propidium iodide and cover with a coverglass.

In an alternative schedule to localize micro FISH signal on chromosome bands, first process the slides by trypsin-Giemsa banding, photograph and destain by sequential washing in 70% ethanol (1 min), 85% ethanol (1 min), Carnoy's fixative 10 min, 3.7% formaldehyde in PBS (10 min) and PBS (5 min twice) before hybridization. For hybridization using fluorochrome-labelled *whole chromosome paints* (WCP, Imagenetics, Naperville IL), include 10 ng µl^{-1} WCP in the hybridization mixture. Examine slide with a Zeiss Axiophot microscope equipped with a dual bandpass fluorescein-rhodamine filter. This method may be used in plants with modifications.

Genomic *in situ* hybridization with non-radioactive probe (GISH)

Isolation of genomic DNA

Surface sterilise 0.5 to 1.0 g fresh leaf material in 2% bleach in deionized water for 10 min. Briefly rinse in water. Crush to fine powder in mortar and pestle, with a little sterilized water and liquid nitrogen. Add 1 ml isolation buffer to form ice slurry (buffer; 2× CTAB, 100 mM Tris-HCl pH 8.0; 1.4 M NaCl; 20 mM EDTA, 2% CTAB [(hexadodecyl (cetyl) trimethyl ammonium chloride]. Continue brisk homogenization.

Add 100 µl of 5% SDS (sodium dodecyl sulphate), followed by 100 µl proteinase K (0.2 mg/ml stock solution) and stir vigorously. Transfer homogenate to a 15 ml centrifuge tube. Rinse pestle and mortar with 1 ml isolation buffer and add to centrifuge tube. Incubate at 37°C for 3 h.

Extract DNA by adding an equal volume of phenol/chloroform to mixture in centrifuge tube. Gently invert several times. Centrifuge at 13,000 rpm for 10 min. Transfer supernatant fluid to a clean centrifuge tube. Discard bottom layer. Repeat steps from extraction of DNA once more.

To centrifuge tube add an equal volume of chloroform to the contents. Gently invert several times, centrifuge at 13,000 rpm for 10 min. Separate out supernatant.

Precipitate the DNA by adding 3 M sodium acetate equal to 1/20 th of total volume of mixture to separated supernatant in tube and then absolute ethanol about two and a half times the present volume of mixture in tube. Leave at −20°C for overnight. Centrifuge at 13,000 rpm for 10 min. Discard supernatant, wash pellet with 70% ethanol 1 ml and drain. Resuspend pellets in TE (pH 8)—500 µl aliquots until dissolved.

Incubate in 10 µg/ml RNase at 37°C for 3 h. Add SDS (5%) equal to 1/20 th of total mixture in tube. Repeat step for precipitation of DNA leaving to reprecipitate for at least 4 h. Centrifuge at 13,000 rpm for 10 min, discard supernatant, wash in 70% ethanol and resuspend pellet in TE pH 8.0, until dissolved.

Store in 50 µl aliquots at −20°C for short term or −70°C for long term.

Assessment of isolated DNA

Prepare 2 µl DNA, 8 µl TE (pH 8.0) and 5 µl loading buffer (40% sucrose, 0.25% bromophenol blue; 1 kb ladder with 30% loading buffer (100 ng/µl). Run on a 0.5% (or 1%) agarose gel in 1× TBE, 60 V, 0.5 mA for 2 h.

For shearing DNA from *Milium montanium*, a shearing strategy included use of hypodermic syringe coupled with brief pulsing (1–2 s) in a microcentrifuge and longer periods (up to a minute) on a vortex mixer. The method is as follows: Vortex 50 µl of genomic DNA for 1 min; pulse briefly; suck up-and-down 25 times using hypodermic apparatus; vortex; pulse; suck up-and-down

25 times; vortex; pulse; repeat several times. For optimal labelling, genomic DNA should be sheared into fragments 10 to 12 kb in length.

Biotinylation of genomic DNA (adapted from manufacturers of the BRL labelling system)

Shear genomic DNA to sizes between 10–12 Kb. Add 2–3 µg DNA to 70 µl water; mix well; add 10 µl 10× dNTP mixture; mix and then add quickly 10 µl 10× enzyme mixture from the freezer. Mix and pulse in benchtop centrifuge. Incubate at 16°C for 1 h. Add 10 µl stop buffer, making a total of 100 µl. Add 60–90 µg of sonicated salmon sperm DNA, which reduces non-specific binding and helps to bring down biotinylated DNA. Add 3 M sodium acetate equal to 1/10 th volume in tube (11 µl). Mix well. Add 300 µl cold absolute ethanol. Close tube tightly and invert gently several times.

Precipitate the DNA at −20°C for about 1 h or at −70°C for 20–30 min. Centrifuge at 15,100 rpm for 10 min. Discard supernatant. Wash pellet with 70% ethanol (500 µl). Invert tube several times. Dry pellet gently under vacuum for 15–20 min. Dissolve pellet in 21 µl of TE (pH 8.0), keeping overnight at room temperature.

Test biotinylation

Prepare buffers 1, 2 and 3 as instructed by the manufacturers BRL. Pipette 1 µl of the 21 µl "Labelled" DNA in TE onto nitrocellulose strip and dry at 80°C under vacuum for 1–2 h. Make up 100 µl of buffer 2 and heat 50 µl to 40°C. Wash strip in buffer 1 for one min for rehydration; then incubate in buffer 2 (3% BSA in buffer 1) for 20 min at 40°C. Gently blot strip between filter paper; dry at 80°C for 15–20 min; remove and store at room temperature overnight.

Rehydrate in buffer 2 for 10 min. Incubate in 2 µg/ml BRL streptavidin (2 µl in 998 µl) in buffer 1 for 10 min with gentle shaking. Make it up to 1 ml and wash strip in petri dish. Wash strip three times in 40 ml buffer 1, for 3 min each time with gentle agitation. Incubate in 1 µg/ml BRL biotin AP (1 µl in 999 µl) in buffer 1 for 10 min, agitating gently. Wash twice in 40 ml buffer 1 for 3 min each. Wash in buffer 3 for 3 min twice.

For detection: To 7.5 ml buffer 3, add 33 µl NBT and 25 µl BCIP solution. Incubate in sealed plastic bag for upto 4 h in dark. Well-labelled DNA turns dark. Wash strips in TE, blot between filter paper and dry in oven for 10 min.

RNAse treatment

Wash slides in 2× SSC for 10 min. Pipette 100 µl of RNAse solution onto slides (100 µg/ml boiled RNAse in 2× SSC); cover with coverglass and incubate in moist chamber with 2× SSC in incubator at 37°C for 1 h. Carefully remove coverglass and place slide in metal rack. Wash slides thrice in 2× SSC, keeping for 5 min in each. Dehydrate slides through ethanol series of 70% and absolute on ice, keeping twice for 5 min in each. Dry slides at approx 45°C, usually for a few hours.

Hybridization mixture
Prepare probe mixture containing per slide (20 µl) the following: DNA (in TE, pH 8.0), 100–200 ng; deionized water (20% vol); formamide (Fisons electrophoresis grade), 50% vol; 50% dextran sulphate (20% vol); 3 M NaCl (10% vol). Mix and leave on ice. Add dextran last. Denature in water bath at 85°C for 15 min. After 13 min, add salt, vortex briefly to mix and return tube to water bath. Transfer tubes immediately to ice to quench for 2 min. Pipette 20 µl of probe mixture onto denatured slide. Cover with plastic coverslip. Place slide in moist chamber, seal with masking tape and incubate overnight at 37°C.

Chromosome denaturation and hybridization
Transfer slides after RNase treatment as described earlier, to 70% formamide in 2× SSC and keep in water bath (68–72°C) for 2 min. Dehydrate slides through ethanol series kept at −20°C (70%, 70%, absolute, absolute), keeping 2 min in each. Airdry slides, add hybridization mixture prepared as given in the previous step and leave to hybridize.

Post-hybridization washes
Remove coverglass and keep slides in 2× SSC for 10 min at 42°C. Wash slides in 50% formamide in 2× SSC for 10 min at 42°C. Keep in 2× SSC for 10 min at 42°C, followed by 5 min in 2× SSC at room temperature on an orbital shaker.

Detection
Prepare: BN buffer (0.1 M sodium bicarbonate, 0.05% v/v, Nonidet p-40 Sigma, pH 8.0) or BT buffer replacing Nonidet with Tween 20; 5% (w/v) BSA (bovine serum albumin) in BN buffer; 5 µg/ml F-avidin (Vector Labs) in 5% BSA-BN; 10 µg/ml F-avidin (Vector Labs) in 5% BSA-BN; 5% (v/v) normal goat serum (NGS) in BN buffer; 25 µg/ml propidium iodide in phosphate-buffered saline (PBS); 1–2 µg/ml DAPI in McIlvaine's buffer (pH 7).

Wash slides in BN for 10 min. Block with 5% BSA-BN (10 µl/slide) for 5 min *without coverglass*. Incubate slides in 5 µg/ml F-avidin in 5% BSA-BN (50–100 µl) *under coverglass* for 1 h at 37°C in moist chamber. Remove coverglass. Wash slides twice in BN at 40°C, for 5 min each. Wash in BN at room temperature for 5 min. Block with 5% NGS (100 µl/slide) for 5 min *without coverglass*. Incubate slides *under coverglass* in 25 µg/ml biotinylated anti-avidin in 5% NGS-BN (50–100 µl/slide) for 1 h at 37°C in moist chamber. Remove coverglass and wash in BN twice at 40°C and once at room temperature, keeping 5 min in each.

Incubate slides in 10 µg/ml F-avidin in 5% BSA-BN under coverglass for 1 h at 37°C. Remove coverglass and wash slides twice in BN at 37°C. Incubate slides in 1–2 µg/ml propidium iodide in PBS for 10 min in dark. Rinse in BN buffer (30–60 s). Incubate slides in 2 µg/ml DAPI (100 µl under coverglass) for 30 min in dark. Remove coverglass, rinse in BN (30 s) and then distilled water. Dry slightly. Mount under 22×22 mm coverglass with BN: Citifluor (Agar Scientific) mountant (1:1, v/v). Scan under UV light (365 nm), locate cells and photograph FITC fluorescence under blue light (450–490 nm). Usually half of automatic exposure time gives good results.

The method, with slight alterations, has been used for localization of parental genomes in a wide hybrid between *Secale africanum* and *Hordeum chilense* (Schwarzacher *et al.*, 1989) and genetic divergence in *Gibasis* (Parokomy *et al.*, 1992).

In situ **hybridization with automated chromosome denaturation**

A programmable temperature controller was modified for denaturation of the chromosome spread preparations on slides prior to DNA–DNA hybridization (Make: Cambio from Genesys Instruments, Cambridge, after Heslop–Harrison *et al.*, 1994).

Slide preparation and pretreatment for DNA–DNA
Dry chromosome preparations in an oven at 35 to 60°C overnight. Add 200 µl of × 100 µg/ml (w/v) RNase A[*] in 2× SSC (0.3 M NaCl, 0.03 M Na citrate pH 7.0), cover and incubate for 1 h at 37°C in a humid chamber. Wash slides thrice for 5 min each in 2× SSC at room temperature. Transfer slides to 0.01 M HCl for 2 min and add 200 µl pepsin[**] to each slide, place a coverglass and incubate for 10 min at 37°C. Transfer to water and wash twice in 2× SSC. Transfer slides in 4% (w/v) freshly depolymerized paraformaldehyde[***] in water and incubate for 10 min at room temperature. Wash thrice for 5 min each in 2× SSC at room temperature. Dehydrate through successive grades of 70, 90 and 100% ethanol, keeping 3 min in each. Airdry at room temperature.

Denaturation, hybridization and post-hybridization washes
Saturate pad at bottom of temperature controller with 2× SSC. Set at 80 to 85°C. Denature the hybridization mix at 70°C for 10 min. Transfer to ice for 5 min. Add 30 to 40 µl of denatured hybridization mix to each slide and cover with plastic coverglass. Transfer to preheated humid chamber. Denature at 80°C for 10 min and slowly cool to 37°C. It can be done automatically using the programmable temperature controller.

Keep the chamber in an oven at 37°C overnight for hybridization. Float coverglass off in 2× SSC at 42°C. Wash slides twice for 5 min each with 20%

---

[*]Stock RNase A; dissolve 10 mg/ml of DNase-free RNase in 10 mM Tris-HCl, pH 7.5, 15 mM NaCl. Boil for 15 min, cool, store frozen. For use, dilute 1:100 in 2× SSC.
[**]Stock Pepsin : 500 µg/ml in 0.01 M HCl, stored at −20°C. For use, dilute 5 µg/ml in 0.01 M HCl.
[***]Depolymerised paraformaldehyde: Add 2 g of paraformaldehyde to 40 ml water, heat to 60 to 80°C, clear with 10 ml 0.1 M NaOH, adjust to 50 ml.

Flow diagram for Fluorescence in situ hybridization
(from Dr. U. C. Lavania, Lucknow, Personal communication)

Preparation of clean air dried chromosome spreads (free from cell wall/cytoplasm/ RNA/other cell debris)	Fluoro-labelling of Probe (DNA sequence/genomic DNA : with dUTP, either directly conjugated with fluorochromes or linked with haptens) by PCR/Nick Translation

Fixation   : in 4% paraformaldehyde (pH 8) at 60°C, 10 min.,
Dehydration : in ethanol series

Denaturation of probe and material

Hybridization solution (total hybridization mix required for one slide = 40 µl)

Forma mide (50%)	Dextran sulphate (10%)	2 × SSC	SDS (10%)	sheared Salmon sperm DNA (5 µg)	fluoro-labelled probe(s) 0.1 µg each	H$_2$O to balance

Denaturation (70°C) and hybridization (at 37°C overnight) through gradual cooling

Post-hybridization and stringent washing (a critical step)
(i) in 2 × SSC at 30°C, (ii) 0.1 × SSC at 42°C,
(iii) in 2 × SSC at 42°C, (iv) in 4 × SSC at 30°C in 2% detergent

Detection
(i) Direct   : for probes directly labelled with fluorochromes
(sensitivity is limited, good for detection of middle-to-high copy number probes)

(ii) Indirect : via immunocytochemical attachment of fluorescent markers (e.g. digoxigenin; and biotin labelled probes could be visualised by anti-digoxigenin conjugated to fluorescein/rhodamine; and avidin conjugated to fluorescein/texas red/rhodamine/ coumarin, respectively).
Appropriate combination of above would facilitate simultaneous detection of two or more sequences

Counterstaining : with UV fluorescing dyes (e.g. DAPI/propidium iodide at 2 µg/ml)
Mounting   : in antifade/vectashield/neutral mounting medium
Observation   : as epifluorescence through compatible excitation filters

(v/v) formamide in 0.1× SSC at 42°C. Wash four to six times more as follows: thrice in 2× SSC for 3 min each at 42°C; cool for 5 min; thrice in 2× SSC for 3 min each at room temperature.

Visualization of chromosomes and labelling sites
Add 100 μl of 2 μg/ml (w/v) DAPI (4',6-diamidino-2-phenylindole) in McIlvaine's buffer (9 mM citric acid, 80 mM disodiumhydrogen phosphate) to each slide, apply a coverglass and stain for 5 to 10 min at room temperature.

Remove coverglass, drain, add 100 μl of propidium iodide (PI, 5–10 μg/ml) in 2× SSC per slide, replace coverglass and stain for 5 to 10 min at room temperature. Wash in 4× SSC/Tween and use 10 μ antifade mountant (AF-2, Citifluor, London, UK). Place thin glass coverglass over specimen on slide, blot out excess antifade under filter paper. DAPI stains DNA and fluoresces blue under ultraviolet excitation. Propidium iodide stains DNA and is used as counter stain for FITC. Both are excited by blue light but FITC emits green and PI orange-red light. Since FITC level overlays PI counterstain, the sites of label appear yellow and the unlabelled chromatin red-orange.

DAPI stock solution; 100 μg/ml in water; propidium iodide stock solution: 100 μg/ml in water; both stored at −20°C.

For studying fluorescence, epifluorescence microscope is needed (suggested filters Zeiss 09. Leitz 12/3 for FITC and PI; Zeiss 02, Leitz A for DAPI). *For photography*, suggested colour print films are: Fujicolor Super HG 400 and Kodak Ektar 1000.

DNA–DNA hybridization methods are generally based on those of Pinkel *et al.* (1986) for mammalian chromosomes but have been modified for plants almost at each step.

Genome *in situ* hybridization in sectioned plant nuclei

Fix root-tips for 1 h at about 20°C in freshly prepared 1% (v/v) glutaraldehyde (diluted from Taab vacuum-distilled 10 ml sealed ampoules) and 0.25% (v/v) saturated aqueous picric acid solution in phosphate buffer (0.05 M Na_2HPO_2; 0.05 M KH_2PO_4). Dehydrate through a graded ethanol series, embed in LR white (medium) resin and polymerize at 65°C for 16 h.

[**] *Suggested material*; Secale cereale, root-tips pretreated in colchicine, fixed in acetic acid–ethanol (1 : 3); softened by maceration with enzymes and squashed in 45% acetic acid.
[***] *DNA probe used*: Clone pSc119.2, contains a 120 bp tandemly repeated DNA sequence isolated from rye. It was labelled with digoxigenin-11-dUTP (Boehringer-Mannheim) by nick translation (Leitch *et al.*, 1991). Biotin-labelled probes can also be used (Schwarzacher *et al.*, 1989).

Cut ultrathin (0.1 µm, gold) and thin (0.25 µm, blue) sections with a Reichert Ultracut for electron microscopy and light microsopy respectively. Expand all sections in knife trough containing 1% benzyl alcohol in water. Pick up ultrathin sections on 200 gold mesh grids and transfer thin sections to glass slides coated with poly-L-lysine solution (Sigma). Incubate slides overnight at 60°C prior to hybridization. The subsequent steps are similar to those for *in situ* hybridization of squashes (after Leitch *et al.*, 1990).

In situ hybridization with biotin-labelled probe and enzymatic detection

Pretreat fresh root-tips in ice water for 24 h and fix in glacial acetic acid: 95% ethanol (1 : 3) for 2 to 5 days. Squash in 1% acetic–carmine and store at $-70°C$.

Nick translate probe pSc199 (Bedbrook *et al.*, 1980) with kit obtained from Enzo Biochem Inc (325 Hudson Street, New York, NY 10013) and label with biotinylated UTP (Deteck I-hrp kit from Enzo Biochem Inc).

For nick translation, mix 400 pg of DNase 1; 12 units of DNA polymerase 1, 1 µg of clone DNA and the nucleotides as given in the kit in a 50 µl reaction mixture of 0.5 M Tris (pH 7.5) in 5 mM $MgCl_2$. Keep two and half hours at 18°C and pass through Sephadex G-50 spin column to separate the nucleotides from the labelled DNA.

For in situ hybridization, place 0.4 µg of labelled pSc 119 in a mixture of 50% formamide, 10% dextran, and 30 µg of sheared carrier DNA in 2× SSC (0.6 M NaCl, 0.06 M sodium citrate). Denature the DNA at 85°C for 10 min. Place the slides in 70% formamide in 2× SSC at 70°C for 2.5 min. Rapidly dehydrate in an ethanol series (70, 95, 100%) at $-20°C$. Apply 20 µl of probe mixture to each slide. Cover with a coverglass. Incubate slides in moist chamber at 38°C for 6 h. Remove coverglass. Rinse slides in 2× SSC at room temperature for 5 min; at 37°C for 10 min and again at room temperature for 5 min. Rinse in 0.1% Triton X-100 in PBS (0.13 M NaCl; 0.007 M dibasic sodium phosphate and 0.003 M monobasic sodium phosphate) and then briefly in PBS alone. Drain the slide, *without* drying. Place 120 µl of a complex of streptavidin-biotinylated horse-radish peroxidase on each slide and coverglass. Incubate at 37°C for 30 min. Remove coverglass, rinse slide in 2× SSC for 5 min and then in 0.1% Triton X-100 in PBS for 2 min, both at room temperature. Rinse briefly in PBS, place 500 ml of a solution of 0.05% diaminobenzidine tetrahydrochloride (DAB) and 1% hydrogen peroxide on the slide for 5 min. Rinse slide with PBS and stain immediately with 2% Giemsa for 1 min. Air dry overnight and mount in Permount. Observe hybridization sites under bright field optics to distinguish hybridization sites from Giemsa bands. Photograph under both bright field and phase contrast optics in a Zeiss Photosystem III technical pan 2415.

Minor modifications have been used for different plants. For ISH of *Triticum aestivum* cv. Chinese, denature slides in 70% formamide/2× SSC at 70°C for 2 min, dehydrate in an ethanol series at −20°C. Denature labelled pScT7 DNA at 100°C for 10 min. Hybridize in a mixture containing 50% formamide, 2× SSC, 10% dextran sulphate, salmon sperm carrier DNA at 500 µg/ml and biotin-labelled pScT7 at 4 µg/ml for 6 h in a moist chamber at 37°C (Mukai *et al.*, 1990).

Direct fluorochrome-labelled DNA probes for direct FISH to chromosomes (after Schwarzacher and Heslop–Harrison, 1993)

Make chromosome preparations as described earlier.

Label the probe DNA with direct fluorochrome-labelled nucleotides (Amersham, UK or Boehringer Mannheim, Germany). Incorporation of fluorochrome labels into the probe should be tested by putting a small drop of probe on a microscope slide and observing it under the epi-fluorescence microscope using the correct filter block. Steps are carried out at room temperature unless otherwise indicated.

Dry the chromosome spreads in an incubator at 37°C overnight. Add 200 µl of RNase A, cover, and incubate for 1 h at 37°C in a humid chamber. Wash the slides in 2× SSC three times for 5 min each. Place slides in 0.01 M HCl for 2 min. Add 200 µl of pepsin, cover, and incubate for 10 min at 37°C.

Stop the pepsin reaction by placing the slides in water for 2 min and then wash in 2× SSC twice for 5 min each. Place slides into freshly depolymerized paraformaldehyde and incubate for 10 min. Wash slides in 2× SSC three times for 5 min each. Dehydrate slides 3 min each in 70, 90, and 100% ethanol and then airdry.

Prepare the humid chamber and place in a water bath to raise the internal temperature to 90°C. Leave to equilibrate for at least 15 min.

Denature the hybridization mixture prepared as 20 µl formamide (final conc. 50%), 4 µl 20× SSC (2×), 8 µl dextran sulphate (10%), 0.5 µl SDS, 25–100 ng probe with 1–5 µg salmon sperm DNA (H₂O final vol 40 µl) at 70°C for 10 min and then transfer to ice for 5 min. Add 40 µl denatured hybridization mixture to each slide and cover with a plastic coverglass. Ensure no bubbles are trapped. Quickly place the slides in the preheated humid chamber and incubate for 10 min at 90°C.

Transfer the humid chamber to a 37°C incubator or water bath and leave slides to hybridise overnight. After hybridization carefully remove coverglass and place slides in a Coplin jar containing 2× SSC at 42°C. Pour off 2× SSC and replace with stringent wash solution at 42°C. Incubate for 10 min shaking gently. Replace stringent wash solution once during incubation. Wash slides in 2× SSC three times for 3 min at 42°C, then three times for 3 min at room temperature.

Add either 100 µl of DAPI or 100 µl of PI per slide; cover with a plastic coverglass and incubate for 10 min. Wash briefly in 2× SSC, and drain. Apply

1–2 drops of antifade on the wet slide, cover with a glass coverglass. Blot excess antifade from the slide with filter paper.

Slides can be viewed immediately, but the signal stabilizes after storing for a few days in the dark at 4°C. Slides can be kept at 4°C for up to 1 year. Photograph areas of interest.

METHODS FOR IDENTIFICATION AND MAPPING OF CHROMOSOME SEGMENTS AND GENE LOCI

Mapping of single-copy DNA by *in situ* hybridization

In situ hybridization of single-copy DNA may be achieved using radiolabelled or biotinylated probes (after Levi and Mattei, 1995). The procedure involves:

Hybridization consists of applying the denatured probe in hybridization solution on to the denatured DNA on the chromosome slide. The slide is then covered with a coverglass, transferred into a moist chamber, and incubated at 37°C overnight.

Post-hybridization washes are performed to remove probe—DNA hybrids which are not perfectly homologous. Such hybrids are less stable than perfectly matched hybrids and can be dissociated by washing the slides.

Detection of hybridized probe through autoradiography with radiolabelled probes. Immunofluorescence is normally used with biotinylated probes.

Hybridization with a radiolabelled probe
Prepare 100 ml of hybridization solution by mixing the following: deionized formamide, 50 ml (5 gm of mixed lead iron exchange resin to 100 ml formamide); 20× SSC 10 ml; 100× Denhardt's solution 1 ml (2 g Ficoll 400; 2 g polyvinyl pyrrolidone (360); 2 g BSA in 100 ml water, store at −20°C); 10% SDS, 1 ml; 1 M sodium phosphate buffer pH 6.8, 4 ml.

At this step, filter through a Millipore 0.22 μm filter. Then add: 50% dextran sulphate, 20 ml; double-distilled H_2O, 14 ml. Adjust the pH to 7.0, if necessary, using 1 M HCl. The hybridization solution may be stored in aliquots at −20°C. Dissolve the probe in the hybridization solution. As a guide, probe concentrations of 25 ng/ml and 100 ng/ml are required for probes with sizes of 2000 bp and 500 bp, respectively.

Before hybridization, denature the radiolabelled probe by heating at 70°C for 10 min and then quickly chill in ice. Put 50 μl of denatured probe mixture onto each slide. Cover with a coverglass (24 × 40 mm) and incubate at 42°C in a sealed sandwich box containing 50% formamide, 2× SSC for 16–18 h.

Dip the slides in 50% formamide, 2× SSC at 39°C to remove the coverglass. Rinse the slides successively in: 50% formamide, 2× SSC at 39°C, 3 times for

10 min each time; 2× SSC at 39°C, 3× 10 min; 2× SSC at room temperature, 3 times for 10 min each time. Wash for 1 h in 0.1× SSC at room temperature and then for 1 h in 0.1 × SSC at 4°C.

Dehydrate the slides in 70%, 90%, and absolute ethanol and air-dry. The next steps are carried out in a darkroom with a dark-red safelight (Wratten series 2) from a safe distance.

Handle the NTB-2 emulsion with plastic or ceramic spoons and melt a portion of emulsion in a disposable plastic tube at 42°C in a waterbath. Dilute the emulsion 1:1 with 2% glycerol, mix gently, and transfer to a dipping tube at 42°C. Dip each slide for 1 s in the diluted emulsion. Allow the slides to dry in an upright position for 2 h at room temperature. Put the slides in light-tight boxes containing silica gel desiccant and seal the boxes with black insulating tape. Place the boxes at 4°C for the desired exposure time, away from source of radiation.

Develop the slides in Kodak Dektol solution diluted 1:1 (v/v) with distilled water, at 20°C for 2 min. Rinse briefly in distilled water (30 s) and then fix in Kodak fixer for 5–10 min. Wash 2–3 times in distilled water (5 min each) and air-dry.

To avoid errors in the location of silver grains resulting from any slippage of the emulsion during the chromosome banding procedure, first stain the slides with 4% buffered Giemsa solution (pH 6.8) for 5 min, and photograph the chromosomes. Then perform chromosome banding.

For chromosome replication banding, stain the slides with 150 µg/ml Hoechst 33258 in 2× SSC for 30 min, in the dark. During this step, the Giemsa-stained chromosomes will completely fade away. Rinse the slides in distilled water. Mount in 2× SSC under coverglass. Expose the slides to UV light at 48°C for 60 min. Rinse in distilled water. Stain with 6% Giemsa in water for 6 min. Rinse in distilled water and air-dry. Re-photograph the chromosomes.

Hybridization with biotinylated probe
Lyophilize 75–750 ng of biotinylated probe in a microcentrifuge tube. Dissolve the lyophilized probe in 15 µl of hybridization buffer to give a final concentration of 5–50 ng/µl. Allow to dissolve for 1 h at 37°C. Denature the probe by heating at 75°C for 5 min in a waterbath. Immediately, transfer the mixture to ice and leave for 5 min.

Hybridize the probe to the slides as described in previous schedules. Cover 15 µl of probe with an 18 × 18 mm coverglass. Take the slides out of the moist hybridization chamber and transfer to 50% formamide, 2× SSC pre-warmed at 45°C and agitate for 5 min in a shaking waterbath. The coverglasses are removed.

Repeat this wash twice more. Transfer the slides to 2× SSC, pre-warmed at 45°C, and agitate for 5 min in a shaking waterbath. Repeat twice. Transfer the slide to PN buffer. If needed, store at 4°C in PN.

For detection of biotinylated probe Block the non-specific hybridization sites by incubating the slides for 10 min at room temperature in PNM buffer. Drain the slides briefly and apply 100 µl of 5 µg/ml avidin-FITC in PNM buffer to each slide. Replace the coverglass and incubate the slides at 37°C for 30 min.

Remove the coverglass in PN buffer. Rinse the slides briefly three times for 2 min each time in PN buffer.

For signal amplification, treat the slides with 100 µl of 5 µg/ml biotinylated goat anti-avidin D in PNM buffer. Replace the coverglass and incubate for 30 min at 37°C. Wash the slides 3 times for 2 min each in PN buffer. Block by incubation for 10 min in PNM buffer. Add another layer of avidin-FITC for amplification as before. Wash the slides again 3 times for 2 min each in PN buffer.

Drain the slides. Apply 40 µl of propidium iodide diluted 0.5 µg/ml in antifade solution pH 11.0 under a 24 mm × 40 mm thin coverglass. Store at 4°C in dark. Visualize and photograph the hybridization signals on the chromosomes. With propidium iodide diluted in antifade solution pH 11.0, BrdU-substituted chromosomes appear banded.

Single copy gene from cell culture in plants

Maintain cell suspension cultures of parsley in dark and subculture every 7 days. *For chromosome preparation*, treat three day-old cultures with aphidocoline (7.5 µg/ml) and 5-bromodeoxyuridine (BrdU, 5×10^{-4} M) for 24 h, wash and reincubate for 6 h with fresh medium containing deoxythymidine (3×10^{-5} M). Treat for 2 h with colchicine (0.00625%). Isolate protoplasts as follows: Collect colchicine-treated cells by centrifuging at $100 \times g$ for 5 min. Reduce supernatant to one-fourth of the original volume. Add an equal volume of enzyme solution (2% cellulysin, 1% macerase in 0.7 M mannitol, 10 mM MES, pH 5.6). Continue protoplast isolation for 1 h at 25°C in the dark on a rotary shaker (50 rpm). Filter the protoplasts through 80 µm nylon mesh and wash twice with a solution containing 100 mM glycine, 2.5 mM $CaCl_2$ and 5% glucose at pH 6.0. Hypotonise in 75 mM KCl solution for 10 min and fix overnight in Farmer's solution (glacial acetic acid–ethanol 1 : 3). Change fixatives till a clear supernatant is obtained. Spread the suspension on ice-cold wet slides and air dry at room temperature.

The steps needed for *in situ* hybridization are similar to those described earlier, using the relevant probes (after Huang *et al.*, 1988).

Mapping of low copy DNA sequence in plants

Prepare chromosome spreads from fixed root-tips after digesting with cellulase and pectolyase. Nick translate the cDNA clone pSc503 corresponding to a rye-γ-secaline gene, containing a *Hind*III insert of 900 bp in a pUC8 vector (supplied by PR Shewry, Rothamsted Experiment Station, England) with biotin-11- dUTP as follows:

Add one and one half of plasmid DNA to a 50 µl mixture of 50 mM Tris-HCl, pH 7.5 and 5 mM MgCl$_2$/30 µM each of dATP, dGTP, dCTP (Pharmacia) and biotin-11-dUTP (Enzo Diagnostics), also containing 20 ng of DNase I (Sigma) and 12 units of DNA polymerase (BRL). Incubate at 15°C for 2 h. Terminate labelling by adding 5 µl of 0.2 M EDTA (pH 8.0). Purify labelled probe from unincorporated nucleotides by passing through a Sephadex G-50 spin column. Evaluate incorporation of biotin-11-dUTP by means of dot blot using a BRL streptavidin alkaline phosphatase-detection system. Hybridize biotinylated probes to chromosome preparations following instructions given by Enzo Diagnostics. Incubate preparations in RNase A (1 µg/ml) in 2× SSC solution at 37°C for 45 min. Keep slides in 70% formamide/2× SSC at 70°C for 3.5 min. Immediately dehydrate in a series of ethanol washes (70, 90, 100%), at −20°C for 5 min each. Air dry and place in plastic high humidity chambers. Prepare hybridization mixture as 50% (v/v) formamide/10% (w/v) dextran sulphate/2× SSC with salmon sperm DNA at 0.1 mg/ml. Denature in a boiling water bath for 10 min and store on ice. Add a measured amount of the mixture in a droplet to the chromosome preparation on a slide, cover with a coverglass. Keep the chamber at 80°C for 10 min to denature the preparations and the probe. Transfer to 37°C and incubate overnight.

After hybridization overnight, rinse off extra probe with 2× SSC, and keep successively in fresh changes of 2× SSC for 5 min at room temperature, for 10 min at 37°C and again 5 min at room temperature. Wash in 0.1% Triton/PBS or 4 min and then in PBS alone for 5 min at room temperature. Use primary and secondary antibodies to amplify the signals from the hybridized biotinylated probes. Apply goat antibiotin IgG (Sigma) at a concentration of 30 µg/ml and incubate slides at 37°C for 30 min. Remove unbound antibody by washing thrice in PBS, for 1 min each at room temperature. Next apply biotin-conjugated rabbit anti-goat IgG (Sigma) in a 1:400 dilution of the supplier's stock. Incubate and wash as for goat anti-biotin treatment. Incubate the slides with 0.7% secondary antibody horseradish peroxidase (BRL) for 30 min at 37°C. Streptavidin horse radish peroxidase conjugate was used as the reporter molecule. Wash slides in 2× SSC for 5 min at room temperature and then in 0.1% Triton/PBS for 4 min. To visualize the hybridization complex, add 0.05% diaminobenzidine hydrochloride (BRL) and 0.03% hydrogen peroxide (Sigma) in PBS. Keep for 5 min at room temp. Rinse with PBS and counterstain in 2% Giemsa for one min. Measure armlengths of chromosomes from a high resolution TV screen (Gustafsson *et al.*, 1990).

Two colour mapping of plant DNA sequences

Prepare chromosome squashes of plant root-tip (after Leitch *et al.*, 1992) in 45% acetic acid.

DNA probes

pSc 119.2 is a 120 bp tandem repeat unit of DNA isolated from *Secale cereale* and labelled with digoxigenin dUTP (Boehringer–Mannheim). pTa71 is a ribosomal DNA sequence isolated from wheat *Triticum aestivum* labelled with biotin-11-dUTP.

Mix the probes immediately before use (each to a final concentration of 5 μg/ml) in a solution of 50% (v/v) formamide, 10% (w/v) dextran sulphate, 0.1% (w/v) SDS (sodium dodecyl sulphate) and 2× SSC. Incubate slides in 100 μg/ml DNase-free RNase in 2× SSC for 1 h at 37°C, wash twice in 2× SSC for 10 min at room temperature, dehydrate in a graded ethanol series and air dry. Denature the probe mixture at 70°C for 10 min, load onto the slide preparation and cover with a plastic coverglass. Place the slides in a humid chamber and denature chromosomes and probe together at 90°C for 10 min.

DNA–DNA hybridization

Incubate slides overnight at 37°C. Then wash in 2× SSC twice for 5 min each at 40°C. Wash thoroughly in 50% (v/v) formamide in 2× SSC for 10 min at 40°C to allow DNA sequences with more than 85% homology to the probe to remain hybridized. Wash slides in 2× SSC twice for 5 min at 40°C and then twice for 5 min each at room temperature.

Detection of hybridization

Biotin is detected with Texas Red-labelled avidin and digoxigenin with sheep antidigoxigenin-fluorescein (Boehringer–Mannheim) simultaneously. Transfer slides to detection buffer (4× SSC, 0.2% (v/v) Tween 20) for 5 min, treat with 5% (w/v) BSA (bovine serum albumin) in detection buffer for 5 min, and incubate in 5 μg/ml Texas Red-labelled avidin and 20 μg/ml an tidigoxigenin-fluorescein in detection buffer containing 5% (w/v) BSA for 1 h at 37°C. Wash the slides afterwards thrice in detection buffer at 37°C, keeping in each for 8 min. The signals from biotin and digoxigenin are amplified. Keep in 5% (v/v) normal goat serum in detection buffer for 5 min for blocking. Then incubate with 25 μg/ml biotinylated anti-avidin D and 10 μg/ml FITC-conjugated rabbit anti-sheep in detection buffer containing 5% (w/v) normal goat serum for 1 h at 37°C. Wash in detection buffer three times at 37°C, keeping 8 min each time, reincubate the slides with TexasRed-labelled avidin as before. Counterstain the slides with 2 μg/ml DAPI in McIlvaine's citrate buffer. Mount in antifade solution (AFI, Citifluor) to reduce fading of fluorescence. Examine with a Zeiss epifluorescence microscope with filter sets 02, 09 and 12.

The sites of pSc 119.2 hybridization show green fluorescence near the ends of the chromosomes while pTA71 hybridization sites fluoresce orange/red.

Differentiation of bean heterochromatin (after Fuchs *et al.*, 1994)

Different types of heterochromatin have been distinguished on the basis of their replication behaviour and the patterns of heterochromatic bands as also the presence of satellite or highly repetitive single sequence DNA. A 59 bp repeated *FokI* element had been recorded in the field bean (*Vicia faba*). Giemsa bands may be subdivided into two groups, one completely cleaved by *FokI* and the other more resistant.

Obtain chromosome suspensions from synchronized root tip meristems. Either drop the suspensions on slides for fluorescent *in situ* hybridization or use for flow sorting of chromosomes by a FACStar Plus.

Amplify the *FokI* repeat from genomic DNA of *Vicia faba* by PCR with a sequence-specific primer pair (5'-GCCTTGTCATTATGGAAG-GTAGTCTG-3' and 5'-CCATCCATTGGAGTAACAAAAACTTCG-3'), comprising positions 54–20 and 51–25 of the repeat. Reamplify the PCR products for labelling with Bio-11-UTP, using the following protocol:

Perform primary *FokI* repeat amplification, using 40 pg genomic DNA in a final volume of 50 µl, in IX *Taq* polymerase buffer (Promega), 1 mM MgCl$_2$, 0.2 µM primer, 0.2 mM of each d NTP. Denature DNA at 94°C for 10 min; cool to 85°C and add 2.5 U *Taq* DNA polymerase. Follow 30 cycles at 55°C for 30 s, 67°C for 30 s and 94°C for 30 s. For labelling, use 2 µl of the primary amplification volume in a two-step symmetric PCR. Perform the first 30 amplification cycles in a final volume of 100 µl containing 1× *Taq* polymerase buffer; 1 mM MgCl$_2$; 0.22 µM of one the *Fok* primers; and 0.2 mM dNTPs, except dTTP. Add dTTP(150 µM) and biotin-11-dUTP (150 µM). Denature. Add 5 U *Taq* polymerase and follow 30 cycles at 55°C for 30 s, 67°C for 50 s and 94°C for 30 s. For the second amplification, add 0.1 µm of the missing second *Fok* sequence specific primer and 5 U *Taq* polymerase. Follow 10 cycles at 55°C for 50 s, 72°C for 50 s, 94°C for 50 s and a final extension step at 72°C for 5 min.

Amplify DNA of a specific flow sorted chromosome (III of line EF) by PCR, using degenerated oligonucleotide primers. Label for FISH using 1× *Taq* polymerase buffer; 2.5 mM MgCl$_2$; 0.7 µM 6-MW primer; 0.2 mM dNTPs, except dTTP. For labelling, add 150 µM each of dTTP and Biotin-11-dUTP. After denaturation, follow 25 cycles at 55°C for 1 min; 72°C for 1.5 min, with an extension of 1 s per cycle; 94°C for 1 min and final extension at 72°C for 10 min.

In situ hybridization with biotinylated *FokI* sequence repeats produces distinct chromosome-specific signal patterns after detection with streptavidin-FITC and amplification with biotinylated antistreptavidin.

Localization of geminivirus-related DNA sequences
(after Kenton *et al.*, 1995)

Nicotiana tabacum nuclear genome carries approximately 25 multiple direct repeats of a geminivirus-related DNA (GRD) sequence. This might have arisen by illegitimate recombination, following geminivirus infection, during *Nicotiana* evolution.

GRD 3 repeat sequence used as probe was originally isolated from an EMBL4 bacteriophage lambda genomic library of *N. tabacum*. This 14 kb fragment was subcloned into Bluescript vector to yield the plasmid pPC121, carrying 14 direct repeats of GRD 3. Plasmid pTa71 contains a 9 kb insert that includes the 25S, 18S and 5.8S ribosomal genes and part of the intergenic spacer from wheat.

Linearize plasmids with the appropriate restriction enzyme and label by nick translation; either immediately or after separating the insert from the plasmid through 1% agarose and purifying it with Geneclean II (Stratech scientific). Label linearized pTa71 with a 1 : 1 mixture of Fluorored (Rhodamine-4-dUTP, Amersham) and Fluorogreen (fluorescein-11-dUTP). Label insert from pPC121 with biotin-14-dATP (Bethesda Res. lab.). For use as a probe in sequential *in situ* hybridization, mechanically shear total genomic DNA from *N. sylvestris* and label with biotin-14-dATP.

Prepare chromosome spreads from root-tip metaphases or from meiotic protoplasts at diakinesis in flower-buds of *N. tabacum*, as described elsewhere in this book. For simultaneous or sequential *in situ* hybridization, use several probes at concentrations of 5 µl/ml for pTa71 and 10 µg/ml for the insert from pPC121. Hybridize in 50% formamide at 37°C and give stringent wash in 50% formamide at 42°C. Detect biotin-labelled probes with avidin conjugated to fluorescein isothiocyanate (FITC, Vector Lab) after one amplification with biotinylated anti-avidin D (Vector lab). Visualize probes labelled with a 1 : 1 mixture of FluoroRed and FluoroGreen, without further amplification. Counterstain all slides with DAPI. Observe hybridization signals with single band-pass filters and record using multiple photographic exposures. Photograph with Kodak Ektachrome daylight transparency film (800/1600 ASA).

This work is used to map by ISH a viral DNA sequence integrated into a plant genome.

In situ identification of chromosome (after Bennett *et al.*, 1995)

Zingeria biebersteiniana (Claus) P. Smirnov (Gramineae), is one of five known angiosperms with only four chromosomes in somatic cells ($2n = 4$, $n = 2$). FISH has been used at the initial steps to construct a physical gene map of this species.

Schedule

Germinate seeds on moist filter paper in petri dishes in dark at 23°C. After germination, transplant seedlings into potting compost. Remove fresh root-tips and pretreat in either 0.05% (w/v) colchicine for 4 h at room temperature or iced water at 1°C for 22–24 h and fix in acetic acid–ethanol (1 : 3) at 4°C overnight.

Prepare chromosome spreads following incubation in enzyme mixture for 45–60 min and squash in 45% acetic acid (Bennett *et al*., 1992).

For fluorochrome C-banding, treat slides in 45% acetic acid at 60°C for 10 min, wash in running tap water for 15 min, and then distilled water for 1 min. Dehydrate in absolute ethanol at room temperature for 10 min. Dry in an oven at 42°C for at least 30 min. Denature in freshly prepared sat. aqueous solution of barium hydroxide at 30–35°C for 10 min and wash again as above.

Incubate slides in 2× SSC (0.3 M NaCl, 0.03 M sodium citrate, pH 7) at 60°C for 90 min. Wash and dry as above. Finally, stain slides in either 0.02% (w/v) Hoechst 33258 in absolute ethanol for 3 min or 1–2 μg/ml DAPI in McIlvaine's buffer in dark for 30 min. Mount the former in glycerol-distilled water and the latter in Citifluor PBS–McIlvaine's buffer (both 1 : 1, v/v) respectively. Observe fluorescence by excitation with UV light (365 nm) using a Zeiss epifluorescence microscope and filter set 02.

rDNA probes used were: The clone pTa71 containing a 9 kb Eco RI fragment isolated from wheat and recloned into pUCI19. The clone pTa794 contains a 410 bp *Bam*HI fragment isolated from wheat and cloned in pBR322. It was further amplified by PCR using *Bam*HI primers. Label both probes with biotin-14-dATP by nick translation (Gibco BRL BioNick Labeling System), according to manufacturer's instructions.

For FISH using single probes: Refix slides in 45% acetic acid; dehydrate in absolute ethanol and air-dry overnight. Treat with 100 μg/ml DNase-free RNase in 2× SSC at 37°C for 1 h; wash in 2× SSC; dehydrate through ascending ethanol series and dry. The probe mix contains 3–5 ng/ μl biotinylated probe DNA; 5% deionized formamide; 10% dextran sulphate; 900 μg/ml sonicated salmon sperm DNA. The later processes of denaturation, hybridization, post-hybridization washes, detection of *in situ* hybrids, counterstaining and visualization are similar to the general method described in the beginning of this chapter.

For Multiple FISH: Incorporate 5 ng/μl rhodamine labelled pTa71 and 5 ng/μl digoxigenin-labelled pTa794 together into the hybridization mix described above. Observe as described in earlier methods.

In situ hybridization in rice

For chromosome preparations, disinfect dehulled rice seeds with 0.75% sodium hypochlorite for 10 min, wash in running water for 30 min; soak in tap water for 26 h at 30°C. Transfer to petri dish lined with cheesecloth and incubate at 30°C for 6 h. Immerse two-thirds of the height of the seeds in 15 or 20 mM 2'-deoxyadenosine for 16 h at 30°C and wash thoroughly with distilled water. Immerse the roots and wash in distilled water. Pretreat root-tips in 1.5 mM 8-oxyquinoline for 2 h at 20°C. Wash in distilled water, keep in a solution of 6% pectinase and 6% cellulase in 75 mM KCl, pH 4.0 for 60 min at 37°C. Wash with distilled water, fix in glacial acetic acid–methanol (1 : 3), smear on a glass slide and flame dry.

Label DNA probe with ^{3}H-deoxynucleotides using nick translation. Specific activity of the probe should be approximately 4×10^7 cpm/µg DNA. Incubate slides with metaphase preparations with 200 µl of 100 µg/ml RNase per slide, cover with a coverglass, at 37°C for 1 h. Denature with 70% formamide, $2 \times$ SSC (0.3 M NaCl, 0.03 M sodium citrate, pH 7.0) at 70°C for 2 min. Dehydrate through ascending ethanol series and airdry.

For in situ hybridization, apply 40 µl of labelled and denatured DNA probe on each slide. Cover with a coverglass and keep in a moist chamber for 8 to 16 h at 37°C. Wash three times, 2 min each, with 50% formamide, $2 \times$ SSC, pH 7.2 at 39–40°C and then five times (2 min each) with $2 \times$ SSC at 39–40°C. Dehydrate by passing the slides through ascending ethanol series, keeping 2 min in each. Air dry. Dip the hybridized slides in Ilford K2 emulsion diluted with 0.5% gelatin (1 : 1). Expose the slides at 5°C for 7 days. Develop in Kodak D19 (diluted twofold with water) solution, fix in a Kodak Fixer at 15°C, stain with 4% Giemsa in phosphate buffer and rinse with tap water.

Mapping of three genomes by multicolour FISH (after Mukai *et al.*, 1993)

The method has been utilized for simultaneously discriminating the three constituent genomes of the allohexaploid wheat, *Triticum aestivum* with different colours.

Protocol

Germinate seeds of *T. aestivum* and prepare chromosomes from root-tips for FISH, following enzyme maceration as described earlier.

For preparation of DNA probe, isolate total genomic DNA from the three putative parents of *T. aestivum*, namely *T. urartu*, *Aegilops speltoides* and *Ae. squarrosa* using the method described in the chapter on Isolation. The probes used are clone pScl19, consisting of highly repeated sequences from rye, *Secale cereale* (3 subunits designated as 1, 2 and 3) and clone pAs1, containing a 1 kb insert of *Ae. squarrosa* repeated sequences.

For labelling of DNA probes, shear total genomic DNA by passing it several times through an 18-gauge hypodermic needle. Label total genomic DNA of *T. urartu* with biotin-16-dUTP and that of *Ae. squarrosa* with digoxigenin-11-dUTP by nick translation or random primer method. Label the insert of pSc119 (pSc119.2) and the whole plastid of pHs1 with biotin-16-dUTP and digoxigenin-11-dUTP, respectively, by random primer method.

For FISH, denature chromosomal DNA in 70% formamide—2× SSC for 2 min at 68°C and dehydrate in an ethanol series at −20°C. The hybridization mix (100 μl for total volume) for FISH consists of 50% formamide; 10% dextran sulphate; 2× SSC; 250 ng of biotin-labelled *T. urartu* genomic DNA; 250 ng of digoxigenin-labelled *Ae. squarrosa* genomic DNA; 2 μg of unlabelled sheared *Ae. speltoides* genomic DNA, and 20 μg of sonicated salmon sperm DNA. For FISH using repeated sequences, the hybridization mix (100 μl) contains, in addition to 50% formamide, 10% dextran sulphate and 2× SSC; 50 ng of biotin-labelled pSc119.2; 200 ng of digoxigenin-labelled pAs1 and 20 μg of sonicated salmon sperm DNA. Denature the mix for 10 min at 100°C and quench immediately in ice for at least 10 min. Apply a 10 μl aliquot of the mix to each slide and allow to hybridize overnight in a moist chamber at 37°C. Then wash slides at room temperature, first in 2× SSC for 15 min, followed by 15 min at 1× SSC and 5 min at 4× SSC. Detect simultaneously biotin with FITC-conjugated avidin and digoxigenin with rhodamine-conjugated sheep anti-digoxigenin Fab fragment (Boehringer, Mannheim). Incubate slides in 10 μg/ml FITC-conjugated avidin and 20 μg/ml rhodamine conjugated anti-digoxigenin in detection buffer containing 4× SSC—1% BSA for 1 h at 37°C. After incubation, wash slides in 4× SSC for 10 min; 0.1% Triton × 100 in 4× SSC for 10 min; 4× SSC for 10 min and 2× SSC for 5 min, all at room temperature. Immediately mount the slides in antifade solution (90% glycerol and 0.1% para-phenylene diamine). Examine with a Nikon fluorescence microscope with a B_2 or G filter and photograph on Kodak EKTAR film ASA 1000.

The hybridization sites of the A genome probe fluoresce yellow while those of the D genome probe fluoresce orange. The B genome chromosomes appeared a faint brown as a result of cross-hybridization of the A and D genome probes.

Physical mapping of the 5S *rDNA* genes (after Fukui *et al.*, 1994)

The material used initially was a variety of japonica rice (*Oryza sativa* L).

Collect root-tips about 1 cm long from kernels germinated at 32°C. Extract total genomic DNA from the leaf tissues of the seedlings by the cetyltrimethyl ammonium bromide (CTAB) method described in the chapter on extraction (II.6). Prepare chromosome spreads by the enzyme maceration/airdrying method as described in the chapter on preparation of chromosome spreads (II.4). For *in situ* hybridization, prepare the biotinylated 5S rDNA from interphase nuclei by direct cloning and direct labelling. Enzymatically macerate fixed root-tips on polyester membrane disks. Cut out fragments bearing about 100 nuclei and transfer individually to 500 μl microtubes; treat with a proteinase K solution (1 mg/ml; Wako Pure Chemical Industries,

Osaka) and subject to PCR with 70% substitution of biotin-11-dUTP for dTTP (Sigma).

To clone the DNA fragment containing 5S rDNA, carry out PCR, using 2.5 units of *Taq* DNA polymerase (Promega); 0.5 µg total DNA from the plant to be studied and 1 µM each of a pair of primers, 5'-GATCCCATC-AGAACTCC-3' and 5'-GGTGCTTTAGTGCTGGTAT-3'. These are hybridized to the two strands of the 5S rRNA coding sequence. Ligate the amplified fragments to a linear pCRII vector using a TA cloning Kit (Invitrogen) and introduce by transformation into In V × F cells (Invitrogen). Carry out DNA sequencing by the dideoxynucleotide chain termination method, using sequence ver. 2.0 (U.S. Biochemicals).

This method can be effectively used in mapping the 5S rDNA locus in different cultivars of rice.

Localization of rDNA loci by FISH (after Fukui *et al.*, 1994 in *Oryza sativa*)

Germinate seeds. Excise root-tips 1–2 cm long, fix in acetic acid–ethanol (1 : 1). Store at −20°C for 7 days before examination.

Coat glass slides to be used for FISH with a 0.1% poly-L-lysine solution (Sigma). Treat root-tips for 15–30 min in a decompression chamber, macerate in enzyme mixture in 1.5 ml Eppendorf tube at 37°C for 60–90 min according to the procedure described under preparation of metaphase spreads (Chapter II.4). Flame-dry the preparations.

Treat the chromosome spreads on glass slides with an enzymatic mixture (2% cellulase Onozuka RS, Yakult Honsha, Tokyo; 1.5% macrozyme R-200, Yakult Honsha and 0.3% pectolyase Y-23, Seishin Pharmaceutical Ltd. Tokyo; 1 mM EDTA, pH 4.2), in 2× SSC at 37°C for 30 min. Then treat with 1 mg/ml proteinase K (Wako Pure Chemical, Osaka) at 37°C for 30 min. Wash in 45% acetic acid for 5 min. Dehydrate progressively through a 70, 95 and 99% ethanol series for 10 min each. Air dry. Treat with 100 µg/ml RNase A (Sigma) in 2× SSC at 37°C for 60 min.

The rDNA probe is 3.8 kb long and covers most of the coding regions of the rRNA genes and flanking spacer regions. Label the probe by a random primer labelling method with biotin-dUTP as instructed by the manufacturer. Drop a 15 µl aliquot of the hybridization mix containing 100 ng of biotinylated-rDNA in 50% formamide/2× SSC on a glass slide. Cover with a coverglass, seal with liquid Arabian gum and air-dry. Heat the covered slide on a thermal cycler at 70°C for 6 min and 37°C for 18 h.

Remove coverglass. Wash slides with 2 × SSC thrice and once with 4× SSC at 37°C for 10 min each. Drop a 70 µl aliquot of FITC-avidin conjugate (0.1 mg/ml, Boehringer–Mannheim) onto the glass slide. Incubate at 37°C for 60 min. Rinse with BT buffer (0.1 M sodium hydrogen carbonate, 0.05% Tween 20, pH 8.3) thrice at 40°C for 10 min each. Drop a 70 µl biotinylated anti-avidin solution (1%, Vector Lab., Calif, U.S.A.) onto the glass slides. Incubate at 37°C for 30 min. Wash briefly with BT buffer. Apply a 70 µl fluorescein-avidin solution (1%, Vector Lab) to each slide and again incubate at 37°C for 30 min. Wash thoroughly with BT buffer thrice at 40°C for 10 min each.

Blocking is carried out three times before probe hybridization and before the immunological reaction with 5% bovine or goat serum albumin in BT buffer at 37°C for 5 min. This method can be used in conjunction with imaging method.

Localization of seed protein genes by FISH
(after Fuchs and Schubert, 1995)

In *Vicia faba*, field bean, three genes coding for storage proteins of the legumin or 115 class have been mapped physically by FISH.

Prepare chromosome suspensions from synchronized root tip meristems of seedlings. Fix in 6% formaldehyde for 15 min at 4°C and isolate the chromosomes according to the method of Schubert *et al.* (1993) given in Chapter II.6. Drop chromosome suspension on ice-cold slides and air dry.

Label DNA with Bio-16-UTP (Boehringer, Mannheim) by nicktranslation as described earlier and purify using Sephadex G-50. Derive specific probes for the different proteins from different genomic fragments for hybridization. Pretreat slides in 100 µg/ml DNase free RNase at 37°C for 45 min. Wash in 2× SSC. Treat with proteinase K (4 µg/ml for 15 min at 37°C). Post-fix chromosomes by treating with 4% paraformaldehyde for 10 min. Wash slides in 2× SSC, dehydrate by ethanol and air-dry. On each slide, mix 300 ng of labelled probe DNA with 5 µg each of salmon sperm DNA and *t*RNA from *Escherichia coli*. Denature in hybridization buffer (50% formamide, 10% dextran sulphate, 2× SSC) for 5 min at 75°C. Drop the probe onto prewarmed slides. Denature probes and chromosomes together on a Zytotherm (Schutron) for 10 min at 80°C. Slowly cool the slides to 37°C. Transfer to a wet chamber at 37°C for overnight hybridization. After hybridization, wash three times in 50% formamide at 42°C and 0.2× SSC at 60°C for 5 min each.

Detect signals with FITC after two rounds of amplification with biotinylated antistreptavidin as described earlier. Counterstain with propidium iodide (PI). Visualize signals in a Zeiss Axioskop with the following filter combinations: 487915 (PI); 487910 (FITC) and 487909 (FITC/PI).

High resolution physical mapping by FISH to extended DNA fibres
(after Fransz *et al.*, 1996 in *Arabidopsis thaliana*)

For isolation of nuclei, grind 2 g of young leaves to a fine powder in a mortar and pestle in liquid nitrogen. Transfer powder to a 50 ml centrifuge tube containing 20 ml ice-cold nucleus isolation buffer (NIB, containing 10 mM

Tris-HCl, pH 9.5; 10 mM EDTA; 100 mM KCl; 0.5 M sucrose; 4 mM spermidine; 10 mM spermine and 0.1% (v/v) 2-mercaptoethanol. Mix gently for 5 min. Filtrate homogenate through consecutive nylon mesh filters (170, 120, 120, 5 and 20 μm). Add about 1/20 volume of 10% (v/v) Triton X-100 in NIB to the filtrate. Spin down the nuclei at 2000g for 10 min at 4°C and resuspend in 200 μl NIB at a concentration of about 5×10^5 per ml. Mix this suspension with an equal volume of glycerol and store at -20°C.

For preparation of extended DNA fibres, centrifuge 50 μg of the nuclear suspension at room temperature in a microfuge at 3600 rpm of 5 min. Gently resuspend the pellet in 50 μl PBS (10 mM sodium phosphate, pH 7.0 + 140 mM NaCl). Pipette two 1 μl droplets of suspension onto one end of a clean object slide and air-dry. Disrupt the nuclei for 5 min at room temperature in 10 μl of STE1 lysis buffer (0.5% SDS, 50 mM EDTA, 100 mM Tris pH 7.0), or in STE buffer containing 5 mM EDTA instead (STE2). Stretch DNA fibres by tilting the glass slide and allowing the buffer to float downwards in 4 to 10 s. Air dry. Immerse slides in acetic–ethanol (1 : 3) for 2 min. Air-dry. Bake at 60°C for 30 min and store in a dry box at room temperature.

Alternative method   Resuspend the nuclear pellet from the microfuge run in acetic acid–ethanol (1 : 3). Pipette two droplets of 1 μl suspension on a microscope slide and fix by slowly dropping successively, before evaporation of the buffer, 10 μl each of ethanol 100%, 70% and 50%. Air dry. Lyse the nuclei by carefully adding 100 μl of an alkaline ethanol mixture (50 mM NaOH, 30% ethanol) at one end of the slide. Spread the mixture evenly over the slide with a coverglass. Incubate for 40 to 60 s or longer. Fix the material by pipetting 100 ml ethanol onto the lysed nuclei, while tilting the slide.

For labelling the probe, use either biotin-16-dUTP (Boehringer) or digoxigenin-11-UTP (Boehringer) for a standard nick translation assay, together with a 50-fold salmon sperm DNA, dissolved in 50% deionized formamide in 2× SSC, 50 mM sodium phosphate, pH 7.0 and 10% (w/v) dextran sulphate.

Hybridize the DNA probes (0.1–1.0 g μl^{-1} of probe for each target) to the extended DNA fibre preparations in a mixture of salmon sperm DNA (50× probe DNA concentration); 10% (w/v) sodium dextran sulphate; 50% (v/v) deionized formamide; 2× SSC; 50 mM sodium phosphate, pH 7.0 in a final volume of 20 μl. Cover the slide with a coverglass. Incubate at 80°C for 2 min to denature both probe and target DNA. Incubate the slides in a moist chamber for 16–20 h at 37°C. Wash after hybridization as usual in 50% formamide, 2× SSC, pH 7.0 for 3–5 min at 42°C; rinse briefly in 2× SSC at 42°C and finally in 0.1× SSC, pH 7.0, thrice for 5 min each at 55–58°C.

For *visualization of hybridized probes*, pre-incubate slides in 4 M buffer containing 5% (w/v) non-fat dry milk in 4× SSC, pH 7.0 at 37°C for 30 min and rinse in 4T buffer (4× SSC, 0.05% (v/v) Tween 20) for 2 min. Carry out the next incubation steps at 37°C for 30 min. Follow each step by washing three times for 5 min each at room temperature. In the first detection step, incubate with Texas red-conjugated avidin $(5 \mu l^{-1}$, Vector Lab) in 4 M and then wash for 5 min in 4T buffer and 2×5 min in TNT (100 mM Tris HCl pH 7.5, 150 mM NaCl, 0.05% Tween 20). In the second detection step, use biotinylated goat anti-avidin $(5 \mu g^{-1}$, Vector Lab) and mouse anti-digoxigenin (0.2 μg/ml, Boehringer) in TNB (100 mM Tris HCl pH 7.5; 150 mM NaCl, 0.5% blocking reagent, Boehringer). For third detection step, use Texas Red-avidin and FITC-conjugated rabbit anti-mouse (1 : 1000, Sigma) in TNB. For the fourth detection step, use FITC-conjugated goat anti-rabbit (1 : 1000, Sigma) in TNB.

Subsequently dehydrate the slides in an ethanol series of 70, 90 and 100%, keeping 2 min in each. Air dry and mount in Vectashield (Vector Lab) with 1.0 μg/ml DAPI as counterstain. Observe hybridization signal under Zeiss Axioplan microscope equipped with epifluorescence illumination and Plan Neofluar optics. Directly photograph selected images on a 400 ISO colour negative film using single band pass excitation filters (01, 09 and 14 for DAPI, FITC and TRITC/Texas Red fluorescence) or a triple band filter block with separated excitation filters.

Primed *in situ* hybridization (PRINS)

PRINS is a very useful method for detecting nucleic acid sequences *in situ* in cells fixed to a solid support or in a suspension. It can be used for high copy sequences on mitotic chromosomes (after Abbo *et al.*, 1993a).

Germinate seeds, treat with ice-cold water for 24 h, fix in acetic acid–ethanol (1 : 3). Prepare slides according to the schedule described earlier, pretreating with RNase and paraformaldehyde (Heslop-Harrison *et al.*, 1991).

PRINS: Prepare primer DNA for PRINS from plasmid DNA carrying the pTa71 9 kb *Eco*RI fragment of wheat ribosomal DNA. Digest plasmid DNA with *Sau*3AI to generate fragments smaller than 2 kb in size. Use about 0.5 g of digested plasmid DNA per slide.

For PRINS procedure, denature chromosomes in 70% formamide in $2 \times$ SSC at 70°C for 3.5 min. Dehydrate the slides in an ethanol series, air dry for about 30 min and preheat for 15 min at 65°C in a humid chamber. Denature probe (primer) DNA by boiling for 7 min and add immediately to nick translation buffer (NT) at 65°C. Apply about 0.5 μg of digested plasmid DNA to the slides in 40 μl of NT buffer. Transfer the closed chamber to a 60°C water bath for annealing. After 45 min, rinse slides briefly in NT buffer at 65°C, drain quickly and place in a humid chamber floating in a 42°C water bath.

Apply 40 μl of the labelling mixture (100 μM of dATP, dCTP, dGTP; 75 μm of dTTP; 25 μM of dig-dUTP (Boehringer, Mannheim) and 5 units of Klenow DNA polymerase in NT buffer). Incubate slides at 42°C for 1 h and briefly wash in stop buffer (0.5 M NaCl, 0.05 M EDTA, pH 8.0) at 65°C.

Incorporate digoxigenin-dUTP with fluorescein-anti-digoxigenin following the usual schedule described earlier (Heslop-Harrison *et al.*, 1991).

PRINS to detect clustered low copy sequences on plant chromosomes
(after Abbo *et al.*, 1993b)

Short oligonucleotide primers permit greater penetration into the debris and the highly condensed chromatin present in plant chromosome preparations. As a result, the sensitivity of *in situ* preparations is increased and clustered, low copy sequences can be detected.

Prepare plant material as mentioned in the earlier schedule. Synthesize the oligonucleotides on a Applied Biosystems oligonucleotide synthesizer (model 380 A). The test initially was carried out on the *Hor* 2 locus of barley. The primers designated were Hor 2.1 (5'-ATG-TAAAGTGAATAAGGT-3') and Hor 2.3 (5'-TACCTTGCATGGGTTTAG-3').

Pretreat chromosome spreads with RNase (1 μg/ml in 2× SSC) for 1 h at 37°C. To prevent generation of random nicks, include a nick-filling step. In this step, incubate slides for 2 h in a humid chamber at 37°C with 40 μm each of dATP, dCTP, dGTP, TTP and 0.1 unit of Klenow DNA polymerase per ml in NT buffer. Finally treat with 4% paraformaldehyde as described by Heslop-Harrison *et al.* (1991).

For PRINS, denature chromosomes in 70% formamide in 2× SSC at 70°C for 3.5 min. Dehydrate slides in an ethanol series, air-dry for 30 min, preheat to the annealing temperature (32°C or 42°C) for 20 min in a humid chamber. Apply 40 μl of probe mixture to each slide. The mixture contains between 0 and 500 ng of each B-hordein-specific primer, 10% dextran sulphate, 1% SDS, 0.5 μg of heat denatured salmon sperm DNA per μl, and 2× SSC.

Anneal the primers for 16 h. Wash slides twice for 10 min each in 2× SSC at annealing temperature. Then wash slides twice for 10 min each in NT buffer and then at annealing temperature.

Add 40 μl of labelling mixture per slide. It consists of 75 μM dATP, dCTP, dGTP and 20 μM TTP; 30 μM 12-fluorescein-dUTP or 4-rhodamine dUTP and 0.1 unit of Klenow DNA polymerase per μl prepared in NT buffer. Carry out DNA polymerization for 40 min at the annealing temperature (32°C or 42°C). Place slides in stop buffer (0.5 M NaCl/0.05 M EDTA) and incubate at 65°C for 5 min. Stain chromosomes with DAPI and PI (when fluorescein is used as label). Treat the spread with a drop of antifade mounting agent under a glass coverglass and observe by fluorescence microscopy.

Direct labelling by rapid FISH (after Reader *et al.*, 1994)

In a shortened method, modifications include omission of formamide from hybridization and stringency washing.

Fluorolabel total genomic DNA from eye for use as a probe in nick translation. Mix the fluorochrome-labelled nucleotide mix, 1 mM Fluorored (Amersham) in a 1 : 1 ratio with 0.05 mM dTTP at a 2.5 : 1 ratio. Use 3.5 μl of each mix in nick translation reaction. Incubate the mix containing Fluoro-green at 15°C for 3.5 h. Stop the reaction with EDTA. Add immediately unlabelled total genomic wheat DNA at 70 times the concentration of the probe DNA, thus combining treatments with blocking and carrier DNA just before ethanol precipitation of the probe. Dissolve the precipitated pellet in 20 μl 10 mM Tris, 1 mM EDTA. Probe can be stored at −20°C upto 3 months.

Treat chromosome preparations at 37°C with 5 μg/ml DNase-free RNase (Sigma 7875) in 2× SSC for 3 min. Wash twice in 2× SSC, 2 min each. Wash twice more in 2 mM CaCl$_2$, 20 mM Tris HCl, pH 7.5 (2 min). Treat for 10 min with 2 mM CaCl$_2$, 20 mM Tris HCl, pH 7.5; 1 ng ml^{-1} proteinase K. Wash twice (2 min each) with 20 mM Tris HCl, 20 mM EDTA, pH 7.5. Cool slides to room temperature. Store in 0.1× SSC. Prepare 4% aqueous paraformaldehyde by heating with constant stirring to 80°C, adding IM NaOH (0.05% v/v), adjusting to pH 7.5 with 10× PBS (1–2% v/v) and cooling to room temperature.

Incubate slides for 10 min. Wash three times at room temperature in 2× SSC (2 min each). Denature in 70% formamide, 2× SSC (70°C for 2 min). Dehydrate by sequential incubation in increasing concentrations of ice-cold ethanol (70%, 90% and 100%). Air-dry for 15–20 min at room temp.

For FISH, mix labelled probe to a final concentration of 1.25 μg/ml in 10% dextran sulphate, 2× SSC, 0.25% SDS and denature by boiling for 5 min. Incubate on ice for 5 min. Add 40 μl of probe to each slide and apply coverglass (cut from polythene). Incubate the slides in a humid chamber at 65°C for 2 h. Remove coverglasses. Wash slides for 5 min each twice in 2× SSC at 65°C. Incubate again at 37°C for 20 min. Cool at room temperature. Then rinse in 4× SSC, Tween 20 and stain for 1.5 min with 0.25 μg/ml DAPI under plastic coverglass. Mount in Vectashield under glass coverglass.

This rapid technique is completed within a single day and can be successfully applied to both mitotic and meiotic chromosomes and both to freshly prepared slides and to slides which had been stored at −20°C for upto 12 months. In working with mitotic chromosomes, digestion with proteinase K can be omitted and total genomic DNA or cloned nuclear rDNA used as probe.

Direct chromosome mapping of genes by *in situ* PCR
(after Mukai and Appels, 1996)

In situ PCR combines the sensitivity of PCR with the cytological localization of nucleic acid sequences.

Schedule for Secale cereale: Prepare root-tip squashes following pretreatment and enzymatic maceration. Synthesize three pairs of oligonucleotides to be used as primers for PCR, based on sequences reported for wheat *Nor-BI*, rye *Nor-RI* and rye *5s-Rrna-RI* (rye 5s). Denature chromosomal DNA in root-tip squash preparation by incubating the slides in 70% formamide/ 2× SSC at 70°C for 2 min; rinse for 5 min each in 70% ethanol at −20°C and 90% and 100% ethanol at room temperature. Air dry the slides. PCR reaction mixture (100 µl) contains each primer at 300 nM; 10 mM Tris-HCl (pH 8.3), 50 mM KCl; 3.0 mM $MgCl_2$; 200 µM of each dNTP with 30% of the dTTP replaced by digoxigenin-11-dUTP and 10 units of *Taq* polymerase. Apply 20 µl of the reaction mixture to the denatured slides and cover with coverglass, previously cleaned with acid/ethanol. Seal slide with rubber cement. Perform *in situ* PCR using an FTS-320 Thermal Sequencer. The programme includes initial annealing at 55°C for 4 min and an initial extension at 72°C for 10 min (one cycle), followed by 20 cycles of denaturation at 94°C for 1 min, annealing at 55°C for 1 min and extension at 72°C for 3 min.

After completing the temperature programme, remove coverglass. Wash slides in 70% ethanol/0.05% SDS, 2× SSC and 4× SSC for 5 min each at room temperature. Detect *in situ* PCR products incorporating digoxigenin- labelled nucleotides by immunochemistry with a fluorescein-conjugated anti-digoxigenin anti body as follows: Incubate the slides for 1 h at 37°C with 1 : 10 diluted, fluorescein-conjugated anti-digoxigenin Fab fragments (stock solution 200 µg/ml in 4× SSC/1% BSA). Then wash in 4× SSC for 10 min; 0.1% Triton × 100 in 4× SSC for 10 min; 4× SSC for 10 min; and 2× SSC for 5 min, all at room temperature in dark. Mount slides in an antifade solution (90% glycerol/0.1× PBS/1.25% 1,4-diazabicyclo [2.2.2] octane, DABCO), containing propidium iodide (1 µg/ml). Observe through fluorescence microscope with a blue filter.

In situ localization of yeast artificial chromosome sequences on metaphase chromosomes of higher plants
(after Fuchs *et al.*, 1996)

For localization of short low or single copy sequence, a target sequence of at least 10 kb per locus is needed. It has been proposed to use large insert clones available from libraries constructed in yeast artificial chromosomes (YAC). YACs were hybridized *in situ* onto tomato metaphase chromosomes.

For isolation of chromosome specific YACs, screen the YAC library of tomato by PCR, with a primer pair derived from a chromosome-specific RFLP marker. Sequence the cloned marker CT277 (900 bp), localized on chromosome 2 of the molecular linkage map of the tomato genome, from both ends with an A.L.F. sequencer (Pharmacia), according to standard procedures. Designate a primer pair of this marker for PCR amplification as (CAT CTC ACA GTG TTT TGC AG and GAT GCT GGC ATT GTT GCT TC).

The YAC library, consisting of more than 36,000 clones, was organized in 380 DNA pools containing 96 single yeast clones each. Screen the DNA pools by PCR with the CT277-specific primer pair. Analyse amplified fragments by separation on 1% agarose gels. Subject the 96 yeast clones of those DNA pools that revealed marker specific fragments directly to a second PCR step. For lysis of cells, add 5 µl of a 1 mg/ml solution of zymolyase 100 T (ICN) to 5 µl of the glycerol stock cultures. Incubate for 30 min at 37°C. Run PCR in a total volume of 50 µl. Isolate single positive clones on selective media. Extract yeast chromosomes according to Carle and Olson (1985). Determine the size of the respective YACs by pulsed gel electrophoresis, blotting and hybridization with the marker. Isolate YAC DNA from agarose gels after electrophoresis by adsorption to glass milk with a Gene Clean Kit (Bio 101) .

For *in situ* hybridization, prepare metaphase spreads from fresh root-tips of tomato and a dihaploid potato. Treat root-tips with 0.05% colchicine for 2 h; fix in acetic acid–ethanol (1 : 3); digest in 1% pectinase and 1% cellulase (in 0.01 M citric acid–sodium citrate buffer, pH 4.5–4.8) for 30 min at 37°C and squash in 45% acetic acid. Either use immediately for FISH or store in glycerol at 4°C.

Label YAC-DNA with biotin-16-dUTP either by PCR after amplification with degenerated primers or directly by nick translation using a kit (Amersham) according to the instructions of the manufacturer. Perform FISH as described by Fuchs and Schubert (1995) and also given earlier in this chapter.

Evaluate metaphases using a Zeiss Axioskop with appropriate filter combinations. Capture fluorochrome images separately with a cooled CCD Camera (Photometrics), pseudocolor and merge and if needed, process on the computer (Adope Photoshop, Gene Join). Print complete images on a Phaser II (SDX, Tectronix).

METHODS FOR IMMUNOLOGICAL DETECTION

In the study of chromosome structure, immunological techniques have lately been widely used. The objective is to utilize antibody for localization of antigens at the cytological level. In chromosome analysis, identification of DNA sequences *in situ*, and the localization of antigenic chromosomal proteins have involved immunological methods. The DNA localization has been dealt with extensively in the preceding part of this chapter, where FISH and other associated techniques have been covered. The following discussion would principally deal with localization of chromosomal proteins.

The immunofluorescent techniques need as a prerequisite, the maintenance of chromosomal integrity without hampering its antigenic property and the accessibility of the applied antibody into the chromosomal antigen. However, since accessibility involves cell permeability which may cause loss of antigen and fixation may result in loss of antigenic activity of chromosomal components, suitable pretreatment and fixation schedules are crucial for antigen localization (Jeppeson, 1994 and Heyting *et al.*, 1994). In recent techniques these limitations have been overcome through refinements as far as possible (see Gosden, 1994).

Advances

In chromosome structure, since histones and non-histones play an important role in functioning, immunofluorescence techniques have been applied to detect and monitor modifications in protein compounds during development and differentiation. The role of histone proteins in nucleosomal assembly is well established. However, the importance of non-histone proteins in the fundamental core of chromosome scaffold, in the centromere structure as well as in the regulation of heterochromatin condensation, has been demonstrated through immunofluorescence method as well. Similarly, the existence of Topoisomerase II as a component of chromosome scaffold has been shown. The application of antibodies raised against histones exhibits homogeneous labelling, indicating the continuity of nucleosomal units. With modified histone epitops non-uniform labelling has been noted, indicating chromatin variation.

The immunoserological approach involving the use of sera containing autoantibodies from autoimmune patients has helped in localizing different types of centromeric protein. The combination of PRINS with antibody technique has helped to localize specific DNA sequences associated with specific proteins. A few sample techniques are included in the following paragraphs.

Antibody method for replicating bands in plants

Treat seedling (two-day old) with 1 µg/ml 5-bromodeoxyuridine (BrdU) containing 1 µg/ml fluorodeoxyuridine (FdU) for 30 min in dark at 22°C. Rinse with 10 mM thymidine containing 1 µg/ml FdU in dark. Incubate the seedlings in dark on wet filter paper after rinsing thrice with deionized water. The period of incubation depends on the species, varying from 4 h for *Crepis capillaris* to 10 h for *Allium fistulosum*. Add 0.02% colchicine to the seedling for the final hour of incubation.

Fix seedlings in acetic acid–methanol (1:3) for 1–7 days at −20°C. Excise the root-tips, hydrate, macerate with enzymes on slides and flame dry. For denaturation, incubate the slides in 1 NHCl at 60°C for 10 min. Wash gently with deionized water. Incubate in a phosphate-buffered saline (PBS, pH 7.2), containing 0.1% Tween 20 at 0°C for 5 min. For antibody treatment, coat the slides with 5000-fold diluted anti-BrdU antibody (mouse IgG, IMT Co.). Incubate at room temperature for 30 min in a moist chamber, rinse in cold PBS twice for 5 min each. Coat the slides with a biotin-conjugated anti-mouse antibody (Vector Laboratories), obtained by mixing avidin-DH and biotinylated horseradish peroxidase H. Incubate at room temperature for 30 min in a moist chamber. Rinse the slides twice in cold PBS, keeping in each for 5 min.

Visualize the peroxidase with a solution of 1 mg/ml 3.3'diaminobenzidine-4HCl (DAB, Dottite); 0.4 mg/ml $NiCl_2$ and 0.02% H_2O_2 in 100 mM Trisbuffer (pH 7.2) at room temperature for 7 min in a moist chamber. Rinse the slides in distilled water twice for 5 min each. Counterstain the chromosomes with 0.1% neutral Red (Chroma-Gesellschaft Schmid GmbH) for 15–30 min. Rinse gently with distilled water, air dry, dip in xylene for 10 min and amount in Eukitt (Kinder Germany).

The positively stained bands indicate the replication sites labelled during the S phase and SCE patterns.

Kinetochore visualization with CREST serum

Germinate pollen grains. After 7 h, harvest the pollen tubes, fix and process. Expose overnight at 4°C to a x15 dilution of CREST serum E.K. in phosphate-buffered saline containing 3% bovine serum albumin and 0.02% sodium azide, followed by a x50 dilution of a fluorescein-conjugated goat antihuman secondary antibody (Sigma) for 1.5 h.

For double localization, apply the CREST serum and fluorescein-linked secondary probe first and follow with an antitubulin protocol, employing a rhodamine-conjugated antimouse secondary antibody (Sigma). Mount the pollen tubes in a solution containing 1 mg/ml phenylenediamine (Sigma), 10 μg/ml Hoechst 33258 (Sigma), 50% glycerol, and 0.1 M Tris (pH 9.3) and view on a Universal microscope (Carl-Zeiss), equipped with Olympus 60 and 100x DApo UV objectives for epifluorescence.

Stain pollen tubes with 100 μg/ml mithramycin A (Sigma) in deionized water and view on a laser scanning confocal microscope (MRC 500, BioRad Instruments, Cambridge, MA) using fluorescein excitation and a 60x planapochromat objective.

Kinetochores are shown by CREST antibodies to be paired and single fluorescent dots, located at the ends of microtubule bundles previously identified as kinetochore fibres (Palevitz, 1990).

Immunostaining by antibodies of acetylated histone H4 variants
(after Houben et al., 1996a, 1997)

Metaphase chromosomes of *Vicia faba* were exposed to antibodies recognizing defined acetylated isoforms of histone H4. After indirect immunostaining with antibodies directed against H4 acetylated on lysines 5, 8 and 12 respectively, the entire chromosome complement was labelled.

The *antibodies* are polyclonal rabbit antisera and recognize specifically histone H4 acetylated at lysine positions 5 (R 41/5), 8 (R 12/8), 12 (R 20/12), 16 (R 14/16) and non-acetylated H4 (RSU) respectively. Monoclonal anti-DNA antibodies (Boehringer) can be used as controls directed against single and double stranded DNA.

In *Vicia faba* (Houben *et al.*, 1996b). Prepare a pure suspension of metaphase chromosomes of *Vicia faba* and interphase nuclei from synchronized root meristems of seedlings. For indirect immunofluorescence, drop the suspension on slides and cover with coverglass. Remove coverglass by freezing in liquid nitrogen and transfer slide immediately to phosphate buffered saline (PBS, pH 7.3). Incubate the chromosomes for 1 h at 4°C in primary sera (diluted 1 : 400 in PBS) in a humidified chamber, wash in PBS and incubate for 1 h at 4°C with fluorescein-isothiocyanate (FITC)-conjugated anti-rabbit IgG (Dianova). Wash thrice in PBS and post fix for 10 min in PBS containing 4% (v/v) formaldehyde. Observe fluorescently labelled chromosomes using a microscope with epifluorescence optics. Photograph with Kodak Ektachrom 400 films.

Immunostaining in brachycome dichromosomatica

Treat root-tips overnight in ice water and fix in ice-cold, freshly prepared 12% (w/v) formaldehyde solution containing 10 mM Tris HCl (pH 7.5); 10 mM EDTA and 100 mM NaCl. Wash for 20 min in 10 mM citric acid-sodium citrate buffer (pH 4.5) at room temperature; digest at 37°C for 25 min with 2.5% pectinase, 2.5% cellulase Onozuka R-10 and 2.5% pectolyase Y-23 (w/v) dissolved in citric acid–sodium citrate buffer. Wash root-tips for 10 min in 50 mM phosphate buffer containing 5 mM $MgSO_4$ at room temperature. Squash on glass slides in phosphate buffer (Houben *et al.*, 1997).

Remove coverglass by freezing in liquid nitrogen; transfer slides immediately into phosphate buffered saline (PBS, pH 7.4). Preincubate slides for 30 min with 3% bovine serum albumin (BSA); 5% non fat milk powder (w/v) and 0.2% Tween 20 (v/v) in PBS at room temperature.

Then incubate chromosome preparations for 1 h at 4°C in primary sera (diluted 1 : 200 in PBS) in a humidified chamber, wash in PBS and incubate at 4°C for 1 h with FITC-conjugated anti-rabbit IgG (Dianova) diluted 1 : 40 in PBS with 3% BSA (w/v), 5% (w/v) non fat dry milk and 0.2% Tween 20. Wash three times in PBS. Counterstain with DAPI and mount in antifade. The slides must never become dry throughout the procedure. Observe fluorescence as in the previous material.

C-banding with DAPI involved the replacement of Giemsa staining by DAPI.

Fix root-tips, pretreated in ice-water as described earlier, for several days in acetic acid–ethanol (1:3). Wash in enzyme buffer (0.01 M citric acid–sodium citrate, pH 4.5) for 30 min at room temperature treat with 1% (w/v) cellulase Onozuka R10 and 1% (w/v) pectinase in the same buffer at 37°C for 22 min; wash in enzyme buffer; transfer to 45% acetic acid and squash on glass slides.

For C-banding, treat preparations on slides with 45% acetic acid at 60°C; wash for 15 min with distilled water; dehydrate in absolute ethanol and dry at 42°C for 30 min. Treat slides with saturated barium hydroxide solution at 30°C for 10 min and wash as above. Then incubate in 2× SSC at 60°C for 90 min, wash in water and air dry. Finally, stain with DAPI and mount in antifade.

For *DNA late replication pattern through autoradiography*, incubate excised roots with ^{3}H-thymidine (5 µCi/ml) for 30 min, wash thoroughly and grow for 5 h in water. Then treat with ice water overnight, fix in acetic acid–ethanol (1:3) and stain in Feulgen. Squash. Remove coverglass. Dry the slide and coat with photographic emulsion (EMI, Amersham). Expose at 4°C for 4–7 days in dark. Develop silver grains in Kodak DI9B developer, fix in photographic fixative and wash extensively. Observe under light microscope.

Replication pattern by indirect immunofluorescence
(after Yanagisawa *et al.*, 1993)

The method utilizes indirect immunofluorescence to detect the incorporation of BrdU in order to analyze DNA replication pattern in very small chromosomes.

Transfer freshly germinating seeds to aqeous aphidicolin (Wako solution 10 µg/ml), known to prevent DNA replication from entering into a new replicon and keep for 16 h. Remove seeds from the solution, wash in distilled water for 15 min and soak in aq. 5-Bromodeoxyuridine (BrdU, Sigma) soln (50 µg/ml) for 30 min.

Remove root-tips, wash in distilled water for 15 min and place on wet filter paper. Keep for 8 h and then fix in acetic acid–methanol (1:3). Prepare chromosome squashes, following enzyme maceration, as described in the previous schedule.

For *indirect immunofluorescence staining*, immerse chromosome preparations on slides in an alkaline solution (70 mM NaOH: ethanol pH 7.3, v/v) at room temp for 5 min and then wash in PBS (pH 7.4) for 5 min. Place a few drops of a mouse monoclonal antibody against BrdU (mouse-anti-BrdU, Partec AG and Beeton Dickinson), diluted 100 times with PBS containing 0.5% Tween 20 (Bio-Rad) and 0.5% bovine serum albumin (BSA, Sigma), on the slides. Cover slide with parafilm, incubate in a humid chamber at 37°C for 1 h. Wash in PBS three times, 10 min each, at room temperature. Then place a few drops of FITC-conjugated goat antibody against mouse immunoglobin G (goat-anti-mouse IgG FITC-conjugated, Tago), diluted 1:100

with PBS containing 0.5% Tween 20 and 1% neutral goat serum (Cappel), on each slide. Cover with parafilm, incubate in a humid chamber at 37°C for 1 h. Wash twice in PBS and once in distilled water, each for 10 min at room temperature. Counterstain with aq. propidium iodide (PI) solution (1 μg/ml) and keep in stain for 24 h in dark at 4°C. Place a drop of McIlvaine's buffer (6 mM citric acid, 80 mM dibasic sodium phosphate, pH 7.0) on each slide and cover with a coverglass. Photograph fluorescent images with Bar G excitation filters and negative colour films (Super HG ISO 400 and 1600, Fuji).

Isolation of centromeric DNA by immuno-adsorption
(after Houben et al., 1996b)

Isolate chromosomes, as described elsewhere, from root-tip meristems. Centrifuge and resuspend in 200 μl of ice-cold 1× PBS (pH 7.3), containing 400 mM NaCl, 1 mM MgCl$_2$, 10 mm PMSF (phenylmethylsulfonyl fluoride), 5 mM EDTA, 0.1% Triton ×100, 0.1% Nonidet P-40 and 2% BSA. Sonicate the suspension 5 times for 30 s each until the size of the fragments, was less than 0.5 μm, as seen under the microscope.

Load tosyl-activated magnetic Dynabeads M-280 (Dynal) with purified anti-kinetochore anti-body 5/1, according to manufacturer's instructions. Incubate together the antibody-loaded magnetic beads (100 μl) and the sonicated chromosome suspension (200 μl) for 2 h under gentle rolling. Pipette off unbound chromosome fragments. Keep beads with bound fragments at the wall of the Eppendorf tube by means of a magnetic particle concentrator. Wash five times, 10 min each, at room temperature in 200 μl 1× PBS with 0.1% Tween 20 and 2% BSA. Wash beads again in 1× PBS, 0.1% Tween 20 for 10 min at room temperature. Treat with 0.2 mg proteinase K in 0.5 ml of 10 mM tris-HCl (pH 8.0), 10 mM NaCl and 0.1% SDS for 8 h at 50°C. Extract DNA with phenol/chloroform and precipitate with ethanol. Resuspend precipitate in 5 μl TE and amplify DNA by DOP-PCR.

STATUS AND ADVANCES

The in situ hybridization has been of immense help in gene mapping and in the study of phylogeny, evolution and origin of biodiversity, though complex DNA sequences are involved in the process. As far as the study of allopolyploids or amphidiploids is concerned, its contribution in the genome analysis is well known (Bennett, 1996; Jellen et al., 1994). However, it has been claimed that at the initial phase of amphidiploidy, sterility may occur out of the non-interaction between one set of genome with the other genome and cytoplasm. In order to overcome this barrier, arising out of disturbed nucleocytoplasmic interaction, often intergenomic translocations may occur in polyploids. Such translocations have been demonstrated in species of *Triticum* (Jiang and Gill, 1994). The origin of biodiversity and stabilization of complex genotypes through polyploidy, as well as species specific trans-location, could also be confirmed through *in situ* hybridization. Such inter-genomic translocation in species of *Nicotiana* (Kenton *et al.*, 1993) has also been demonstrated through genome *in situ* hybridization.

The *in situ* hybridization has further revealed that widely different genomes have contributed to the origin of biodiversity through interkingdom transfer. Lately, it has been shown that gemini related viruses (GRD sequences) are present in the genome of *Nicotiana*. Their presence in *T.* genome donor of *N. tabacum* before the origin of amphidiploidy has been suggested (Bejarano *et al.*, 1996; Leitch *et al.*, 1997). Further, this technique has served as a good indicator of stable integration of an alien gene sequence artificially introduced into the genome. This is specially useful where the donor and recipient genomes do not show marked differences in karyotypes (de Jong *et al.*, 1996; vide Leitch *et al.*, 1997).

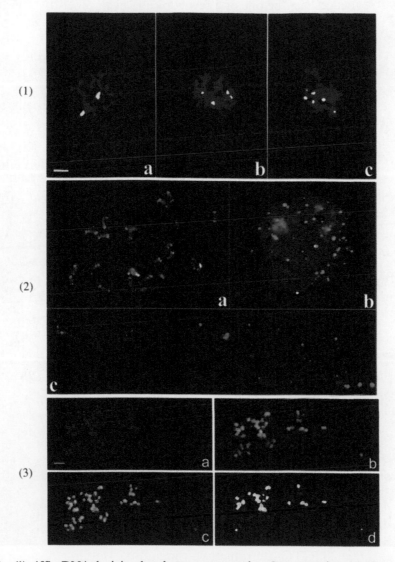

Plate 1 (1) 45S rDNA loci in the three representative *Oryza* species. (a) *O. sativa* spp. *japonica*, (b) *O. sativa* spp. *javanica*, (c) *O. punctata*. *O. sativa* spp. *indica* has the same number of the loci as *javanica*. Bar indicates 3 μm. TAG (1994). (2) Multi-colour FISH using rice A genome specific tandem repeated sequence (TrsA, red) and telomere sequences (green). (a) IR 36 has 6 pairs of TrsA sites at the distal ends of the long arms. (b) Interphase mapping of TrsA and telomere sequences. (c) TrsA and telomere signals on the extended DNA fiber. Plant Mol. Biol. (in press). (3) Genomic *in situ* hybridisation allows differential painting of C genome chromosomes among B and C genome chromosomes in *O. punctata* (BBCC). (a) Counterstained chromosome with PI. (b) Fluorescent signals after genomic *in situ* hybridisation. (c) C-genome chromosomes with the more intense signals were marked by red contour lines. (d) Merged image of the fluorescent signals and counterstained images. Yellow signal indicates the chromosomes belonging to C genome. Bar indicates 10 μm. TAG (in press). (Courtesy of Prof. K. Fukui, Hokuriku Agr. Expt. Station, Niigata, Japan.)

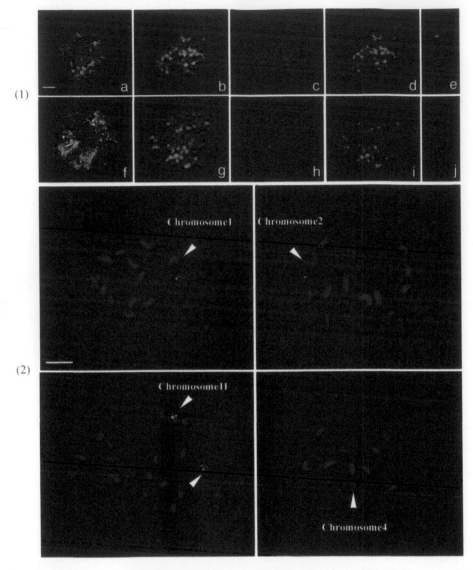

Plate 2 (1) Multicolour FISH reveal three different genomes within the two somatic hybrids, *O. sativa* (AA)+*O. punctata* (BBCC). (a) A genome (red), B genome (green), and C genome (blue) in the nucleus. (b) Chromosomes stained by the same method. (c) and (h) Monochrome detection of A genome chromosomes. (d) Detection of B genome chromosomes. (e) Insertion of the B genome chromosomal fragments to the C genome chromosomes. (f) Discrimination of the three different genomes with red (A genome) and green (C genome) fluorescence in the nucleus. (g) Chromosomes stained by the same method. (i) Monochrome detection of C genome chromosomes. (j) Insertion of the A genome chromosomal fragments to the B genome chromosomes. Bar indicates 5 µm. TAG (in press). (2) Physical mapping of nucleotide sequences with a variety of sizes on rice chromosomes. (a) Yeast artificial chromosome with a 399 kbp rice genomic DNA insert. (b) Bacterial artificial chromosome with a 150 kbp rice genomic DNA insert containing a rice leaf blast resistant gene. (c) Cosmid clone with a 35 kbp containing a rice bacterial leaf blight gene. (d) RFLP clone with a 1.29 kbp. Bar shows 5 µm. *Plant Mol. Biol.* (Submitted.) (Courtesy of Prof. K. Fukui, Hokuriku Agr. Expt. Stn., Niigata, Japan.)

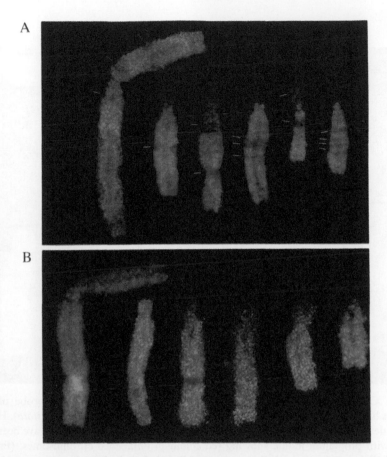

Plate 3   A–B—*in situ* hybridisation of biotinylated DNA and counterstaining with propidium iodide. DNA of chromosomes *III* and *V* respectively of karyotype EF of *Vicia faba* hybridise to chromosomes of karyotype ACB of the same species. The reduced signal density in NOR and some heterochromatic regions characterised tandem repeats are absent from *III* of EF are to be noted. (Courtesy of Professor I Schubert, Plant Genetics Institute, Gatersleben and Springer Verlag.)

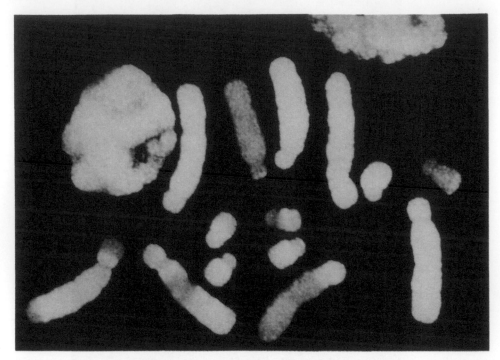

Plate 4 Genome *in situ* Hybridisation (GISH): Total genomic DNA probe of *Gasteria lutzii* was hybridised to a back cross progeny *G. lutzii* × (*G. lutzii* × *Aloe aristata*). Hybridised sites are detected with yellow fluorochrome—fluorescein and chromosomes have been counter-stained with propidium iodide showing red colour in unlabelled chromosomes. (Preparation by Dr C. Takahashi—through the courtesy of Dr's I. Leitch, P.D. Brandham and Professor M.D. Bennett, Jodrell Laboratory, Royal Botanic Gardens, Kew.)

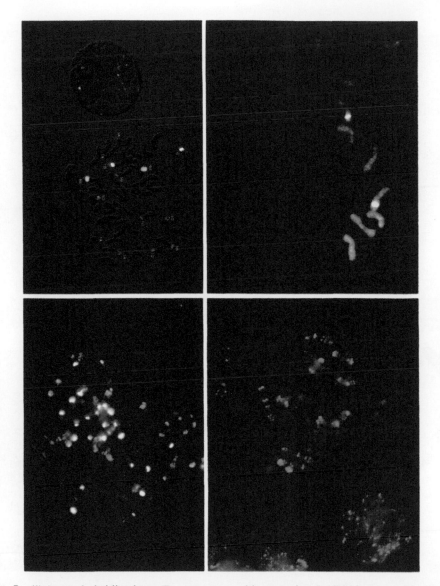

Plate 5 (1) *In situ* hybridisation—*T. aestivum* cv. chinese spring 5s rRNA genes—1A, 1B, 3A, 5B, 5D (Red). 18S—26S rRNA genes—1B, 6B, 1A, 5D (Green). (2) *In situ* PCR—Imperial rye 18s rRNA genes on 1R chromosome—NOR R1 Primer—386 bp. (3) Triticale (8× —Cs wheat and Imperial rye). Seven colour FISH on metaphase cell. Seven DNA sequences pSc 119.2, pSc 74, pAsl, telomere 185.26S rDNA, 5S rRNA and gliadin detection by red, bluish green, green, orange, pink, blue and white colour respectively. (Courtesy of Dr Y. Mukai, Osaka Kyoiku University, R-698-1 Asahigaoka Osaka, Japan). (4) FISH on wheat and rye. (Courtesy of Dr U.C. Lavania, CIMAP, Lucknow.)

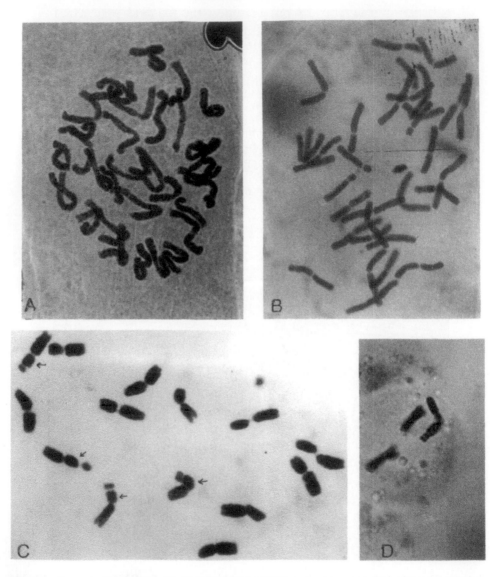

Plate 6 Somatic chromosomes of plants from different species (A) and (B)Natural Auto-tetraploid *Allium tuberosum* (4× = 32) in prophase and metaphase respectively. (C) and (D) Diploid *Hordeum vulgare* (2n = 14) and *Haplopappus gracilis* (2n = 4) respectively.

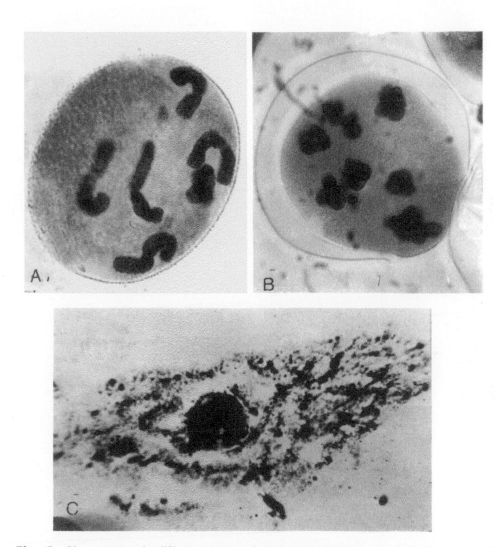

Plate 7 Chromosomes in different organs of plants. (A) Chromosomes in pollen grains of *Tradescantia paludosa* (n = 6). (B) Meiosis showing eight quadrivalents in autotetraploid *A. tubersum*. (C) Endoreplicated polytenic chromosomes in suspensor cells of *Psophocarpus tetragonolobus* (2n = 18). (Plates 6 & 7—Courtesy of Professor Sumitra Sen, Dr Kana Talukder and Dr Aparna Basu, Center of Advanced Study, Department of Botany, University of Calcutta.)

SECTION—III

MOLECULAR PATTERN ANALYSIS AND GENE MAPPING IN CHROMOSOMES

The manipulation and engineering of chromosomes, at the gene level, need a thorough understanding of and experience in the localization, identification and isolation of genes, and a study of their complexity at the molecular level.

The study of polymorphism of gene sequences and mapping of genes are gaining importance in the analysis of biodiversity and evolution. The molecular documentation of genotypes, through fingerprinting of sequences, has an immense potential.

The genes can be localized by the preparation of a genetic map. It can be achieved by crossing a series of isogenic lines, where different lines are homozygous for almost all the genes, attained through continued inbreeding. The preparation of such a map becomes easy if there are several blocks with linked genes corresponding to chromosome numbers. In the absence of isogenic lines, various markers are to be used, which are homozygous for different alleles in the parents.

The morphological markers are obtained through the analysis of vegetative and reproductive characters of the parents and the offspring. Other markers involve the use of isozyme patterns as well as molecular probes such as RAPD, microsatellites, AFLPs and RFLPs. Since repeat DNAs may be homologous to different loci, unique sequence probes are better used in RFLP analysis. This map may involve a combination of all the markers. Genetic maps of different species are often based on different polymorphic marker loci.

The genetic map with restriction enzymes is based on the principle of locating specific cleavage sites of different endonucleases which recognize short DNA sequences and cleave at specific sites. The identification and mapping of cleavage sites yield the restriction map. It delineates short linear sequences cleaved by specific enzymes, the distance being measured in terms of DNA base pairs. This method permits measurement of the distance between closely related genes and thus maps their relative position.

Genetical maps of several species of agricultural importance, such as barley, wheat, tomato, potato, sorghum, alfalfa, pearl millet and bean have been prepared for special reasons, such as utilization for resistance against biotic factors as well as analyzing the hybridization pattern, as in sunflower (Adam Blordon *et al.*, 1995; Rieseberg *et al.*, 1993).

CHAPTER III.1

SEQUENCE ANALYSIS AND MAPPING OF
GENES AND PREPARATION OF PROBES

In the molecular analysis and mapping of genes and genomes, the most convenient method is the use of specific gene probes and nucleic acid hybridization. The preliminary information on the size and sequences of a genomic segment can be obtained through Southern blotting with requisite probes, and cloned DNAs. The principle of southern blotting is the transfer of fractionated DNA sequences from the gel to nitrocellulose filter, followed by identification of sequences through hybridization with specific probes. This method permits a preliminary understanding of the number of gene

Table 1 Some restriction enzymes and their cleaving sequences

Microorganism	Enzyme abbreviation	Sequence $5' \rightarrow 3'$ $3' \rightarrow 5'$
Bacillus amyloliquefaciens H	*Bam*HI	G↓GATCC CCTAG↑G
Escherichia coli RY 13	*Eco*RI	G↓AATTC CTTAA↑G
Haemophilus aegyptius	*Hae*II	PuGCGC↓Py Py↑CGCGPu
Haemophilus aegyptius	*Hae*III	GG↓CC CC↑GG
Haemophilus haemolyticus	*Hha*I	GCG↓C C↑GCG
Haemophilus influenzae Rd	*Hind*II	GTPy↓PuAC CAPu↑PyTG
Haemophilus influenzae Rd	*Hind*III	A↓AGCTT TTCGA↑A
Haemophilus parainfluenzae	*Hpa*I	GTT↓AAC CAA↑TTG
Haemophilus parainfluenzae	*Hpa*II	C↓CGG GGC↑C
Providencia stuartii 164	P_{st}I	CTGCA↓G G↑ACGTC
Streptomyces albus G	*Sa*II	G↓TCGAC CAGCT↑G
Xanthomonas oryzae	*Xor*II	CGATC↓G

copies to be utilized for genomic cloning and primary characterization. The extent to which the entire genomic clone can be conveniently prepared depends to a great extent on the size of the gene and the genome. Small genes can easily be sequenced but for larger genes, sequences of exons and regulator elements are essential. Moreover, delimiting sites of exons, and introns and the transcription initiation elements need to be identified. Sequence specific probes, nuclease S mapping and primary extension method are needed to achieve these objectives. For efficiency and accuracy in nucleic acid hybridization, a battery of protocols has been developed (for review vide Hall, 1995).

RESTRICTION FRAGMENTS

For cutting the DNA into small segments, restriction enzymes are used which are normally present in the organisms. The organisms use these enzymes for destruction of foreign DNA molecules. These are endonucleases which recognize and cut specific DNA motifs, principally of 4–6 bases. These enzymes act on phosphodiester bonds and the organism protects its own DNA from restriction cuts by methylation at specific sites.

The technique is quite simple. It involves the extraction of DNA, digestion with restriction enzymes and separation of the fragments according to size through gel electrophoresis. The later steps include identification of sequences and loci through transfer to a membrane and hybridization with known probes. The method is known as "Southern Blotting". The probes used for analysis of restriction fragment length polymorphism are locus and species-specific, permitting identification of codominant markers as well. Through the choice of appropriate gels, the DNA fragments, ranging from a few nucleotides to several megabases, can be analysed. However, for analysis of gene structures and short range mapping, the needed resolution is from a few hundred base pairs to several kilobases. Such resolution is conveniently achieved using a low percentage namely, 0.3–0.7% of agarose gels. The viscosity of both polyacrylamide and agarose permits DNA molecules to separate according to size and electrophoretic pattern.

The basic principle is the inverse relationship between the log of the molecular weight of the fragment with the distance the molecule is to transverse in the gel. The gel can be viewed under UV light with the help of fluorescent ethidium bromide staining. Following electrophoresis, prior to hybridization, it is necessary to treat the fragments with acid for depurination and alkali for denaturation. The next step is the transfer of the gel to nitrocellulose or nylon membrane for hybridization with the selected probe for Southern blot.

TRANSFER MEMBRANES

The membrane used for transfer normally should have very low background. Nitrocellulose is satisfactory from that standpoint but it is fragile, does not bind covalently with DNA and has a low capacity for retention of smaller DNA fragments. On the other hand, nylon membrane, though having a high background, yet is strong, has high DNA binding capacity and can retain small fragments as small as 50 nucleotides. Moreover, it permits reprobing several times.

The buffers needed for transfer are different for nitrocellulose and nylon membranes. For nitrocellulose membranes, alkaline buffer is not suitable because of inability of DNA to bind at high pH and instability of nitrocellulose at long alkaline exposure. A suitable transfer buffer is $20 \times SSC$ solution. With nylon membranes, a wide range of buffers can be used, including concentrated NaOH soln. Nylon membrane is well suited for autoradiographic analysis but its background effect is a limitation in its use in detection methods based on chemiluminescence.

Of the different methods of transfer, namely the capillary, vacuum blotting and electroblotting, the former is the least expensive and simple. However, for large DNA fragments where upward capillary movement is slow, partial depurination is necessary for degrading high mol.wt. DNA to 1–2 kb fragments.

There are also methods for downward transfer, which are rather rapid including alkaline transfer. For vacuum blotting, specialized blotters are supplied by different agencies with matching protocols for rapid transfer. Electroblotting has a limited application but it is very effective with polyacrylamide gels (Hall, 1995).

IMMOBILIZATION

The next step is immobilization to prevent loss of DNA. With nitrocellulose, vacuum baking at 80°C and with uncharged nylon membrane, UV treatment are adopted to secure crosslinking. For the latter, commercially available self calibrating UV-cross linker is recommended. With *positively charged* nylon membrane, no special immobilization step is necessary for DNA blot, since covalent linkages are formed in alkaline transfer.

REQUISITES FOR SOUTHERN BLOTTING

For Southern blotting, preceded by gel electrophoresis, a number of precautionary steps have been recommended against shearing, incomplete dissolution and poor enzyme access (see Hall, 1995). These include: proper handling of high molecular weight DNA using cut pipette tips; keeping DNA along with addition of restriction buffer with continuous stirring in ice for 1–2 h; stirring DNA at 4°C for 2–3 min after addition of restriction enzyme, incubation for 1 h at appropriate temperature and addition of a second aliquot of enzyme to secure complete digestion. The addition of spermidine dihydro-chloride is needed to overcome inhibition of enzyme activity by impurities.

For hybridization with the probe, certain prerequisites are to be fulfilled. These are: the prevention of nonspecific binding through prehybridization with a blocking DNA, the use of a suitable probe for specific hybridization and critical washing of the membrane to destabilize nonspecific hybrids.

The probes to be used for hybridization may be homologous or hetero-logous. For hybridization with homologous probes, that is of similar geno-types, a low ionic strength and high temperature yield specific hybrids. With GC-rich probes, nonspecific hybrids may arise which are to be washed off thoroughly at higher temperature. Very stringent conditions are necessary for hybridization with heterologous probes, which are from different species. The washing temperature is determined through trial and error but in general 1% mismatch results in 1°C reduction in hybrid melting temperature. Therefore, reduction in hybridization temperature for washing should be approximately equivalent to percentage of mismatch expected. The probes may be prepared from different genomic sources but synthetic oligonucleotide probes have distinct advantages. They can be used in excess over the complementary target sequence permitting specific binding. These probes, being single stranded, do not require any denaturation step. As such, oligonucleotide probes are now widely used.

Outlines of techniques for restriction fragment analysis of genomic DNA. Southern transfer and hybridization are given later.

With oligonucleotide probes, the prehybridization and hybridization of Southern blots are performed in $6 \times$ SSC or 1 M sodium buffer. The tempera-ture for hybridization for 1–2 h should be 5°C below the Td of the probe $(Td - 4 (G + C) + 2 (AHT)$ in 1 M Na for 11–20 base oligonucleotides. Initial wash with $6 \times$ SSC at room temperature is followed by 1–2 min at 5°C below the Td and finally 20 min wash at room temperature.

For analysis of DNA fragments larger than 25–30 kbp, standard gel electrophoresis may not be adequate. These fragments can be analyzed through pulse field equipments, which involve periodic changing of the direction of the electric field. In this method, short pulses of electricity are run in two directions and embedding of DNA is done in the form of agarose plugs so that fragmentation is avoided. Field inversion electrophoresis may resolve DNA molecules upto 2 Mbp in size while with CHEF (Contour-Clamped Homogeneous Electric Field), the resolution can go even to a higher level (Maliga *et al.*, 1995). Several fungi have been subjected to this technique and successful maps of individual chromosomes prepared.

PREPARATION OF PROBES

The probes play a very crucial role in gene identification and analysis. They can be obtained through cloning specific nucleic acid sequences in single or double stranded phage vectors in plasmids or cosmids. They can be chemically synthesized oligonucleotides derived from known sequences. Such synthetic sequences do not require genomic DNA and as such are widely used. For gene identification, probes are hybridized with target DNA sequences and high specific activity is necessary to secure strong hybridization signals. As such the probes are labelled with radioactive isotopes or fluorescent compounds for detection. For labelling, two methods are widely used, namely Nick translation and Random primer labelling. The amplification of probes through polymerase chain reaction helps in preparing a large amount of probes from a small sample.

Labelling of probes

By Nick translation: Both radioactive and non radioactive labels are used in Nick translation. The principle is to nick one strand of the double helix with DNAse I which generates free 3' hybroxyl ends in the unlabelled DNA. The *E. coli* DNA polymerase I, by virtue of its 5.3' exonucleolytic activity, removes nucleotides from 5' end of the nick, simultaneously adding new nucleotides to the 3' hydroxyl nick terminus. The incorporation involves all four deoxynucleotides, of which one is labelled and becomes incorporated into the new strand through complementary base pairing. This method yields high specific activity with ^{32}P labelled adenine or cytosine phosphates. The technique is given later.

Random primer labelling is performed on linear single stranded DNA template. Random hexanucleotides are synthesized by automatic DNA synthesizer, containing all four bases in every position. Oligonucleotides are randomly annealed and labelled nucleotides are enzymatically added. Different commercial companies also supply mixtures of random primers with instructions for their use. The primers are needed for DNA synthesis from circular or linear denatured DNA as templates, with the help of Klenow fragment of *E. coli* DNA polymerase. As this enzyme does not have exonuclease activity, the DNA is synthesized exclusively through primer extension. During synthesis the labelled nucleotides also get incorporated in the new strand. This method permits the generation of probes with high specific activity ($10^8/10^9$ cpm/μg), since more than 70% of labelled nucleotides partici-pate in the new strand (Kaufman *et al.*, 1995). Moreover, being devoid of any nick, the probes can be very long. The method of preparation of radioactive probes through random primer method is later given in detail.

TECHNIQUES FOLLOWED

Restriction enzyme digestion of DNA

Materials required include: micro-centrifuge tubes (0.5 and 1.5 ml); auto-pipette (0.5–10 and 1–200 μl); sterile autopipette tips; restriction enzymes and respective 10X assay buffers.

Pipette the following into a sterile micro-centrifuge tube: X μl DNA (10 to 15 μg DNA); 10 μl 10X assay buffer; 90X μl sterile water, making a total of 100 μl. Add restriction endonuclease (2–5 units/μg of DNA) and incubate at 37°C for 6 h. Use 5 μl of the reaction mixture to check for digestion on agarose gel electrophoresis.

DNA gel electrophoresis

Materials required include: electrophoresis buffer –50× TAE; ethidium bro-mide solution-10 mg/ml; electrophoresis grade agarose; 6x loading dye; DNA molecular weight marker; heating block or water bath and horizontal gel electrophoresis apparatus.

50 × TAE-Buffer stock contains 242 g Tris-base; 57.5 ml glacial acetic acid; 10 ml 0.5 M EDTA (pH 8.0); volume made up to 100 ml.

10 mg/ml ethidium bromide stock = 10 mg ethidium bromide powder in 1 ml sterile water and stored in 4°C.

6 × gel loading dye: 0.25% bromophenol blue; 30% glycerol; stored at 4°C.

Preparation of the gel
With an adequate volume of 1× electrophoresis buffer (from 50× TAE), fill the electrophoresis tank. Add electrophoresis grade agarose to electrophore-sis buffer. Melt the agarose in a micro-wave oven or on hot-plate and swirl gently to ensure mixing. Gel concentration typically varies from 0.8 to 1.5% agarose. Cool the solution to 60°C and if desired, add ethidium bromide to a final concentration of 0.5 μl/ml and mix thoroughly (Gels that are run in the absence of ethidium bromide can be stained by placing the gel in a dilute solution of ethidium bromide 0.5 μg/ml in water) with gentle agitation for 10 to 30 min. Seal the gel casting platform. Pour the melted agarose and insert the gel comb, making sure that no air bubbles are trapped.

Loading and running the gel

After the gel has set, remove the tape from the ends of the gel platform and withdraw the gel comb. Place the gel casting platform containing the solidified agarose gel in the electrophoresis tank. Add sufficient electrophoresis buffer to cover the gel to a depth of 1 mm. Make sure no air pockets are trapped within the wells. Mix the DNA samples with appropriate volumes of 6X loading dye. Close the lid of the electrophoresis tank and connect the leads in such a way so that the DNA will migrate into the gel towards the anode. Monitor the progress of the separation by the migration of the dye. Turn off the power supply when bromo phenol blue dye has migrated sufficiently to the bottom. Observe on a UV transilluminator.

Photograph the gel and then process for the Southern transfer and hybridization.

Southern transfer and hybridization

Setting of gel for Southern transfer

Digest the desired amount of DNA with the appropriate restriction enzyme. Load the DNA onto the gel. Run the gel and then process for the transfer.

Materials required include: 0.2 N HCl; Denaturing solution (1.5 M NaCl, 0.5 M NaOH); Neutralizing solution (1 M Tris-HCl pH 7.4, 1.5 M NaCl); 20× SSC.

After electrophoresis, stain and photograph the gel. Trim away the unused area and cut off the bottom left-hand corner of the gel to ensure orientation. Transfer to a glass baking dish. Place the gel in 0.2 N HCl for 10 min. Rinse with deionized water twice. Denature the DNA by soaking the gel in several volumes of 1.5 M NaCl, 0.5 N NaOH with constant gentle agitation for 45 min by changing the solution once. Rinse the gel briefly in de-ionized water, and then neutralise by soaking in several volumes of 1 M Tris (pH 7.4), 1.5 M NaCl at room temperature for 45 min by changing the buffer once with constant gentle agitation.

While the gel is in the neutralization solution, wrap a piece of Whatman 3 MM paper around a piece of plexiglass to form a support that is longer and wider than the gel. Place the support inside a large baking dish, fill with transfer buffer (10× SSC) until the level of the liquid reaches almost to the top of the support. Smooth out all air bubbles with a glass rod.

Using a clean blade or scalpel, cut a piece of nitrocellulose or nylon membrane filter about 1 mm larger than the gel in both dimensions. Use gloves and blunt ended forceps. Float the nitrocellulose or nylon filter on the surface of deionized water until it wets completely from beneath and then immerse it in transfer buffer for 5 min. Cut a corner from the filter to match the corner cut from the gel.

Remove the gel from the neutralization solution and invert it. Place it on the support so that it is centered on the wet 3 MM paper without air bubbles. Surround, but do not cover, the gel with Saran wrap, as a barrier to prevent liquid from flowing directly from the reservoir to paper towels placed on top of the gel.

Place the wet membrane filter on top of the gel so that the cut corners are aligned without air bubbles between the filter and the gel. Wet 2 pieces of 3 MM paper (same size as the gel) in 2× SSC and place on top of the wet nitrocellulose filter. Cut a stack of paper towels (5–8 cm high) just smaller than the 3 MM papers. Place the towel on the 3 MM papers. Put a glass plate on top of the stack and weigh down with a 500 g weight. Allow the transfer of DNA to proceed for 8–24 h. Replace the paper towels as they become wet.

Remove the paper towels and 3 MM papers above the gel. Mark the positions of the gel slots on the filter. Peel the gel from the membrane filter. Discard the filter after checking the transfer. Soak the membrane filter in 6×SSC for 5 min at room temperature.

Remove the membrane filter from 6× SSC and allow the excess fluid to drain. Fix the DNA to the membrane filter by UV cross linking. Hybridise the DNA immobilized on the membrane filter to a ^{32}P-labelled probe (for preparation of probe, see next pages).

Pre-hybridization and hybridization
Materials required include:

Pre-hybridization solution containing: 50% formamide (stock 100% formamide); 6× SSPE (from 20× SSPE); 5× Denhardt's reagent (from 50 × stock); 0.5% SDS (from 10% SDS stock); Carrier DNA (100 µg/ml from stock 10 mg/ml)

Hybridization solution containing pre-hybridization solution and labelled probe.

Preparation of Denhardt's reagent (50×): Ficoll (400)—5 g; Polyvinyl-pyrolidone—5 g; BSA (Pentax-V fraction)—5 g; dissolved in water to make up a volume of 500 ml.

Preparation of Salmon sperm carrier DNA: Weigh DNA (10 mg) and add 1 ml sterile water. Allow it to stand at 37°C for a period of 3 h. Using an 18 gauge needle, syringe the DNA solution in and out, till it becomes uniformly viscous. Place at 95°C water bath for 15 min. Instantly cool on ice and aliquot and store at −20°C.

Add all components of the pre-hybridization solution into a hybridization tube, except Salmon sperm carrier DNA. Place the membrane filter, following UV cross-linking, in the bottle containing the pre-hybridization solution. Wet the filter with the solution so that there are no air bubbles trapped.

Heat the carrier DNA stock in a boiling water bath for 15 min and instantly cool on ice. Add the carrier DNA to the tube containing the filter.

Mix the solution in the tube by rotating it and fix the hybridization bottle in the hybridization incubator.

Rotate at the requisite temperature for 4–6 h. Then heat the probe (labelled DNA) in a boiling water bath and cool on the ice for 5 min. Add the probe into the pre-hybridization solution in the tube and continue hybridization at the requisite temperature.

Washing the hybridized filter and autoradiography

After the requisite period of time, wear gloves and remove the filter from the hybridization tube with a pair of forceps. Submerge immediately into a tray containing $2\times$ SSC and 1% SDS at room temperature. After 5 min, transfer the filter to a fresh tray containing the same fresh mixture and gently agitate for 5 min. Repeat this step once more.

Remove the filter and then subsequently submerge it in $1\times$ SSC and 1% SDS and gently agitate it for 30 min at room temperature.

Wash the filter further as follows *depending on the stringency required*.

High stringency washes for 95% to 100% homology between probe and target sequences: $0.1\times$ SSC, 0.1% SDS at 55°C to 65°C, 3 times for 15 min each.

Medium stringency washes for 95% homology between probe and target sequences: $0.1\times$ SSC and 0.1% SDS at 40–50°C, 3 times for 15 min each.

Low stringency washes for 75% to 85% homology between probe and target sequences: $0.1\times$ SSC and 0.1% SDS at room temp. to 37°C, 3 times for 15 min each.

Place the wet filter between the folds of the Saran wrap and squeeze out the extra fluid. Apply adhesive labels marked with radioactive ink to several asymmetric locations on the Saran wrap. These markers serve to align the autoradiograph with the filter. Cover the labels with smooth tape. Expose the filter to X-ray film (Kodak X-OMAT or equivalent) in a X-ray cassette with intensifying screen. Place the X-ray cassette in −70°C.

For developing the exposed film

Remove the X-ray cassette from −70°C and allow it to warm at room temperature for 5 min. Open the cassette in the dark room, place it in the developer and gently agitate for 5 to 10 min. Remove the film from the developer, rinse with water and place in the fixer. Gently agitate the tray containing the fixer for 10 min. Remove the film and place it in running water for 10–15 min and air dry.

For removal of the radiolabelled probe for reprobing

Efforts should be made to keep the filters wet during the complete process, even during the period of exposure.

For removing probes from nitrocellulose filters, heat $0.05\times$ SSC, 0.01 M EDTA (pH 8.0) to boiling. Remove the fluid from heat and add SDS to a final concentration of 0.1%. Immerse the filter in the hot elution buffer for 15 min. Repeat the step with a fresh batch of boiling elution buffer. Rinse the filter briefly in 0.01× SSC at room temperature. Remove most of the liquid from the filter and place it on a Saran wrap and cover by squeezing. Removal of the probe from the filter can be monitered by exposing it to X-ray film. On developing the X-ray film no signal should be seen.

For removal of probe from nylon membrane, immerse the membrane in 1 mM Tris-HCl (pH 8.0), 1 mM EDTA (pH 8.0) and 0.1×Denhardt's solution for 2 h at 25°C. Repeat the step. Rinse the membrane briefly with 0.1× SSPE at room temperature. Cover with the Saran wrap by squeezing out the extra fluid.

Radioactive labelling methods

Nick translation labelling of dsDNA

One strand of a double-stranded DNA molecule is nicked with *DNase* 1, generating free 3'-hydroxyl ends within the unlabelled DNA. *E. coli* DNA polymerase I removes nucleotides from the 5' side of the nick and simultaneously adds new nucleotides to the 3'-hydroxyl terminus of the nick. During the incorporation of new nucleotides one of four deoxyribonucleotides is radioactively labelled and is incorporated into the new strand by a base complementary to the template. In this way, a high specific activity (10^8 cpm/μg) of labelled DNA can be obtained using ^{32}PdATP or ^{32}PdCTP.

Reagents needed include

10× Nick-Translation Buffer: 500 mM Tris-HCl, pH 7.5; 100 mM MgSO$_4$; 1 mM DTT; 500 μg/ml BSA (Fraction V, Sigma); aliquot and store at −20°C until use.

Unlabelled dNTP Stock Solutions: 1.5 mM each dNTP.

Radioactively labelled dNTP [α-^{32}P] dATP or [α-^{32}P] dCTP (3000 Ci/mmol).

Pancreatic DNase I Solution: DNase I (1 mg/ml) in a solution containing 0.15 M NaCl and 50% glycerol. Aliquot and store at −20°C.

E. coli DNA Polymerase I Solution.

Stop Solution: 0.2 M EDTA, pH 8.0.
1 X TEN Buffer: 10 mM Tris-HCl, pH 8.0; 1 mM EDTA, pH 8.0; 100 mM NaCl.

Sephadex G-50 or Bio-Gel P-60 Powder.

Set up a reaction on ice with 10× Nick translation buffer (5 μl); DNA sample (0.4 to 1 μg in <2 μl); mixture of three unlabelled dNTPs (10 μl); [α-^{32}P] dATP or [α-^{32}P] dCTP (>3000 Ci/mmol; 4 to 7 μl); diluted DNase I (10 ng/ml) (5 μl); *E. coli* DNA polymerase I (2.5 to 5 units) in distilled water to a final volume of 50 µl. The amount of sample DNA can be as low as 2.5 ng in a reaction of 5 to 20 µl.

Incubate the reaction for 1 h at 15°C. Stop the reaction by adding 5 µl of 0.2 M EDTA (pH 8.0) solution. Place on ice. Determine the percentage of [α-^{32}P] dCTP incorporated into the DNA with either of the following methods:

DE-81 filter-binding assay: Dilute 1 µl of the labelled mixture in 99 µl (1 : 100) of 0.2 M EDTA solution. Spot 3 µl of the diluted sample, in duplicate, on Whatman DE-81 circular filters. Dry under a heat lamp. Wash one of the two filters in 50 ml of 0.5 M sodium phosphate buffer (pH 6.8) for 5 min to remove unincorporated cpm. Repeat washing once. Use other filter directly for total cpm in the sample. Add scintillation fluid (about 10 ml) to each tube containing one of the filters. Count the cpm in a scintillation counter according to the instructions.

TCA precipitation: Dilute 1 µl of the labelled reaction in 99 µl (1 : 100) of 0.2 M EDTA solution. Spot 3 µl of the diluted sample on a glass-fiber filter or a nitrocellulose filter. Air dry. Add 3 µl of the same diluted sample into a tube containing 100 µl of 0.1 mg/ml carrier DNA or acetylated BSA and 20 mM EDTA. Add 1.3 ml of ice-cold 10% trichloroacetic acid (TCA) and 1% sodium pyrophosphate to the mixture. Mix well and incubate on ice for 20 to 25 min to precipitate the DNA. Filter the precipitated DNA on a glass-fiber or nitrocellulose filter under vacuum. Wash with 5 ml ice-cold 10% TCA four times under vacuum. Rinse with 5 ml acetone (for glass-fiber filters only) or 5 ml of 95% ethanol. Air dry. Transfer the filters to two cpm counting tubes and add 10 to 15 ml of scintillation fluid to each tube. Count the total cpm and incorporated cpm in a scintillation counter according to the instructions.

Calculate the specific activity of the probe.

$$\text{Theoretical yield (ng) of probe} = \frac{\text{dNTP added} \times 4 \times 330\,\text{ng/nmol}}{\text{specific activity of the labelled dNTP (µCi/nmol)}}$$

$$\text{Percent incorporation} = \frac{\text{cpm incorporated}}{\text{total cpm}} \times 100$$

DNA Synthesized (ng) = percent incorporation $\times$ 0.01 $\times$ theoretical yield.

$$\text{Specific activity (cpm/µg) of probe} = \frac{\text{total cpm incorporated}}{(\text{DNA synthesized} + \text{input DNA})\,(\text{ng}) \times 0.001\,\text{µg/ng}} \times 100$$

Purify the probe by removing unincorporated isotope. Chromatography on Sephadex G-50 spin columns. Sephadex G-50 or Bio-Gel P-60 spin column is very effective for separating labelled DNA from unincorporated radioactive precursor such as [α-^{32}P] dCTP or [α-^{32}P] dATP and oligomers that are retained in the column.

Resuspend 2 to 4 g Sephadex G-50 or Bio-Gel P-60 in 50 to 100 ml of TEN buffer and allow to equilibrate for at least 1 h. Store at 4°C until use.

Insert a small amount of sterile glass wool in the bottom of a 1 ml disposable syringe using the barrel of the syringe to tamp the glass wool in place. Fill the syringe completely with the Sephadex G-50 or Bio-Gel P-60 suspension.

Insert the syringe containing the suspension into a 15-ml disposable plastic tube and place the tube in a swinging-bucket rotor in a bench-top centrifuge. Centrifuge at 1600× g for 4 min at room temperature.

Repeat adding the suspended resin to the syringe and centrifuging at 1600× g for 4 min until the packaged volume reaches 0.9 ml in the syringe and remains unchanged after centrifugation.

Add 100 μl of 1 × TEN buffer to the top of the column and recentrifuge as above. Repeat this step two to three times. The volume of the column should remain unchanged.

Transfer the spin column to a fresh 15-ml disposable tube. Add the labelled DNA sample onto the top of the resin dropwise using a pipette.

Centrifuge at 1600× g for 4 min at room temperature. Remove and discard the column containing unincorporated radioactive label in a radioactive waste container. Carefully transfer the effluent (about 0.1 ml) to a fresh microcentrifuge tube. Cap and store at −20°C until use.

Labelling of DNA by Random Primer Method

Materials required include: purified DNA fragment (>100 ng); 5 μl of 3000 Ci/mmol α-^{32}P dATP or dCTP (50 μCi); sterile micro-centrifuge tubes; Random primer labelling kit (NEB or USB).

Pipette 25 ng of template DNA into a sterile micro-centrifuge tube. Make up the volume to 34 μl. Denature in boiling water bath for 5 min. Quickly place this tube on ice for 5 min. Centrifuge briefly in cold.

Add the following reagents to the DNA in the order listed: 5 μl 10× labelling buffer (includes random octa or hexa deoxyribonucleotides); 6 μl dNTP mixture (2 μl each of dATP, dTTP, and dGTP); 5 μl a^{32}P dCTP (3000 Ci/mmol, 50 μci); 1 μl Klenow fragment (5 units).

Incubate at 37°C for 2 h. Terminate the reaction by adding 5 μl of 0.2 M EDTA.

PREPARATION OF AMPLIFIED DNA BY PCR

The technique, involving polymerase chain reaction, is based on the principle of *in vitro* amplification of selected DNA sequences through simultaneous primer extension of complementary DNA strands. The two primers are short 22–24 oligonucleotides, which match the 5' end of the complementary DNA strands. The amplification is carried out by a thermostable DNA polymerase in presence of four DNA nucleotides. The enzyme, Taq polymerase, is isolated from *Thermus aquaticus*—a thermostable bacterium. The primer acts as template for the amplification of target sequence. Following termination of the reaction, which may yield millions of copies of DNA within a few hours, the fragments can be visualized by staining the DNA after gel electrophoresis.

The reaction thus consists of three principal steps, namely, denaturation at high temperature, annealing at lower temperature and finally elongation at a temperature (approximately 72°C) optimal for the action of thermostable polymerase.

Several factors control the specificity and strength of polymerase chain reaction. These are: the temperature profile of thermal cycles including transition time between different steps; activity and amount of Taq polymerase; concentration of primers, mostly between 0.1 to 0.2 µ; template DNA, Mg and the chemicals like dimethyl sulphoxide or Tween or gelatin which make the reaction more specific. In general (Innis *et al.*, 1990), the melting temperature (Tm) of prime and annealing site can be calculated as 2°C for each AT and 4°C for each GC pair respectively. However, this calculation is based on oligonucleotide hybridization in presence of NaCl and as such may not hold good for PCR reactions (Weising *et al.*, 1995). The annealing temperature is normally set 5°C below Tm. For arbitrary primers, the temperature of 36°C is set annealing temperature. For primers with high GC content, the annealing temperature is quite high.

The entire process needs two synthetic oligomers—the primers, taq polymerase and four nucleotide phosphates functioning on template DNA. On heat denaturation of the template followed by cooling, the primers anneal to the target ends, one in each strand at two different ends of the single stranded templates. Because of primer bonding to complementary strands, the polymerases copy the target sequences. After duplication, the two strands are separated by heating and fresh strands are initiated following cooling and primer annealing. As the 3' ends point towards each other, successive melting and cooling result in exponential increase of strands bound by primers. It is noted that 20 cycles result in amplification of 10^6 copies of the segment.

The wide use of PCR is due chiefly to its capacity to accept different types of primers—ranging from simple, sequence-specific, minisatellite to arbitrary sequences. The oligonucleotides with simple, sequence-repeats or core sequence of minisatellite repeats, are widely used. The core sequence of wild type phage M13 (GAGGG TGG × GG × TCT) and simple sequence such as (CA)$_8$, (CAC)$_5$, (CT)$_8$, (GTG)$_5$, (GACA)$_4$, and (GATA)$_4$ have been utilized in bringing about genotypic distinction in pathogenic fungi (Weising *et al.*, 1995). A large number of variable nucleotide tandem repeats (VNTR) has been utilized in working out genetic polymorphism. Minisatellite and simple sequence primers have come out as a powerful tool in PCR technique for molecular identification of plant taxa at various levels.

Schedule

Materials required

Template DNA (10 ng) for cloned DNA;

Primers (10 µM); (Reverse and forward).

dNTP mix (2 mM stock): dATP dCTP dTTP dGTP (all in equal ratio).

PCR Buffer (10×): Tris-HCl (pH 8.3) 100 mM; KCl 500 mM; MgCl$_2$ 20 mM. MgCl$_2$ concentration varies from 1.5 to 2.5 mM final conc.

Ethanol; 3 M sodium acetate, pH 5.2.

In 0.5 ml eppendorf tube add the following and mix by brief vortexing :
DNA (10 µl); 10×PCR buffer (10 µl); primer (forward) (10 µl, 1 µM final); primer (reverse) (10 µl, 1 µM final); 2 mM dNTP mix (10 µl, 200 µM final); Sterile MQ water 54.5 µl; Taq polymerase (0.5 µl, 2.5 U). Make upto 100 µl. Spin briefly in a microfuge and add 100 µl of sterile mineral oil on top.

Put in a programmable thermocycler using following programme: Initial Denaturation 94°C 5 min, 1 cycle; Denaturation 94°C 2 min, 30 cycle; Annealing 55°C 1 min, 30 cycle; Elongation 72°C 2 min, 30 cycle and final elongation step 72°C 10 min, 1 cycle. After the amplification is complete, remove the eppendorf carefully because the block of machine is hot and put on ice.

Remove mineral oil from top with the help of a pipette. Add equal vol. of phenol (TE saturated, pH 8.0), vortex briefly. Spin in a microfuge for 5 min.

Take upper layer and add equal vol. of chloroform: isoamyl alcohol. Repeat the two earlier steps. Precipitate the upper layer by adding 0.1 vol. of 3 M sodium acetate pH 5.2 and 2 vol. ethanol. Keep at −70° for 30 min. Centrifuge in microfuge for 15 min.

Drain off the supernatant and add 500 µl of 70% ethanol. Vortex briefly and centrifuge for 5 min at room temp. Drain off the supernatant and dry the pellet in vaccuum. Dissolve the pellet in 50 µl sterile water.

Check the DNA amplification on 1% to 1.2% agarose.

TYPES OF PROBES

Microsatellite markers

These molecular markers, otherwise known as simple sequences, are short stretches of DNA containing varying numbers of 2–6 base pair repeat elements such as ATATAT (2×4). Several investigated genomes contain microsatellites showing a high level of polymorphism exceeding even 20 alleles, which differ in the number of repeat elements.

Microsatellites can be analysed through PCR, utilizing specific primers attaching to flanking sequences of microsatellite, followed by gel sequencing. The high polymorphism of microsatellites is due to their high mutation rate.

Microsatellites have been used for DNA Finger Printing as well. One of the limitations of the use of microsatellite markers is the necessity of specific primer sets for each microsatellite. However for a group of allied species, the same marker set may be used. Primers for several species have already been published.

Variable number of tandem repeats, in short VNTRs, are also polyallelic markers derived from locus-specific probes. They have very large minisatellite arrays as ribosomal DNA of nucleolar organizing region.

Random amplified polymorphic markers or RAPDs can help in selective amplification of specific DNA segments. These probes are oligonucleotides synthesized at random, serving as primers at the two ends of specific DNA segments. With the help of such arbitrary primers, DNA segments can be PCR amplified through random selection. The primers are short oligonucleotides often a 10-mer, of random sequence. After amplification in PCR, the analysis is carried out through gel sequencing. It is a very fast process as compared to the use of other probes and has provided enormous data in several crops including soybean, blueberry and others, where it has been utilized in gene mapping as well (Rowland and Levi, 1994).

Amplified fragment length polymorphisms: in short AFLPs are based on selective amplification of a set of restriction fragments derived from a total genome digest. It involves restriction enzyme digestion of DNA, ligation of oligonucleotides and selective amplification of a subset of fragments through PCR, and finally gel electrophoresis of amplified fragments or analysis. Majority of AFLPs are dominant markers of genome and as such it is difficult to detect heterozygotes through AFLP application. The technique requires only a small amount of nucleic acid and the entire process is PCR based. Moreover, no prerequisite information is needed for AFLP markers, as necessary for microsatellites. The technique is useful not only for working out affinities but also introgressive hybridization.

Random Amplified Polymorphic DNA (RAPD)

It is based on *in vitro* amplification of randomly selected oligonucleotide sequences. Amplification takes place by simultaneous primer extension of complementary strands of DNA. In tea *in vitro* assay, the primer acts as a template to the plant DNA, for amplification of its homologue. Primers with 10 nucleotides and at least 50% GC content are used. Amplification is followed by gel separation and staining with ethidium bromide for detection. With arbitrary primers too, very little DNA is needed for polymerase chain reaction.

RAPD fragments as probes may represent single copy, mid repetitive or highly repetitive DNA sequences. In case of highly repetitive sequences, there are multiple bands or smears which are absent in single or low copy sequences. Normally, single copy RAPD probes can be used in restriction fragment length polymorphisms and be utilized as codominant markers. RAPD probes have proved to be very useful in the study of biodiversity, hybridization, gene mapping and genetic map construction.

CHAPTER III.2

GENE CLONING AND PREPARATION OF LIBRARY

In order to identify and characterise the gene, cloning of the gene sequences becomes necessary. For mapping the genes on specific chromosomes and study of gene expression and regulation, cloned genes are essential. In order to characterize a full length genome, the genomic DNA library is prepared through cloning of fragmented DNA sequences. In order to prepare a library, the size of the DNA fragments selected for cloning is of prime importance. The larger the DNA fragments, the smaller will be the number of clones in the library. A library which is mainly a collection of DNA fragments may be prepared out of genomic DNA or complementary DNA sequences. The genomic libraries are obtained through restriction digests of genomic DNA, including non-protein coding inserts as well. The complementary DNA library is the collecting of DNA inserts obtained from reverse transcription of messenger RNA extract.

The vectors chosen for cloning depend to a great extent on the size of the chromosomal fragments to be cloned. There are different types of vectors used for the construction of the DNA library, namely, plasmids, bacterio-phages or phages, cosmids and yeast artificial chromosomes (YAC).

Bacterial plasmids are extrachromosomal elements, capable of independent replication within bacteria, and often confer antibiotic resistance to the host. The phage-lambda, cosmids and YAC have, however, been extensively used in cloning and mapping. The vectors, in general, contain polylinkers with multiple restriction sites for insertion of foreign DNA, frequently flanked by promoters, permitting expression of cloned DNA. In addition, they contain sequences responsible for their stable propagation and some markers such as antibiotic resistance conferring selection value to the host.

CLONING VECTORS AND DIFFERENT LIBRARIES

Cloning through phage and plasmid principally differs in the mode of introduction into the host cell. Transformation and electrophoration are often used for smaller than 20 kb DNA fragments. For larger fragments, such as upto 25 kb, phage vectors are useful as they can be inserted into bacteria through the natural phage infection process. Cosmids can utilize phage infection system and can propagate inside the host cell. They can accept larger DNA fragments ranging between 20 and 50 kb. The cosmids are often tailored to remove most of the phage to accommodate larger inserts. Yeast Artificial Chromosome or YAC can clone very large segments, between 50 and 10,000 kb fragments of foreign DNA. For large genes as well as for single chromosome, YAC vectors as such are very useful. The YACs contain cloned DNA flanked by two vector arms.

The YAC vectors contain a centromere, a telomere, yeast origin of replication (ARS) to ensure replication and segregation of YAC. In addition, there is provision for isolation of short stretches of DNA at two ends of the insert which are of much use in chromosome walking with YAC libraries. Therefore, YAC vectors are tailored in a way so that each carries CEN4 (Centromere), TEL (Telomere), ARS (Autonomous Replicating Sequence) marker gene, and cloning site for selected gene. It is propagated as YAC 4 as bacterial plasmids, linearized by Eco R, and ligated to selected insert. The recombinant YAC vectors are transformed into yeast cells, such as AB1380.

Normally, genomic DNA libraries are used for mapping gene sequences and cloning marker genes linked with certain phenotypic characters. Once a marker sequence near a desired locus is identified through linkage analysis and cloned, overlapping and distinct genomic clones are then isolated to extend further towards the focus of interest. Repeated screening results in contiguous clones spanning a distance from the original marker locus to the desired target gene. This process is otherwise known as "Chromosome Walking".

Complementary DNA libraries are used for isolating the desired gene. The reverse transcription, through reverse transcriptase, can be primed through a random assortment of hexa primers or oligo dT primers to bind to polyA tail. The randomly primed cDNA library does not have inserts at defined positions but oligo-dT primed libraries have inserts starting at 3' end of every message.

The method of cloning gene utilizing *Yeast Artificial Chromosome* is outlined later.

TRANSFORMATION OF *E. COLI* AND ISOLATION OF PLASMID DNA

The bacteria, treated with ice cold solutions of calcium chloride and then briefly heated, can be used to transfect bacteriophage lambda DNA, and to transform plasmid DNA (Sambrook *et al.*, 1989).

$CaCl_2$ treatment induces a transient state of "competence" in the recipient bacteria enabling them to take up DNAs from a variety of sources. Addition of Dimethyl sulphoxide (DMSO) in the final step enhances the rate of transformation as DMSO acts as a carrier molecule for an introduced foreign DNA. Competent cells are usually stored in small aliquots at $-70°C$.

Following the present technique, transformation efficiencies of upto 10^7 transformed colonies per microgram of plasmid DNA can be obtained.

Reagents needed are:

LB (Luria-Bertani) broth; bacto tryptone/peptone, 10 gm; yeast extract, 5 gm; sodium chloride, 10 gm. Adjust pH to 7.0 with NaOH and make upto 1000 ml with water and autoclave.

0.1 M $CaCl_2$: Prepare 10 ml aliquots of 1 M $CaCl_2$ and store at $-20°C$. Dilute when needed. Dimethyl sulphoxide (DMSO).

Technique

Pick out a single colony from a freshly grown plate of *E. coli* cells and transfer to a 100 ml LB broth in a 500 ml conical flask. Grow the bacteria on a rotatory shaker at 200 rpm for 3 h at 37°C. Aseptically transfer the cells into sterile centrifuge tubes and cool on ice for 10 min. Recover cells by centrifuging at 4000 rpm for 10 min at 4°C. Decant the media and stand the tubes in an inverted position till media traces drain away completely. Resuspend cells in 10 ml ice-cold $CaCl_2$ and store on ice for 10 min. Recover cells at 400 rpm for 10 min at 4°C. Decant off the fluid. Resuspend pellet in 2 ml of ice-cold 0.1 M $CaCl_2$ per 50 ml of original culture.

Add 140 µl DMSO/4 ml of resuspended cells, mix gently and store on ice for 15 min. Add an additional 140 µl DMSO and quickly dispense 50 µl or 100 µl aliquots in 0.5 ml microfuge tubes and freeze in liquid nitrogen. Store competent cells at $-70°C$.

TRANSFORMATION AND RECOVERY OF PLASMID CLONES

The competent cells of *E. coli* can be used for bacterial transformation using plasmid DNA clones containing an antibiotic resistance marker. The transformed bacteria can be recovered on an LB plate supplemented with the suitable antibiotic.

The main principle behind bacterial transformation is incubation at about 200 ng of plasmid DNA in competent *E. coli* cells and a heat shock at 42° C. DMSO acts as a carrier molecule for the plasmid DNA. The bacteria, grown in LB medium, slowly recover to express the antibiotic resistance marker gene. When these are plated onto an LB plate, supplemented with the same antibiotic, only the transformed clones with the antibiotic resistance gene can be noted.

Technique

To cooled LB agar medium, add 50 µg/µl antibiotic and pour into 90 mm petriplates to solidify. Dry overnight in 37°C. Take out a frozen aliquot (100 µl) of competent *E. coli* cells and thaw on ice. Add about 200 ng of plasmid DNA and store on ice for 30 min after thorough mixing. Transfer the tube to a 42°C water bath and subject to heat shock by incubating exactly for 90 s. Rapidly transfer to ice bath and chill for 1–2 min. Add 400 µl of sterile LB medium and incubate for 45 min at 37°C (to recover and express the antibiotic resistance marker). Transfer about 100 µl of the cells onto a LB-antibiotic plate, spread evenly with a sterile L-shaped glass rod and leave in the laminar flow for about 10 min.

Invert the plates and incubate at 37°C overnight. Colonies appear in 12–16 h.

ISOLATION OF PLASMID DNA

Plasmids are extra chromosomal closed circular DNA molecules ranging in size from 1 to 200 kb. They are found in bacteria and fungi (like yeasts) behaving as accessory genetic units that replicate and are inherited independently of the host chromosomes. They confer phenotypes like antibiotic resistance, antibiotic production, degradation of complex organic compounds and production of colicins, enterotoxins, restriction and modification enzymes.

Purification of plasmid DNA extract

The extraction and purification of plasmid DNA involve three steps: growth of bacterial culture, harvesting and lysis of bacteria and purification of plasmid DNA.

In the first step a single bacterial colony is grown in LB broth with the appropriate antibiotic at 37°C. The second step involves recovery of bacterial cells at log phase and lysing them with lysozyme aided by detergents like Sodium Dodecyl Sulphate (SDS) in the presence of alkali (NaOH). This treatment disrupts base pairing and causes the linear chromosomal DNA of the host to denature leaving the strands of the closed circular plasmid DNA intact. In the final step an acetate is added to precipitate the proteins along with the linearised chromosomal DNA. The aqueous supernatant is finally treated with ethanol to selectively precipitate the plasmid DNA.

Methods followed

Large scale preparation of plasmid DNA
a. Preparation of bacterial culture: Inoculate 25 ml LB with a single bacterial colony containing appropriate antibiotics (e.g. ampicillin at a final conc. of 100 µg/ml). Incubate in a shaker at 37°C, 200 rpm for 12 h.

Inoculate LB (prewarmed to 37°C) containing the appropriate antibiotic in 4 × 250 ml flasks with 25 ml of overnight culture obtained from the earlier step. Incubate the culture at 37°C with vigorous shaking (300 cycles/min on a shaker), till the OD_{600} of the resulting culture is ~0.4 (~2.5 h). Add 2.5 ml of a solution of chloramphenicol (34 mg/ml in ethanol) for a final concentration of 170 µg/ml. Incubate the culture for a further period of 12–16 h at 37°C with vigorous shaking (300 cycles/min on a shaker).

b. *Isolation of plasmid DNA*: Harvest the bacterial cells from a 1000 ml culture by centrifugation at 4000 rpm. Keep the centrifuge bottle in an inverted position to allow all the supernatant to drain out. Resuspend the bacterial pellet in 100 ml of ice cold STE (0.1M NaCl; 10 mM Tris-HCl, pH 8.0; 1 mM EDTA, pH 8.0).

Collect the bacterial cells by centrifugation as described in the first step. Resuspend the washed bacterial pellet obtained from a litre of culture in 20 ml of solution I (50 mM glucose; 25 mM Tris-HCl, pH 8.0; 10 mM EDTA, pH 8.0). Add 2 ml of a freshly prepared solution of lysozyme (10 mg/ml in 10 mM Tris-HCl, pH 8.0). Add 40 ml of freshly prepared solution II (0.2 N NaOH, fresh 1% SDS). Close the top of the centrifuge bottle and mix the contents thoroughly by gently inverting the bottle several times. Store the bottle at room temperature for 5–10 min.

Add 30 ml of ice-cold solution III (5 mM potassium acetate, 60 ml; glacial acetic acid, 11.5 ml; distilled water, 28.5 ml). Close the centrifuge bottle and mix the contents by shaking. Store on ice for 10 min. A flocculent white precipitate forms. Centrifuge the bacterial lysate at 8000 rpm for 15 min at 4°C in a Sorvall GSA rotor. Filter the supernatant through four layers of autoclaved cheese cloth into a 250 ml centrifuge bottle. Add 0.6 vol of isopropanol, mix well, and store the bottle for 10 min at room temperature.

Recover the nucleic acids by centrifugation at 8000 rpm for 15 min at room temperature in a Sorvall GSA rotor. Decant the supernatant carefully, and invert the open bottle to drain out the last drops of supernatant. Rinse the pellet and the walls of the bottle with 70% ethanol at room temperature. Drain off ethanol, and use a pasteur pipette attached to a vacuum line to remove any beads of liquid that adhere to the walls of the bottle. Invert the bottle on a paper towel for a few min at room temperature to allow the final traces of ethanol to evaporate. The pellet can be dried in vacuum if needed.

Dissolve the pellet of nucleic acids in 3 ml TE (pH 8.0). Purify the plasmid DNA precipitation with polyethylene glycol.

CLONING OF COMPLEMENTARY DNA (cDNA)

The process of complementary DNA cloning and library has been greatly facilitated in recent years owing to the application of multiple enzyme method for simultaneous synthesis of the second strand. Full length *c*DNAs can now be synthesized and highly purified RNAs for strand synthesis are available. Several commercial firms are manufacturing different types of *c*DNA kits designed for cloning with bacteriophage cloning vectors.

In principle, the method involves the use of purified fractionated DNA for synthesis of messenger RNA. The polytailed messenger RNA is denatured in denaturation annealing buffer. In the next step, after the removal or dilution of the buffer, single stranded DNA synthesis is initiated, through reverse transcriptase and all four nucleotide phosphates in the reaction buffer. The single stranded DNA then serves as a template for synthesis of double strand through addition of components at RNase H-RT reaction in the tube without any inactivation. The newly synthesized double strand *c*DNA molecules are then ligated to lambda vector arms involving the use of oligonucleotide linkers which include restriction enzyme recognition site as well. The packaging and cloning of recombinant lambda molecules involve assembling of phage components and take advantage of the normal infection mechanism of *E. coli* host cell. *In vitro* packaging components are also available from several commercial firms. A large number of recombinant clones can be obtained with phage vectors in *E. coli* host. The next step is the screening of the library, normally done for locating low copy DNA sequences. The screening technique in principle involves bringing the phage plaque in contact with the nylon or nitrocellulose membranes. The alkali treatment of filters causes phage lysis and exposes recombinant DNA which becomes denatured and fixed on the filters without affecting viability. The filter is then hybridized with labelled probes. Film exposure reveals the hybridization signal emitted from the recombinant plaque. Through successive plating and plaque separation individual recombinant clones can be purified (Slighton *et al.*, 1993). The method of *c*DNA cloning utilizing phage vector is outlined later.

PREPARATION OF PLANT *c*-DNA LIBRARY THROUGH PHAGE VECTORS (after Glick and Thompson 1993)

*c*DNA synthesis

Denature mRNA and anneal with oligo-dT primer by adding mRNA (1–5 µg) and oligo-(dT) (100–500 ng) into a sterile 1.5 ml microfuge tube, and dry in a Savant SpeedVac. If a directional ds-*c*DNA synthesis approach is being used, add the specific oligo(dT) primer. Resuspend in 50 µl of denaturation–annealing buffer (4.0 mM sodium phosphate pH 7.2, 2.0 mM EDTA; 15% DMSO ultrapure).

Denature mRNA by incubating for 5 min at 65°C. Add 5 μl of 1 M NaCl that has been preheated to 65°C. Slowly cool to 42°C over about 30 min. Remove the annealed mRNA-oligo-(dT) complex from the denaturation–annealing buffer by phenol–chloroform–isoamyl alcohol (1:1:0.04) extraction, followed by ethanol precipitation and vacuum drying.

Initiate ss-cDNA synthesis by resuspending the mRNA-oligo(dT) complex pellet in the following: 2.5 μl, 1 M Tris-HCl (pH 8.3); 5.0 μl, 60 mM MgCl$_2$; 2.5 μl, 1 M NaCl; 5.0 μl, 5.0 mM dNTP mix (5.0 mM each dGTP, dATP, dTTP); 5.0 μl, 2.5 mM dCTP; 5.0 μl, 70 mM DTT; 5–10 μl, [α-^{32}P] dCTP (400 Ci/mmol).

Before adding enzyme, remove two 1 μl aliquots for analysis of ^{32}P incorporation. Add 250 U RNasin and 200 U RT (AMV or M-MLV). Incubate at 42°C for 45 min. Add 5 more units of RT, and incubate at 50°C for 15 min.

Stop the reaction by placing on ice for at least 5 min; then remove two 1 μl aliquots for analysis of incorporated ^{32}P label by TCA precipitation.

Dry the ss-cDNA synthesis reaction sample using a Savant SpeedVac.

Initiate RNase H-RT synthesis of ds-cDNA by resuspending the dried ss-cDNA–mRNA duplex pellet in 10 μl of 0.1 × TE buffer. Then add the following: 10 μl, 1 M Tris-HCl (pH 8.3); 10 μl, 60 mM MgCl$_2$; 10 μl, 70 mM DTT; 5 μl, 1 M NaCl; 10 μl, dNTP mix same as used earlier; 5 μl, 2.5 mM dCTP; 5 to 10 μl, [α-^{32}P] dCTP (400 Ci/mmol).

Add RNase H, 2 U/μg of ss-cDNA–RNA duplex, then add AMV (or M-MLV)-RT, 0.4 U/μl. Adjust to 100 μl by the addition of sterile water. Incubate at 12°C for 1 h; then at 42°C for an additional hour. For the final 15 min, add an additional 2.5 U of RT.

Terminate second-strand synthesis by adding EDTA and SDS to a final concentration of 20 mM and 0.1%, respectively. Extract with phenol–chloroform–isoamyl alcohol (1:1:0.04), with chloroform–isoamyl alcohol (1:0.04), and then with ethanol precipitate. Dry the pellet under vacuum. If linkers (EcoRI or others) are to be added, the ds-cDNA will need to be methylated to protect indigenous sites from digestion. However, if linkers contain the recognition site of a rare cutting restriction enzyme, the step can be avoided.

Linker ligation and ligation of cDNAs with λ vector arms
Before EcoRI linkers are added, the ds-cDNA molecules must be blunt-ended which can be done by a number of general methods, including nuclease (SI or mung bean) digestion or fill-in by *E. coli* polymerase. Prior to starting, the blunt-ended ds-cDNA should be freed of contaminating enzymes and dried under vacuum.

Resuspend the dried ds-cDNA (1 to 2 μg) in 20 μl of EcoRI methylase buffer (50 mM Tris-HCl pH 7.5; 1 mM EDTA; 5 mM DTT).

Table: Commercially available *In vitro* packing mixes

Name	Efficiency	Vendor
Amersham γ packaging kit	$1-2 \times 10^9$ PFU/μg vector	Amersham
DNA packaging kit	2×10^8 PFU/μg vector	Boehringer-Mannheim
SuperScript™	5×10^8 PFU/μg vector	Bethesda Research Labs
Packagene™	2×10^8 PFU/μg vector	Stratagene
Gigapack™ II XL	2×10^9 PFU/μg vector	Stratagene
Gigapack™ Gold	2×10^9 PFU/μg vector	Stratagene

Add 2 μl of 100 mM S-adenosyl-L-methionine and 10 U of *Eco*RI methylase per μg of ds-*c*DNA. Incubate at 37°C for 15 min. Heat-inactivate the *Eco*RI methylase by incubating at 70°C for 10 min. Remove proteins by extracting once with phenol–chloroform–isoamyl alcohol (1 : 1 : 0.04), once by chloroform–isoamyl alcohol (1 : 0.04), followed by ethanol precipitation. Dry the pellet under vacuum.

Resuspend *Eco*RI-methylated blunt-ended ds-*c*DNA pellet (0.1–2 μg) in 5 μl of double-distilled water and add the following: 0.1–2 μl *Eco*RI linkers; 0.5 μl of 10× ligase buffer and 0.5–1.0 μl of T4 ligase (10 U/ μl, Collaborative Research, Inc.). Mix, spin down solution, and incubate at 10–14°C for 24 to 48 h. Heat inactivate the T4 ligase by incubating at 70°C for 10 min, then add *Eco*RI digestion buffer, 2–4 μl of *Eco*RI (20 U/ μl) and incubate at 37°C for 3 h.

Stop the *Eco*RI digestion reaction by heating to 70°C for 10 min, or by extracting once with phenol–chloroform–isoamyl alcohol (1 : 1 : 0.04) and once with chloroform–isoamyl alcohol (1 : 0.04), followed by ethanol precipitation. Dry the pellet under vacuum, then resuspend in 20 μl of sterile TEN buffer. Separate *Eco*RI-linkered ds-*c*DNA molecules from free *Eco*RI linkers using Bio-Rad A-50 M, or equivalent, chromatography. Pool the fractions containing the largest sized ds-*c*DNA molecules, extract once with phenol–chloroform–isoamyl alcohol (1 : 1 : 0.04), once with chloroform–isoamyl alcohol (1 : 0.04), ethanol precipitate, and dry the pellet under vacuum. Resuspend the linkered ds-*c*DNA pellet in TEN buffer to obtain a final concentration of about 10 ng/ μl. Initiate the ligation of the *Eco*RI-linkered ds-*c*DNA with λ vector arms by adding 1 μg (1–2 μl) of phosphatase-treated λ vector arms to a microfuge tube, and then by adding 10–50 ng (1 to 5 μl) of *Eco*RI-linkered ds-*c*DNA. Adjust the reaction volume to between 5 and 10 μl with double-distilled water, mix, spin down, and then heat the sample to 37°C for 5 min. Add 0.5–1 μl of DNA ligase buffer and 1 ml of T4 ligase (10 U/ μl). Adjust the final volume to 10 μl by the addition of sterile, double-distilled water. Mix, spin down, and incubate at 14–16°C for 8–21 h.

Heat-inactivate the ligase by incubating at 70°C for 5 min. Proceed for *in vitro* packaging of recombinant λ-ds-*c*DNA molecules into λ phage particles or store at −20°C.

In vitro packaging of recombinant λ molecules
Take *in vitro* packaging extracts from commercial source and *in vitro* package half of the ligation mix as instructed by the vendor. This generally requires a 2–3 h incubation step for packaging of the recombinant λ molecules into the phage particles. After packaging, add 1 ml of λ dilution buffer (10 mM Tris-HCl, pH 7.4; 10 mM MgSO$_4$; 0.01% gelatin) and 20 μl of chloroform. Mix gently and titer the supernatant solution at dilutions of 10^{-3} and 10^{-4} using the appropriate *E. coli* host. If the ligation and *in vitro* packaging reactions are successful, the titer should be between 10^5 and 10^6 PFU/ml.

SCREENING OF *c*-DNA LIBRARY

Plating the cDNA library: Plate the *c*DNA library from *E. coli* host to get near confluent lysis on at least five 90 mm plates (around 25,000 plaques).

Transfer of plaques onto nylon membrane: Chill the above plates at 4°C for an hour to allow the top agar to harden. Remove from 4°C and keep in a laminar flow bench. In a clean tray put two sheets of 3MM paper and wet with denaturing solution (0.5 M NaOH, 1.5 M NaCl).

In another tray put two sheets of 3 MM paper and wet them with neutralizing solution (0.5 M Tris-HCl, pH 7.2, 3.0 M NaCl, and 1 mM EDTA). In another tray put a sheet of 3 MM paper and wet with 2× SSC. Keep nylon membranes on a 3 MM paper. With the help of a soft lead pencil number the membranes.

Remove the lid of the plates. Hold the nylon membrane with the help of two forceps or disposable gloves and place on the plate avoiding air bubbles. Puncture at three asymmetric places on the nylon membrane through bottom agar with a sterile needle. Mark the positions of the holes on the bottom of the plate with the help of a marker pen. Keep the nylon membrane for a minute.

Transfer the nylon membrane to the tray containing 3 MM paper soaked with denaturing solution with plaques side up and keep for five minutes. Transfer it to the tray containing 3 MM paper soaked with neutralizing solution and keep for five minutes.

Transfer the nylon membrane to the tray containing 3 MM paper soaked in 2× SSC. Keep the nylon membrane with 3 MM paper in UV crosslinker and crosslink the DNA to the membrane. Remove the nylon membrane, dry it and store under vacuum at room temperature or use for prehybridization.

Plaque hybridization: Place the membrane in hybridization bottle with plaques facing the lumen. Add the required volume of prehybridization solution [6× SSC, 0.02% bovine serum albumin, 0.02% Ficoll ($M_1 = 400,000$), 0.02% polyvinyl-pyrrolidone ($M_1 = 36000$) 1% SDS].

Fix the bottle in hybridization incubator and let it rotate at 37°C for at least 2 h. Denature the ^{32}P labelled DNA probe by heating at 100°C for 5 min, chill on ice for 5 min; and add to the prehybridization solution. Fix the bottle back in the incubator and let it rotate at 37°C for 12–16 h.

Washing of nylon membranes and autoradiography: After hybridization, transfer the nylon membranes to a tray containing 300–400 ml of 2× SSC and 0.1% SDS at room temperature and wash by gently agitating the tray. Check the radioactivity left on the membrane and the periphery for background level. If desired, increase the stringency of washing by reducing the salt concentration (0.5× SSC) and raising the temperature to 50°C.

After washing, wrap the membrane in saran wrap and apply radioactive ink at three asymmetric places outside the area of membranes. Place membranes wrapped in saran wrap in the X ray cassette and keep an X-ray film on it in dark or under safe red light covering the radioactive ink and close the cassette. Keep the cassette at −70°C.

Alignment of the film: Develop the film in dark. Align it with the membranes using the marks left by the radioactive ink on the saran wrap. Mark on the X-ray film the positions of puncture marks on filters. Align these puncture marks on X-ray film with the respective plate and pick the positive plaques in SM.

Purification of the clone: Plate the plaques at a lesser density (~500 plaques per plate), pick on nylon, hybridize and pick positive area. Again plate at still lesser density (~100 plaques per plate) and pick single positive plaque. Plate this single positive plaque and confirm it by hybridization.

YAC LIBRARIES

Preparation of cells or tissues for isolation and purification of high-molecular weight DNA

Construction of a YAC library requires the molecular weight of genomic DNA to be as high as possible. The purity and integrity of the isolated DNA are crucial for pulsed-field gel electrophoresis (PFGE) and cloning. To obtain intact genomic DNA, cells or protoplasts are lysed *in situ* in an agarose plug, digested with appropriate restriction enzyme and checked by PFGE or field inversion gel electrophoresis (FIGE).

Phosphate-buffered saline (PBS) contains; NaCl (8 g); KCl (0.2 g); Na_2HPO_4 (1.44 g); KH_2PO_4 (0.24 g). Dissolve well after each addition in 800 ml double distilled water. Adjust pH to 7.4 with 2 N HCl and add double distilled water to 1 l. Autoclave and store at room temperature.

Clorox® solution: 5–10% (v/v) sodium hypochlorite solution or commercial bleach in double distilled water.

CPW-Salt solution (II): KH_2PO_4 (27.2 g); KI (0.16 mg); $CuSO_4$ $5H_2O$ (0.025 mg); KNO_3 (0.101 g); $MgSO_4$ $7H_2O$ (0.246 g).

Enzyme medium: 9% (w/v) mannitol; 3 mM 2-(N-morpholino)-ethanesulphonic acid (MES)-KOH, pH 5.8; 1% (w/v) cellulase; 0.2% (w/v) macerozyme; make up in CPW-salt solution.

Washing medium; 3 mM MES-KOH, pH 5.8; 2% (w/v) KCl. Make up in CPW-salt solution. Autoclave.

Sucrose solution contains: 18% (w/v) sucrose; 3 mM MES-KOH, pH 5.8. Make up in CPW-salt solution. Autoclave.

Cell suspension buffer: 10 mM Tris-HCl, pH 7.6; 100 mM EDTA, pH 8.0; 20 mM NaCl.

For plant tissues protoplasts are preferred instead of entire cells since cell walls are relatively difficult to digest.

Remove six to eight of the youngest, fully expanded leaves from plants grown and place the leaf tissue in a petri dish. Surface-sterilize the leaves by immersing in 5–10% Clorox solution (sodium hypochlorite) for 5–10 min followed by thorough rinsing with 40 ml sterile distilled water four to five times to remove the Clorox.

Add 5–10 ml of sterile enzyme medium, mix, and incubate in the dark at room temperature (24–25°C) for 18–20 h without shaking. Add 15 ml of washing medium and gently shake to loosen the protoplasts from undigested leaf materials. Filter through a nylon mesh (50 μm pore diameter) to remove undigested materials. The protoplasts are in the filtrate.

Centrifuge the protoplasts at 1000× g for 5 min at room temperature and carefully decant the supernatant. Resuspend the protoplasts in 4 ml of washing medium and again centrifuge. Resuspend in 1 ml of washing medium, add 1 ml of 18% sucrose, which will become an underlayer in protoplast suspension and centrifuge at 120× g for 5 min at room temperature.

Carefully transfer the protoplasts from the interface using a wide-bore Pasteur pipette to a clean centrifuge tube and add 1 ml protoplast suspension buffer. Count the protoplasts using a microscope and a haemocytometer.

Centrifuge at 1000× g for 5 min at room temperature and resuspend the protoplasts at approximately 5×10^7 protoplasts/ml in cell suspension buffer.

Isolation of high-molecular weight DNA

Prepare an equal volume of 1% (w/v) low-melting temperature agarose in cell suspension buffer. Melt the agarose in a microwave and allow to cool to 42°C. Warm an equal volume of protoplast suspension (5×10^7 cells/ml) to 42°C and add to the agarose gel mixture (42°C). Mix well to ensure protoplasts are evenly dispersed throughout the agarose.

Add the melted agarose-cell/protoplast mixture to an ice-cold plug former or preformed Plexiglas molds (50–100 μl) or equivalent tubes using a 1-ml pipette. Allow the plugs to harden for 30 min on ice and carefully remove the plugs. Cut the cylindrical plugs into smaller blocks, if necessary. Place the plugs or blocks in 50 volumes of lysis buffer and incubate for 24 h at 50°C with shaking at 60 rpm. Replace the old lysis buffer with fresh lysis buffer and continue to incubate at 50°C for 24 h with shaking at 60 rpm.

Isolation of intact yeast DNA

Harvest yeast cells from a liquid culture by centrifugation at 1000× g for 5 min at 4°C and decant the supernatant. Wash the cell pellet by resuspending the cells in 5 vol double distilled water and centrifuge again. Repeat washing once. Resuspend the cells in 50 mM EDTA buffer (pH 8.0) at approximately 4×10^9 cells/ml on ice.

Prepare an equal volume of 1% (w/v) low-melting temperature agarose gel in double distilled water. Melt in a microwave and allow to cool to 42°C. Warm an equal volume of yeast cell suspension to 42°C and add to an equal volume of agarose gel mixture (42°C). Mix well and pour the mixture into a plug mold on ice. Allow to harden at 0°C for 30 min.

Carefully transfer the plugs into 10 vol of SCEM buffer containing 1 unit/ml of Zymolyase 20-T, and incubate at 37°C for 5 h. Replace SCEM buffer with 10 volumes of DLS buffer and incubate at 50°C for 3 h. Replace the old DLS buffer with fresh DLS buffer and incubate for another 3 h. Rinse the plugs four times with 4 volumes of TE buffer and store at 4°C until use.

Restriction enzyme digestion of DNA in agarose

The DNA purified in agarose plugs or blocks is almost intact and should be partially digested with a rare cutting enzyme used for YAC cloning or be completely digested with an appropriate enzyme for PFGE analysis.

Partial EcoRI restriction enzyme digestion of
genomic DNA for YAC cloning
Rinse plugs containing DNA three times in 50 volumes of IX restriction enzyme buffer lacking Mg^{2+}. Remove the buffer and add one volume of IX

restriction buffer lacking Mg^{2+} with 4 units/μg of *Eco*RI at 4°C for 1 h. Add Mg^{2+} from a stock solution of 100 mM $MgCl_2$ to the desired concentration to initiate *Eco*RI digestion. Immediately incubate at 37°C for 1 h.

Stop the reaction by removing the restriction buffer and adding 10 vol of cold TE buffer. Store at 4°C until use.

Preparation of YAC vectors for cloning

YAC4 is the most widely used yeast artificial chromosome vector that is propagated as bacterial plasmids pYAC4 in *E. coli*.

Obtain an *E. coli* culture containing pYAC4. Linearize pYAC4 with the restriction enzyme *Bam*HI in order to release a HIS3 spacer fragment between the telomeres. The *Bam*HI site is then dephosphorylated to prevent ligation between the telomeres and the HIS3 spacer (see the next protocol).

Extract the linearized and dephosphorylated pYAC4 with phenol/chloroform, chloroform : isoamyl alcohol (24 : 1), precipitate in ethanol, and dissolve the DNA in double distilled water as described for plasmid DNA isolation. Carry out restriction enzyme digestion with *Eco*RI to open the cloning site in the intron of the *SUP*4tRNA gene. Extract, precipitate, and resuspend the DNA in double distilled water. Store at −20°C until use.

Ligation of partially digested genomic DNA insert to pYAC4 vector

Briefly rinse the agarose plug containing partially digested genomic DNA twice with 20 vol of double distilled water followed by 10 vol of IX ligation buffer. Discard the buffer and add the prepared pYAC4 vector to the plug at a vector : insert DNA molar ratio of 40 : 1. The mass of vector is approximately equal to the mass of insert. Melt the plugs in a 68°C water bath for 5 min and transfer to 37°C.

Preheat 2× ligation buffer containing 4000 units/ml DNA T4 to 37°C for 2 min and add 1 vol of the buffer to the gel mixture prepared. Gently mix and allow the ligation reaction to incubate at 37°C for 2–3 h. Transfer the reaction mixture to room temperature and continue to incubate the reaction overnight.

Size fractionation of DNA by CHEF gel or other PFGE

The ligated reaction should be size-fractionated prior to transformation. In 1% (w/v) low-melting point agarose CHEF gel, apply switching conditions to retain fragment above a particular size in a compression zone. By use of switching times of 15 s on the CHEF apparatus, DNA fragments <300 kb are allowed to migrate as a function of their sizes. However, fragments

> 300 kb migrate more slowly in a compression zone without resolution. Electrophoresis is carried out using 0.5× TBE at 10°C.

Agarase hydrolysis of agarose

Agarase treatment can be directly carried out inside the agarose matrix to hydrolyse agarose and to release the recombinant YAC/DNA insert. The solution containing the ligated YAC/DNA and oligosaccharide can be directly transformed into yeast spheroplasts without purification.

Preparation of spheroplasts for transformation

Inoculate a single AB 1380 colony (yeast host) into 200 ml of YPD medium and incubate at 30°C until the culture has reached mid-log growth phase. Harvest the cells by centrifuging at 900× g for 5 min at room temperature. Discard the supernatant and resuspend the cells in 20 ml double distilled water. Centrifuge as before. Rinse. Resuspend the cells in 10 ml of 1 M sorbitol solution, count the cells under the microscope in the haemocytometer, and centrifuge as before.

Resuspend the cells in 5 ml of SCEM. Add Zymolyase 20-T (5–20 units/ 1.5×10^9 cells). Incubate at 30°C for 15 min with gentle shaking at 60 rpm. Centrifuge at 500× g and wash the pellet once in 10 ml of 1 M sorbitol solution. Centrifuge at 500× g and resuspend the cells in YPD medium containing 1 M sorbitol. Allow the cells to recover for 30 min at room temperature.

Add 5 ml of STC and centrifuge at 500 g. Wash the cells in 10 ml of STC buffer, centrifuge and resuspend the cells in 5 ml of STC buffer. Check the spheroplasts with a phase-contrast microscope. The cells at this point are stable and ready for transformation.

Transformation of spheroplasts with recombinant YAC/DNA insert

Add 0.5–1.0 volume of 2 M sorbitol solution to the size-fractionated, agarase-treated liquid mixture and aliquot into 10-ml tubes (15 µl per aliquot). Add 0.1 ml of the protoplast suspension to each of four aliquots, gently mix, and incubate at room temperature for 15 min. Add 1 ml of PEG solution to each tube and incubate at room temperature for 15 min.

Centrifuge at 500× g for 10 min at room temperature and carefully discard the supernatant. Resuspend the cell pellet in 0.15 ml of SOS per tube and incubate at 30°C for 45–60 min.

Add 3 ml of TOP lacking uracil, prewarmed to 40°C, to each tube and transfer to petri dishes containing SORB without uracil. Incubate at 30°C for 3–4 days until transformants appear.

Amplification and storage of the YAC library

The process for triplicate storage of arrayed YAC libraries used for large-scale physical mapping is as follows:

Incubate single colonies in 0.6 ml of YPD medium in Micronics racks (96×1 ml) at 30°C with agitation for 36–40 h. Incubate the cultures in SD medium for colony filters, and PCR pools. Add 0.2 ml of 80% glycerol to each culture and mix well. Transfer 0.2 ml of the cell suspension to each of three microtitre plates. One is used as the master library, the other two as working libraries. Wrap the plates with Saran Wrap and quickly store at −80°C. Cap the Micronics and store at −80°C.

For nonorganized storage of a pooled YAC library used for identifying one or a few of YAC clones:

Scrape off and pool approximately 500 colonies grown on SD medium lacking uracil and tryptophan. Incubate these cells in 500 ml of YPD medium at 30°C for 6–10 h with agitation. Remove about 7×10^8 yeast cells in duplicate to make approximately 20 µg total DNA in 2 ml for 500–1000 PCR reactions. Centrifuge the remaining culture at $900 \times$ g for 10 min at room temperature. Carefully discard the supernatant and resuspend the cell pellet in 20% (v/v) glycerol in YPD medium and store the pooled yeast in aliquots at −80°C. Incubate approximately 2500 colonies from an aliquot on SD medium lacking uracil at 30°C with agitation for 2 days. Make 5 replicates on nylon filters. Two of the colony filters are stored on 3MM Whatman paper soaked with 20% (v/v) glycerol at −80°C. Three colony filters are to be used for colony screening.

Screening of a YAC library

For PCR screening of pooled YACs for exclusion of
a large part of the library: Incubate superpools of 1920 colonies in 20 of 96-well microplates at 30°C for 36 h. Isolate DNA from yeast cells as described earlier.

Carry out PCR reactions using approximately 20–40 ng DNA from a yeast pool in a final volume of 20 µl overlayered with a drop of light mineral oil. The PCR cycles depend on the primer and the size of the PCR products. Analyze the PCR products on 1.0–1.5% (w/v) agarose gels or Seckem gels. Positive pools should have the expected PCR band as compared to the positive control and marker lanes.

Similarly screen 96 colonies included in the positive pool. Store the positive colony pool in 20% (v/v) glycerol at −80°C.

For colony screening on a nylon membrane:
Thaw the glycerol stocks from the positive pool of the PCR screening and incubate in 0.15 ml SD medium without uracil in microtitre plates at 30°C for 3 days. Inoculate a charged nylon membrane with 96 or 384 YAC clones from these cultures. Place the inoculated membranes on SD medium without uracil and tryptophan on 22.5 × 22.5 cm plates and incubate at 30°C for 3 days or until colonies are approximately 2 mm in diameter. Avoid any air bubbles under the membranes.

Individually transfer the membranes onto the 22 × 22 cm 3 MM Whatman paper plates, saturated in SCEM containing 0.5 unit/ml Zymolyase 20-T for at least 30 min. Seal the plates with parafilm™ and incubate at 30°C overnight. Incubate the membrane at room temperature on 3MM Whatman paper saturated as follows: 10% (w/v) SDS for 5 min; 0.5 N NaOH for 10 min. Transfer onto dry 3MM Whatman paper for 5 min; treat with 0.2 M Tris-HCl, pH 7.5, 2× SSC for 3 × 5 min. Air dry the membranes for 2 h or under vacuum for 1 to 2 h at 80°C. Store in aluminum foil or in prehybridization buffer until use.

Prehybridise the membranes at 65°C for 3–6 h in 20 ml per filter prehybridization solution with 7% (w/v) PEG 8000, 10% (w/v) SDS, and 100 µg/ml of sonicated denatured salmon sperm DNA. Hybridise the membrane at 65°C using fresh prehybridization buffer with at least 3×10^5 cpm/ml probe. Wash and expose the membranes as described for Southern blotting. Positive clone(s) appear as black spot(s). Store individual positive clones in 20% (v/v) glycerol at −80°C.

For verification of positive YAC clones
Streak out individual positive clones from the glycerol stock on SD medium lacking uracil and incubate at 30°C for 3 days. Inoculate individual colonies (usually 2–4) in 5 ml of YPD medium and incubate at 30°C overnight with agitation. Harvest the cells and extract the DNA as previously described.

Carry out PFGE, using 2–4 of the agarose plugs prepared, followed by Southern blot hybridization. Check the insert size of the genomic DNA in the positive YAC clones. Prepare 20% (v/v) glycerol stocks of the clones and store at −80°C. These stocks can be used for genomic and physical mapping.

CHAPTER III.3

MAPPING OF SEQUENCES—
CHROMOSOME WALKING AND JUMPING

Chromosome walking involves the localization of gene families within a chromosome domain through the use of overlapping clones. Chromosome walking technique and the mapped markers permit identification of specific loci and isolation of genes for the study of their expression. Large regions of genome can thus be cloned through the isolation of a series of overlapping recombinants. In this method, a segment of non-repetitive DNA isolated from the end of a recombinant can be used as a probe for identification of clones with adjacent sequences. Because of the slowness of the process, the use of cosmids rather than bacteriophage is preferred. In cosmids, foreign DNA of approx. 45 kb can be inserted (Sambrook *et al.*, 1989).

The principle of chromosome walking initially involves the selection of an identified gene from the library, and subcloning of its end segment. The subcloned segment is hybridized with other clones, and on the basis of hybridization of the overlapping end sequence, the adjacent clone is chosen. The end segment of the second clone is hybridized again with clones from the library, and the third adjacent sequence is chosen on the basis of hybridization of the overlapping sequences. The repetition of this process utilizing overlapping restriction sites ultimately leads to mapping all the adjacent genes along the length of the chromosome.

The primary step in cloning genes through chromosome walking is to secure probes within a few hundred kilobases of the desired locus. In order to achieve this objective, meiotic segregation of the mutants is correlated with restriction fragment length polymorphism. After identification of linked RFLPs, cloning of desired genes entails isolation of cloned DNA fragments, bridging the gap between the RFLP marker and the gene. Chromosome walking through small steps, through cosmid or bacteriophage, is time-consuming and laborious.

The later development of the technique for cloning and maintenance of long DNA fragments as inserts in *yeast artificial chromosome vector in yeast cell* has emerged as a powerful tool for chromosome walking. It enables DNA fragments of several hundred kilobases to be cloned. In view of the large size of 600 kb or even more, large genomes can be covered within a few clones. With the YAC system, it is possible to link the genetic map measured in centimorgans to physical map measured in kilobases. In *Arabidopsis*, the conversion of 1 cm for 140 bp has been used to convert the size of YAC inserts to genetic distance (Chang *et al.*, 1988). In higher plants, the difficulty of chromosome walk is due to the heavy amount of repetitive sequence present. *Arabidopsis thalliana*, having small genome and low number of interspersed repeats, is therefore, regarded as a model system.

The YAC vector maintained as circular plasmid in *E. coli* has tailored sequences responsible for centromeric, telomeric, autonomously replicated sequences, selective markers and a cloning site maintained through transformation in yeast cells. In addition to functional elements necessary for maintenance in yeast, transposable elements can be inserted in it if necessary. The generation of YAC clones needs restriction digest of these vectors to release two DNA fragments which become two chromosome arms (Schmid *et al.*, 1992). The next step is to insert high mol. wt. DNA by ligation. Ultimately it is transformed into yeast cells where it remains as a linear artificial chromosome. The YAC clones can have much larger inserts than any other cloning system. The general strategy in the construction of the overlapping YAC library is to hybridise all RFLP markers in the region to at least one of the YAC libraries. All the mapped markers can be identified with corresponding YAC clones. Walk can be initiated to link the clones hybridizing to adjacent markers. This step involves the isolation of end probes from one YAC clone and hybridizing these probes to YAC libraries to identify overlapping clones. The end probes can be generated by inverse PCR and left end rescue.

Yeast artificial chromosome technology permits an increase in genomic cloning size upto 10-fold over cosmids (bellane). Collections of overlapping genomic fragments called "Contigs" can help fingerprinting and cover all parts of the genome. The YAC library has been constructed of *Arabidopsis thalliana* genome to isolate genes by chromosome walking. The hybridization probes could be produced from the ends of YAC insert (Grill and Somerville, 1991). In *Arabidopsis*, the complete library of more than 21,000 YAC with an average insert size of 150 kb contains the entire genome.

In order to test the specificity of transcription, inverse PCR is used to amplify vector-insert junction, using the amplified sequence as the template for RNA polymerase. The non-characterized sequences adjoining the vector sequence can also thus be amplified.

Almost 380 RFLP probes have been used (Hwang *et al.*, 1991; Nam *et al.*, 1989) for correlation of the physical map with the classical genetic map in *Arabidopsis*. In maize, nearly 79,000 clones of an average insert size of 145 kb have been prepared with YAC vector (Edwards *et al.*, 1992). *In vitro* transcripts from phage promoters were successfully employed to secure labelled probes for chromosome walking. RNA probes obtained through *in vitro* transcription of recombinant cosmids carrying promoters of bacteriophage encoded DNA dependent RNA polymerase have also been adopted.

In higher organisms, specially in plants, because of the heavy amount of repetitive sequences present in borders permitting crossover, it is rather difficult to map the genes following chromosome walking. In order to eliminate this difficulty in walking, the "chromosome jumping" approach is adopted so that the genes are brought close to each other.

The method of chromosome jumping, involves selection of the length of the hop depending on the distance to be covered; circularization by ligation of two ends of linear DNA; digestion through restriction enzymes and cloning of small DNA fragments representing junctions, the clones serving as the jumping library. Through successive chromosome jumping, it will be possible to bring the gene close to the marker. This procedure permits isolation, cloning and characterization of genes of unknown function as well.

JUMPING LIBRARY-DOUBLE DIGESTION METHOD

Dilute the DNA to the desired concentration in Tris-EDTA buffer (50 mM Tris, pH 7.4, 1 mM EDTA). Add 100 to 500 fold excess of purified BamHI-cut supF DNA (which has been pre-tested for ligation efficiency). Bring the magnesium concentration to 10 mM by adding $MgCl_2$ solution. Equilibrate the mixture at room temperature for 10 min. Add 1 to 2 units of T4 DNA ligase and incubate at 14°C for 12 h. Add 1 unit of ligase again to the mixture and incubate at 14°C for 12 h. Add 2.5 M potassium acetate and bring the final concentration to 0.6 M.

Add 20 μg of yeast tRNA and then recover the circular DNA by ethanol precipitation at −20°C for 2 h. Spin the DNA pellet in a swinging ultracentrifuge rotor at 23,000 rpm for 30 min. Solubilise the pellet, after washing it with 70% ethanol and drying it, in 100 μl of tris EDTA buffer. Block further ligation by treating the DNA mixture with alkaline phosphatase (which will dephosphorylate the cut ends) or by Klenow enzyme (which will fill the staggered ends).

Extract it with buffer-saturated phenol followed by chloroform–isoamyl (24:1) with shaking. Add ammonium or potassium acetate to a final concentration of 0.6 M. Reprecipitate the DNA with twice the volume of ethanol and keep the mixture on ice for 2 to 12 h.

Pellet the DNA by centrifugation. Resuspend the DNA pellet in 100 μl of TE. Take an aliquot and test the ligation by agarose gel electrophoresis. Digest the circular DNA with EcoRI in RE buffer.

Re-extract the digested DNA with phenol, chloroform–isoamyl (24:1) and then precipitate in 2 volume of ethanol after adding potassium acetate. Resuspend the DNA pellet in TE in as small volume as possible. Digest the charon 3A (λ ch3AΔ lac), where AvaI site has been altered (lambda) vector with EcoRI. Remove the enzyme with phenol extraction. Precipitate the vector DNA with ethanol at 4°C for 2 h. Collect the DNA pellet by centrifugation. Check the packaging efficiency of the vector (both uncut and cut and re-ligated) by plating them on *E. coli supF* strain MC1061.

Dissolve the vector DNA pellet in small volume of TE. Ligate the vector to the genomic insert, using a 4:1 molar excess of vector in order to cut down on the number of clones which have multiple inserts.

Take 0.25 µg of this ligated material to infect the *E. coli* to get an efficiency of 3 to 4×10^8 p.f.u/µg DNA on supF strain MC1061. If the efficiency is low, a three fold increase can be made by adding 1 volume of crude terminase preparation (prepared from strain AZ1069, a temperature induced plasmid-bearing strain) to 3 volume of ligation mixture. Terminase is a phage protein which cleaves ligated DNA at the *cos* site after packaging into the phage head.

Test package by plating the transferred phage on a *supF*+ strain (LE392) and on a *supF*-strain (MC1061). Only the junction fragment bearing supF will be able to form plaques on supF strain. Scale up the packaging reaction, and the library can then be plated on a 150 mm plate at a density of 4×10^4 p.f.u. per plate.

Make a nylon blot of the plaque. Hybridise the blot with radioactive supF DNA. Wash the membrane at low stringency and expose the membrane on a Kodak X-ray film. Pick up the positive clones, number them and grow them, separately in 1 ml each of MC1061 (1 OD at 600 nm) cell culture, with vigorous shaking at 37°C until the turbidity of cell culture disappears. Isolate the phage (chimeric one) DNA by standard procedure and store it at 4°C until use. For long-term preservation keep it in ethanol (as DNA precipitate) at −20°C. After *Eco*RI digestion, genome inserts can be isolated.

Note: Once clones are plaque purified, minilysate DNA can be prepared by plating the DNA on LE 392 rather than MC1061, and then washing the plaques with SM buffer.

For routine analysis, subclone the *Eco*RI insert in pBR[322] plasmid-cut by *Eco*RI and phosphorylated by alkaline phosphatase. The resultant ligated DNA (by ligase) is transformed into a *E. coli* lacz amber mutant (CARD-15). Pick up the purple colonies (transformed colonies) after growing the cells on an agar plate.

With the cloning strategy as outlined above, the fragments at the end of the jump are of length 300–5000 bp, at *Eco*RI sites in the genome. If one uses this jumping fragment as a probe to go back in the same jumping library, it is quite likely that the clones derived will travel back in the direction of the probe, rather than proceeding forward along the chromosome. There are two solutions to this problem. The first is to use a complementary library which uses a different restriction enzyme such as Hind III. By alternating between a *Eco*RI/*Mbo*I and *Hind*III/*Mbo*I libraries, it is more likely that one can continue to move in the same direction. Alternatively, it is reasonable to use the jump clone as a probe to obtain a larger genomic sequence from a standard genomic library, made with a rare-cutter enzyme such as *Not*I or with a six-cutter enzyme. This will provide a large number of probes to screen the jumping library in a sequential manner.

JUMPING LIBRARY WITH RARE ENZYME DIGESTION METHOD

The steps in the library construction are similar to mentioned above except that a rare-cutter restriction enzyme such as *Not*I will be used. The prominent differences are:

1 µg of genomic DNA is sufficient to construct the library. Carry out the digestion with excess (10 to 20 units/µg) of enzyme. Check the digestion with pulse field gel electrophoresis.

Cut the suppressor tRNA gene (supF) with *Not*I and dephosphorylate it by alkaline phosphatase. Add 400 molar excess of cut supF to >100 kb genomic DNA fragment. At 0.1 µg/ml a 2000 kb fragment at this ratio has 90% chance of circularization.

After precipitation, re-cut the genomic circles with *Bam*HI and clone the fragment into λCh3A Δ lac as described before. Test the jumping clone on a pulse-field blot. Subclone the separate halves of the clone (after digesting it with the enzyme) in pBR^{322} or bluescript vector.

LINKING LIBRARIES

Linking clones permit crossing over a rare restriction site and then proceed to the next jump. Linking libraries are easier to construct and would contain very large DNA fragments as genomic inserts.

Cut the genomic DNA by partial digestion with *Mbo*I in the reaction buffer supplied by the company. Run the fragments in 0.8% agarose by standard electrophoresis. Elute the 15 kb to 20 kb fragments in TE buffer and purify with phenol extraction. Dilute the molecules to a concentration of 1 µg/ml with deionised water. Add 200-fold molar excess of purified supF gene with *Bam*HI ends—in the reaction buffer described above. Ligation is done as mentioned before. Extract the circular DNAs with buffer-saturated phenol and precipitate with ethanol. Dissolve the DNA pellet in a small volume of TE and then digest with excess of *Not*I (10–20 units/µg). Phenol-extract the digested DNA and reprecipitate. Ligate the *Not*I cut DNA molecules into the appropriate λ arm A 10:1 ratio of excess of vector arms to clonable inserts. Package the DNA as described earlier.

Plate on a *sup*F-host such as MC1061, and store the plaques in SM buffer if necessary. Otherwise, isolate the phage DNAs and subclone them in a bluescript vector at alternative enzyme sites, such as XhoI and *Sfi*I, present within the supF gene internal to *Bam*HI. *Xho*I and *Sfi*I behave as rare-cutting enzymes.

FINGERPRINTING OF CLONES FOR CHROMOSOME WALKING

The essential feature of the fingerprinting reaction is that it should produce, on average, a suitable number of labelled fragments, separable on a high resolution polyacrylamide gel (40–2000 bp), and then be matched pairwise. The final result is generally assisted by a software programme such as CONTIG9 and with a scanner, at AUT mode (Sulston *et al.*, 1988).

Double digestion method

Prepare the DNA by growing phage streaks with 50 ml host cells in 4 ml of top agar. Scrap up each plaque with a toothpick into 2 ml of cy (per litre: 10 g casamino acids, 5 g of yeast extract, 3 g NaCl, 2 g KCl, pH 7.0) + 10 mM MgCl$_2$ + 1/200 dilution of saturated (OD at 600 nm = 1.0) host cells. Shake at 37°C for 10 h when the suspension becomes clear. Isolate the phage DNA by standard procedure.

Dissolve the DNA in 30 μl TE. Follow double- or single-cutting protocol.

Double cutting protocol

Digest the cloned DNA with Hind III. End-label the digested fragments in 96-well microtitre plate. To each well, add dry 40 μCi of 400 Ci/mole [^{32}P] dATP (in ethanol) or 80 μCi of 800 Ci/mole [^{32}P] dATP (aqueous).

Add 160 μl H$_2$O, 40 μl of 10× medium salt restriction buffer (500 mM NaCl + 100 mM Tris-HCl, pH 7.4 + 100 mM MgCl$_2$ + 10 mM dithiothreitol), 4 μl RNase A (10 mg/ml), 10 μl of 0.5 mM ddGTP (to drive the fill-in reaction), 4 μl (40 units) of Hind III, 4 μl (40 units) of AMV reverse transcriptase and 2 μl (50–100 ng) DNA (*Hind*III-cut) to each well.

Spin briefly at 2000 rpm and then seal the plate with a parafilm. Incubate at 37°C for 45 min and then for 25 min at 63°C (to 'kill' the enzymes). Aliquot 2 μl of the following mix onto the side of each well: 200 μl of H$_2$O, 20 μl of 10× medium salt buffer, 80 units (30 units/μl) of Sau3A.

Spin briefly and reseal. Incubate at 37°C for 2 h. Add 4 μl of formamide/dye/EDTA to each well to stop the reaction. Denature the DNA fragments at 80°C for 10 min before gel loading.

Single cutting method

To each well of the 96-well microtitre plate add 1 µl of digestion mix (10 mM Tris-HCl, pH 7.4 + 10 mM MgCl₂ + 50 mM NaCl + 250 units/ml of Hinf I). Seal the plate with parafilm and incubate at 37°C for 1 h. To each well add 2 µl of labelling mix (10 mM Tris-HCl, pH 7.4 + 10 mM MgCl₂ + 6 mM DTT + 0.1 mM each of dATP, dGTP, dTTP + 0.2–0.4 µCi of [³²P] dCTP (3000 Ci/mole) + 0.1 unit of Klenow). Centrifuge briefly (1000 rpm), seal with parafilm. Incubate at room temp for 10 min. Add 7 µl of formamide/dye/EDTA mix. Denature at 80°C for 10 min before gel loading.

Alternative method after Gibson (1987)

Add 2 µl of DNA (100–200 ng) in each well. Add 2 µl of digestion mix (6 units of HinfI + 0.4 µg Rnase + 20 mM Tris. HCl, pH 8 + 7 mM MgCl₂ + 10 mM DTT) to the side of each well. Centrifuge briefly, seal and incubate at 37°C for 1 h. Dispense 2 µl of labelling mix to each well (0.2 units/ml of AMV reverse transcriptase + 80 nci/µl of [³⁵S] dATP + 40 µM each of dCTP, dGTP, dTTP).

Incubate for 15 min at 37°C. Dispense 4 µl of formamide/dye/EDTA mix. Denature at 80°C for 10 min before gel loading.

T-DNA PLASMID RESCUE

Insertional mutagenesis with the T-DNA of the cointegrate Tiplasmid (such as 3850 : 1003) is a recent gene tagging strategy (Behringer Medford, 1992). The insertion of the T-DNA in genomic DNA provides the opportunity to isolate T-DNA borders along with flanking DNA. This is done by forming chimeric plasmids between the inserted T-DNA and the adjacent plant DNA. Sequences from PBR322 provide an origin of replicon and the B- lactamase gene allows for selection with penicillin/ampicillin/carbenicillin. Two types of replicating plasmids, internal and external, can be formed by the restriction digested genomic DNA with a T-DNA insertion. Internal plasmids are composed entirely of T-DNA. As they do not have any origin of replication and the autobiotic resistance gene, they are not maintained in *E. coli*. External plasmids contain either the left or the right border of T-DNA and the flanking plant genomic DNA. These plasmids could be rescued by high efficiency transformation of *E. coli*, and then could be identified by analyzing restriction enzyme digests of the plasmid DNA, isolated from colonies grown in presence of carbenicillin (10 µg/ml of the medium).

Protocol — Rescue of T-DNA plasmids in transgenics
(from Prof R. K. Chaudhuri)

Prepare *E. coli* strain DH5α as described earlier. Add 1–2 µg transformed plant DNA, 2 µl of restriction buffer, 18 µl H_2O and 10 units of restriction enzyme (e.g. salI for left border or *Eco*RI for right border). Incubate at 37°C for 2 h. Bring the volume to 10 µl with distilled water. Add 100 µl phenol: chloroform (1:1) and vortex. Add 1/10 volume of 3.0 M Na-acetate. Add 2 vol (~400 µl) of absolute alcohol. Precipitate the DNA at −20°C for at least 2 h. Centrifuge 10 min at 4°C in a microfuge for 20 min. Resuspend the pellet in 15 µl TE. Add 2 µl 10 mM ATP, 2 µl of 10 × ligation buffer and 1 µl (8 U) of T4DNA ligase. Incubate at 14°C for 12–24 h. Store the reaction mixture at −20°C until further use. Thaw the competent cells (which are stored in 10% glycerol at −80°C) on ice. Mix 1–2 µl of the ligation reaction (with 100–200 ng plant DNA) with 200 µl of competent cells.

For *transformation* any method could be followed. In the *electroporation* method: Transfer the suspension to an ice-cold electroporation cuvette (0.2 cm electrode gap, Bio-Rad Lab). Adjust the settings of the electroporator (e.g. Gene Pulser, Bio-Rad) to provide an electric field of 12.5 kV/cm and an exponential pulse decay time constant of 8.2 min, by adjusting the capacitance (C) and resistance (R) to 21 m farads and 400 ohms, respectively. Electroporate the cells. Quickly transfer the cells to 10 ml warm S.O.C. medium (2% bactotryptone, 0.5% (w/v) yeast extract, 10 mM NaCl, 25 mM KCl, 10 mM $MgCl_2$, 10 mM $MgSO_4$, 20 mM glucose). Culture in a 100 ml Erlenmeyer flask at 37°C with rapid shaking. After 1 h, spin down the cells in a 15 ml conical bottom centrifuge tube with a clinical centrifuge. Resuspend the cells in 0.3 to 0.4 ml S.O.C. and plate these onto three 100 × 15 mm LB carb plates. Incubate the plates at least 20 h at 37°C to allow colonies to grow. Store plates at −4°C.

Carbencillin resistant colony analysis: Prepare plasmids from colonies. Select streak colonies for analysis on a fresh LB plate for reference. For analysis of left or right border rescue, digest the plasmid by the designated restriction endonuclease in the appropriate buffer. Designated RE means enzyme recognition at the left or right border of the co-integrated plasmid used. Stop reactions with 1/10 (v/v) 10 × loading dye (0.25% bromophenol blue, 0.25% xylene cyanol FF, 10 mM EDTA, 15% Ficoll). Electrophorese the digest on a 7% (w/v) agarose gel in 1 X TBE running buffer containing 0.5 µg ethidium bromide per ml. If necessary, perform southern hybridization of fractionated DNA fragments to identify the plant DNA using plant DNA as the probe and a gene of interest as another probe.

CHROMOSOME CRAWLING

This procedure is often applied to identify chromosomal sequences adjacent to a chromosome segment with known DNA sequences. The method takes advantage of polymerase chain reaction in an inverse direction otherwise termed as *Inverse PCR*. The procedure involves the complete digestion of DNA with a restriction enzyme which does not have any site of cleavage within the target region. The restriction fragment including the target should not exceed 2–3 kb in length. Then the DNA is diluted and ligated to form a circle. The template DNA is then linearized by cleaving within the target sequence with appropriate enzyme. The sequences flanking the target are then amplified with the aid of oligonucleotide primers complementary to 5' termini of the target. Ultimately, a double stranded linear DNA results with the head to tail arrangement of flanking sequences outside the target DNA molecule. The restriction site used originally to cleave the genomic DNA marks the junction of upstream and downstream sequences. This method of identifying chromosome sequences is termed as *Chromosome Crawling*—an aspect of *Chromosome walking*.

CHAPTER III.4

TRANSPOSON-MEDIATED TAGGING AND ISOLATION OF GENES

The transposable element in the DNA sequence is capable of moving throughout the genome due to its inbuilt capacity of excision and reintegration. Most of the transposons have inverted repeats at the two ends, with insertion sites as well. The activation of a transposon may be generated by specific elements as in *Ac* locus in maize mutation or by genomic stress, and its movement facilitated by the enzyme transposase coded in the system. The activation may generate mutation, change in methylation status and chromosome breakage. Autonomous transposable element may be an index of genetic instability, as in tissue culture. The mobility of transposons within a chromosome or between the chromosomes may alter gene function due to insertion. It may lead to duplication of sequences at the target site as well. Nucleotide alteration may arise out of endonuclease excision, effect on the reading frame and alterations of the template and change in protein coding pattern. Genetic variability may arise out of imprecise excision (Duncan, 1997). The movement of transposon can be detected in culture as in TyI elements in yeast where shift in location has been recorded.

The insertion sequences play an important role in evolution. Such transposable insertions may even lead to recessive mutations, as recorded in "Wrinkled Seed Gene" in pea, arising out of dysfunctioning of the enzyme coding for branched starch.

Lately, activation of transposons has also been demonstrated in non-parental systems, like the transposable elements of maize. Moreover, stability of inserted elements through removal of sequences responsible for mobility has been achieved. The wide applicability of transposons has led to the development of techniques for transposon tagging needed for gene isolation and transfer.

METHODS FOR TRANSPOSON TAGGING

(Personal communication from Prof R.K. Chaudhuri, Botany Dept., Calcutta University)

One of the possible approaches towards identification of new genes is by DNA homology—with the conserved nucleotide sequence as the homotype (probe). Another method involves a screening for gene expression patterns using panels of monoclonal antibodies raised against specific parts/metabolite of the organism.

Third approach is the 'enhancer trapping technique' where a transposon is mobilized, tagged with a marker gene, such as lacZ, GUS, luciferase, antibiotic resistance gene etc.—to study the gene expression pattern or developmental pattern of the organism. The inverted repeats of the transposon border sequences help to mobilise the transposon-element, with the activity of transposase, if that transposase is expressed in the organism. The result is a series of insertional mutagenesis by the element, within the life cycle of that organism.

In this protocol, transposon tagging experiment is designed with autonomous maize element *Ac* (activator):

For tagging

Select one prominent mutant from a series induced by EMS. Hybridize with a transformant with an *Ac* element—tagged with a marker gene, in an expression vector. Use leaf-disk transformation method.

Vector construction

pTiB6S3-SE plasmid could be purchased from manufacturers. This plasmid is an expression vector plasmid where the promoter is derived from CaMV 35S gene as well as from NOS gene of the Ti-plasmid. A 4.5 kb *Bam*HI-*Eco*RI fragment contains TL-DNA and a 6.0 kb *Eco*RI-*Bam*HI fragment of TR-DNA provided homology for replacement of most of greater than 80% of the TL-DNA, and the entire TR-DNA with the 1.2 kb Tn 903 segment carrying bacterial kanamycin resistance. The resulting plasmid pTiB6S3-SE carries only the TL left border. *Ac* element could be fused to the plasmid flanked by two *Hind*III sites, at the *Hind*III site of the plasmid. After that, the ATG codon of the transposase gene of the transposon element is modified by single strand site-directed mutagenesis. The chimeric plasmid, if introduced in a plant cell, is mobilized if transposase activity remains in the host cell.

The Monsanto expression plasmid vector has unique restriction sites, at the junction of Ti border and NOS-NPTII-NOS and Spc/StrR, such as *Hind*III, *Xho*I, *Bgl*II, *Xba*I, *Eco*RV, *Cla*I and *Eco*RI sites.

TRANSFORMATION OF *AGROBACTERIUM* BY CHIMERIC PLASMID

By Tripartite mating method

Grow fresh overnight culture in LB with appropriate antibiotics.

Agrobacterium pTiB6S3-SE *strain* Cm, 25 µg/ml; Km, 50 µg/ml; 30°C; *E. Coli*/pRK2013 helper strain Km, 50 µg/ml 37°C; it contains a mobilization gene for T-DNA. *E. coli*/pMON200 plasmid strain Spc. 100 µg/ml 37°C.

Combine 100 µl of each of the Agrobacterium strain and pMON200 plasmid strain without pRK2013 strain as a control for mating efficiency and to test the selection plates.

Pellet the cells by centrifugation at 4000 rpm for 4 min. Pipette off 25 µl aliquot from each mating mix onto a freshly poured LB agar plate. Incubate for overnight at 28–30°C. Collect cells from the mating mix with a loop and resuspend by vortexing in a 1.5 ml LB. Pellet the cells by centrifugation (4000 rpm for a few min). Pour off the supernatant and resuspend the cells in 100 µl of LB. Spread all of the cells on an LB agar plate containing Cm (25 µg/ml), Km (50 µg/ml) and Spc. (100 µg/ml). Incubate at 28–30°C for 48 h. Pick the antibiotic resistant colonies. Pick several colonies and restreak on plates containing selection antibiotic-containing agar plates at 28–30°C for overnight. Pick single colonies to 2 µl of LB containing selection antibiotics and shake at 8–30°C for 48 h.

After tripartite mating, the cointegrate plasmid (a hybrid T-DNA) is a pTi :: pMON200 construct where :: indicates the fusion. The two border T-DNA sequences remain at inverse orientation in that construct, and the *Ac* element is placed within that border sequences.

PLANT TRANSFORMATION

Agrobacterium-mediated transformation involves co-cultivation with protoplasts.

LEAF-DISK TRANSFORMATION

Sterilise seeds, germinate; harvest leaves and cut into small pieces. Pre-culture explants for 1–2 days upside down on MS104 medium (MSO, 1 µg/ml NAA, 1.0 µg/ml BA), inoculate by immersing in *A. tumefaciens* culture. Prepare nurse cell culture; place explants upside down on nurse culture plates; transfer explants to MS selection media containing antibiotics, kanamycin and hygromycin at the desired concentration.

Resistant calli are found to proliferate within 2–3 weeks. Transfer the resistant calli at 3 weeks interval; when good growth is noticed they are placed in the regenerated medium containing antibiotics. Transfer to rooting medium when defined shoots are visible from callus culture; place the plantlets in sterile soil to grow. Generate independent single locus transformants carrying Hyg :: AC or SPT :: AC or both.

Map *Ac* element location, relative to the chemical mutants, by making parental crosses of these transformants, or by using earlier map location of the chemical mutants.

LUCIFERASE ASSAY

In vitro method

Grind small pieces of transgenic plant tissue (frozen in liquid nitrogen) in a 1.5 effendorf tube, kept in liquid nitrogen, in 20 µl extraction buffer (0.1 M potassium phosphate, pH 7.5, 1 mM dithiothreitol (DTT). Add 100 µl of extraction buffer to the tube and vortex it. Mix 10–20 µl of extract with 100 µl of luciferase assay buffer (36 mM glycylglycine buffer, pH 7.8, 20 mM $MgCl_2$, 12 mM ATP, 1 mg/ml BSA) in a luminometer tube or cuvette.

Insert the tube or cuvette in a luminometer and initiate the luciferase reaction by injecting 50 µl of 0.4 mM luciferin into the sample tube. Record peak light output with a strip chart recorder or peak height detector. Peak emission is reached about 25 min after injection.

Express activity in light unit (LU) per amount fresh weight sample or per amount of protein.

In vivo method

Grow transgenic plants by hydroponics, in sand or in agar, in sterile condition, so that the supporting medium can be cleanly removed. Wash away the supporting medium from the roots with double distilled sterile water. Briefly blot the roots and dry.

Immerse the roots in a vessel containing 0.4 mM luciferin in water or in growth medium for at least for 1 h, depending on the size of the plant. For quick resolution, cut small pieces of leaf/flower etc., and determine light output in a luminometer in a luciferase activity measurement solution.

For X-ray film photography, mount the plant pieces/section or the whole plantlet on a surface of plastic wrap mounted on a cardboard frame. Gently position the explanted section or plantlet so that it lies flat against the plastic film. Hold the plant material in place by a piece of foam or sponge rubber and a 1″ (6.4 mm) press board.

Expose the plant material to X-ray film (Kodak OG) in the dark until a suitable exposure is obtained. Typical exposures are from 2–30 min, to produce a good light image from transgenic plants bearing luciferase driven by CaMV 35S promoter. Select the positive transformants for DNA isolation in the scheduled buffer and grind the plant parts in the vial immersed in liquid nitrogen. Take 0.5–10 µg of DNA for further analysis after digesting with restriction endonucleases.

HYGROMYCIN ASSAY

In vitro method

Grow transgenic plants in agar or in hydroponic media, in sterile condition, with the antibiotic hygromycin. Clean the supporting medium by washing with distilled water. Briefly blot the plant and dry. If necessary, grind the plant tissues, in presence of liquid nitrogen, in Effendorf tube. The hygromycin phosphotransferase activity is determined as follows:

Extraction buffer contains: 62.5 mM Tris-HCl (pH 6.8), 10% (v/v) glycerol, 5% (v/v) β-mercaptoethanol, before use add 0.125 mg/ml leupeptin. Reaction buffer contains 67 mM Tris-malate, pH 7.1 with maleic acid, 42 mM $MgCl_2$, 400 mM NH_4Cl. Reaction mixture contains 987 µl reaction buffer; 1.4 µl 22 mM hygromycin, 1 µl 10 mM ATP, 10 µl 1M NaF, 0.4 µl [^{32}P] ATP (10 µCi/µl).

Before performing the assay, soak the blotting or the phosphocellulose filter paper in cold AT solution for 30 min to minimise the background count. Collect tissue samples (ca. 400 mg) in microcentrifuge tubes on ice. Add 100 µl of ice cold extraction buffer and homogenise the tissues with a glass pestle. Take control samples from a plant known to express hygromycin phosphotransferase activity and from a wild type plant where there is no such activity. Spin the samples for 10 min in microcentrifuge tubes, in cold (4°C) and collect the supernatant in a second tube. Measure the protein content. Prewarm the sample to 37°C and start the reaction by adding an equal vol of prewarmed reaction mixture. Mix the mixture by vortexing for a few seconds. Take 4 µl sample at 1 min interval from the reaction tube and pipette directly to pretreated filter paper. Add excess of unlabelled ATP on the paper. Dry the paper, wash it with hot (80°C) phosphate buffer (10 mM sodium phosphate, pH 7.5) for 5 min and twice the same buffer at room temperature for 10 min. Dry the paper and expose to X-ray film for overnight in order to locate the radioactive spots. Cut the spots and measure the radioactivity with a scintillation counter. Plot the enzyme activity as pmol of bound P/min/mg protein (U/mg).

CHAPTER III.5

DNA FINGERPRINTING

Each individual has a unique DNA pattern. The demonstration of this uniqueness on the basis of DNA sequence difference forms the basis of DNA fingerprinting. Digestion of DNA with specific enzymes, electrophoresis, blotting and detection, following hybridization with specific oligonucleotide probes, have made DNA fingerprinting a convenient technique.

The analysis of eukaryotic genome has been greatly facilitated through Restriction Fragment Length Polymorphism technique—otherwise termed as RFLP technique. But RFLP technique is diallelic, involving two loci, whereas/multiallelic or multilocus RFLP probes form the basis of DNA fingerprinting pattern. The DNA differences or polymorphism of the genome can be detected by highly variable number of short repeat sequences arranged in tandem sequence (10–60 bps) known as mini or microsatellites. A DNA polycore probe of such minisatellite can detect, simultaneously, through hybridization, a large number of dispersed variable loci containing tandem repeats. Such a complex fingerprint pattern becomes very specific to individuals. Amplification through PCR permits analysis of minisatellite variant repeats. Short tandem repeat microsatellites have made the test very sensitive, allowing DNA typing of degraded systems as well.

Initially the work was concentrated on Southern blot hybridization. But as synthetic oligonucleotide probes are short and single stranded and can have very high polymorphic information, their use has facilitated DNA fingerprinting research. Such probes, complementary to short tandem repeats, can be synthesized, as the information is available on Data Bank. In addition to VNTR, random amplification of polymorphic DNA termed RAPD, which may have a ten base long chain, is proving to be of much use in the search for variations.

Several short tandem repeats have been recorded in a few plant species, also at mono-, tri- and tetranucleotide levels. In general, it has been recorded that in plant species the most frequently observed class of STRs with AT sequences are ATA. That should serve as a general guide in development of plant STRs, but detailed information on individual species which is species-specific, is needed for the frequency and type. Information about tetranucleotide repeat STRs is yet to be collected. A separate search is also needed for organelle DNA sequences as they do not follow the Mendelian law.

In general in plants, short tandem repeats of 6 bp or more are quite common. Of the different type of STRs, tri- and tetranucleotides are abundant.

STRs of mono-, di- and tetranucelotide repeats are located in non-coding regions. Of the trinucleotide STRs, those containing GC base pairs are found in coding regions whereas AT rich ones are in non-coding regions. The interspersed STRs are quite common. More than 50% of the trinucleotide repeats occur in coding regions (*Hordeum, Oryza, Triticum, Zea, Brassica, Glycine, Lycopersicum, Petunia, Pisum, Solanum* and *Nicotiana*).

The numbers of STRs are also species specific. In *Petunia*, it is one in every 11 kb; in *Brassica*, in every 25.4 kb, whereas in *Arabidopsis, Zea* and *Hordeum*, it is one in every 42, 58 and 156 kb respectively. Rice has a variety of STRs but only AT type of STR is found in *Hordeum*. STRs occur 3 times more frequently in dicotyledons (1 in every 31.2 kb) than in monocotyledons (one in 64.6 kb). In dicots, 15% contain GC whereas in monocots 50% are with GC bp.

Because of their high informativeness and ease of analysis, STRs in plants can be of much use in linkage mapping. The STRs can be significantly useful in genetics, breeding, systematics, strain identification and documentation of germplasm. The importance of fingerprinting in research can hardly be overestimated.

Initially, multiple hypervariable DNA loci were located in man. A family of tandem repeated DNA sequences termed as "Minisatellites" was noted to have a common GC rich core sequence of 10–15 bp in a tandem repeat of four or 33 bp motif in the intron of the myoglobin gene. Hybridization of restriction fragments of human DNA with probe derived from the core sequence showed variable minisatellites, forming individual-specific DNA fingerprint. In fact, the existence of variable number of tandem repeats formed the basis of polymorphism.

Both mini and simple sequence microsatellites are located in different chromosome loci, including telomeres and centromeres. The telomeric repeats located at chromosome ends are rather unusual. They are simple sequences with clear function. There is a marked base symmetry, one strand being GA-rich and the other CT-rich, repeated several times with a single strand 5′ overhang in each chromosome. The end segment is synthesized by a special type of polymerase-telomerase, which utilizes RNA as the template.

The PCR based DNA fingerprinting takes advantage of *in vitro* amplification of DNA through thermostable enzymes and primers, both tandem repeat and arbitrary. The basis of the PCR technique has been discussed earlier which involves three steps, namely denaturation, annealing and prolongation. The minisatellite core sequence as well as simple sequence oligonucleotide repeats which were used as hybridization probes, are now effectively employed as PCR primers for DNA fingerprinting (Lieckfeldt *et al.*, 1993).

The other method is to amplify DNA using arbitrarily chosen primer, prepared through synthesis or obtained from commercial firms. For unknown DNA fragments, only one arbitrary primer is used. However, for amplification product, two identical or almost identical target sequences near each other are needed, one site on one strand and the other site on the other strand in the opposing direction. In general, DNA requirement for fingerprinting using PCR methodology is very low. Even a small piece of endosperm, without causing disruption of viability, can yield sufficient material for 60 RAPD reactions (Weising *et al.*, 1994).

METHOD FOR DNA FINGERPRINTING

The standard DNA fingerprint involves the hybridization of restriction-digested genomic DNA with labelled multiallelic, multilocus probes.

The steps involved include: Complete digestion of genomic DNA with an appropriate restriction enzyme; electrophoretic separation of the restriction fragments, usually performed on agarose gels; denaturation and blotting of the separated DNA fragments onto a membrane (alternatively, drying of the gel matrix on a gel dryer); hybridization of the membrane (or the dried gel) to (non) radioactively labeled multilocus probes; detection of hybridizing fragments (i.e. fingerprints) by autoradiography or by nonradioactive approaches.

Gel hybridization with radioactive probes

Solutions: Hybridization buffer: 5× SSPE, 5× Denhardt's solution, 0.1% SDS, 10 µg/ml fragmented and denatured *E. coli* DNA; sterilise by filtration. Probe: [^{32}P]-labelled oligonucleotide, add to an appropriate amount of hybridization buffer at a concentration of 0.5 pmol/ml. 6× SSC (washing solution): 0.9 M NaCl, 0.09 M sodium citrate, pH 7.0.

Stock solutions: 20× SSPE: 3 M NaCl, 0.2 M sodium phosphate buffer, pH 7.4, 0.02 M EDTA. 100× Denhardt's solution: 2% PVP-40, 2% BSA, 2% Ficoll. Sterilise by filtration, and store in aliquots at $-20°$C.

Schedule: Remove plastic wrap and soak the dried gel in a tray filled with distilled water. After a few minutes, the backing filter paper detaches. Remaining pieces of filter paper should be carefully wiped off the gel using gloves.

Transfer the gel into a new tray filled with 6× SSC, and incubate for 5 min. Wind the gel onto a 10-ml disposable pipette, transfer it into a hybridization tube filled with 6× SSC, and unroll it to the inner wall of the tube. Pour off the 6× SSC, and replace with 10 ml of hybridization buffer including the labelled probe.

Hybridize for 3 h to overnight at T_m −5°C. Tubes should be carefully closed, to avoid contamination and/or loss of probe, after a few minutes and after 1 h (when heat has built up inside the tube). After hybridization, decant the probe into a 50-ml Falcon™ tube. The probe may be reused several times. Store at −20°C.

Fill the hybridization tube up to one half with 6× SSC, close it, and wash off most of the unbound probe by shaking. Decant the washing solution to the radioactive waste, remove the gel from the tube, wearing gloves, and transfer it to a tray filled with 6× SSC. Wash the gel in this tray for 3 × 30 min in 6× SSC at room temperature. Transfer the gel to another tray containing 6× SSC prewarmed to hybridization temperature. Wash for 1 to 2 min (stringent "hot wash"). Neither the exact duration nor the temperature of this hot wash is too critical, and similar fingerprint patterns occur over a wide range of washing temperatures. Transfer the gel to 6× SSC at room temperature. It is now ready for autoradiography.

Blot hybridization with nonradioactive probes

Incubate the membrane in blocking solution in a tray at room temperature for 1 h.

Transfer the membrane to a hybridization tube as described above (DNA side facing inward), add 10 ml of hybridization buffer excluding the probe, and prehybridize overnight at T_m −5°C. Decant prehybridization buffer, and replace by 10 ml of hybridization buffer (including the probe at a concentration of 5 pmol/ml) or, alternatively, just add the probe to the prehybridization buffer.

Proceed as described above for gel hybridization with radioactive probes. The membrane is now ready for nonradioactive signal detection.

SIGNAL DETECTION

Both radioactive and chemiluminescent signals can be detected by exposing the gel/membrane to an X-ray film.

Autoradiographic detection

After the final washing place gel/membrane on a sheet of plastic wrap, and drain excess liquid with filter paper. Cover the gel/membrane with plastic wrap. Remove large air bubbles. Inclusion of a piece of tape between the upper and lower sheets of plastic wrap facilitates future unpacking before reusing the gel/membrane. Evaluate signal strength using a hand monitor. Transfer the gel into an X-ray cassette, go to the darkroom, and place a sheet of X-ray film between gel/membrane and intensifying screen. If strong signals are measured, the exposure can be done at room temperature without screens. In this case, exposure time has to be prolonged by a factor of about three. If screens were used, transfer the cassette into a $-80°C$ freezer.

After several hours to several days (depending on signal strength), remove the cassette from the freezer, and let it warm up to room temperature. Open the cassette in the darkroom, and develop the X-ray film as follows: 5 min in X-ray developer (single-coated film may require more time); 30 s in water containing a few drops of acetic acid, 3 min in X-ray fixer (until the film turns transparent); 30 min in running water. The film is then allowed to dry. Handle wet X-ray films carefully since they are very sensitive to scraping. X-ray developing and fixing reagents are commercially available. Automatic X-ray processing machines are also available, but generally not required for a smaller laboratory.

Fingerprint polyacrylamide gel electrophoresis: Fractionation of labelled fragments

Essentially it is similar to denatured gel electrophoresis for sequencing, with some precautions taken to reduce distortion. The procedure is similar to sequencing gel preparation and running except that 4% polyacrylamide gel in TBE buffer is used and a 3 mm thick metal (aluminium) is clamped to the gel plate, to conduct the heat uniformly, by clips.

Load 3 μl of each sample, interspersed every six lanes with 1 μl marker. Mix 35 μl of water + 5 μl of 10× medium salt-buffer + 2.5 μl of λ DNA + 10 units of *Sau* 3A. Incubate at 37°C for 1 h. Add 4 μl of 800 Ci/mol [^{35}S] dATP + 2 μl of 10 mM dGTP + 2.5 μl of 10 mM dTTP + 10 units of AMV reverse transcriptase and incubate at 37°C for 30 min.

Electrophorese for 1.75 h at 30 W until the dye (blue) is about 1 cm from the bottom of the gel. Separate the plate, wash with water and then fix in 10% acetic acid for 15 min. Wash in tap water for 30 min. Dry on an 80°C hot plate for 30 min. Autoradiograph for 2 days without an intensifying screen. Photograph if necessary.

Chemiluminescent detection

When hybridization to a digoxigenated probe and washing steps are completed, incubate membrane for 1 h at room temperature in blocking solution in a tray. This prevents unspecific binding of the antibody to the membrane. Rinse membrane shortly in solution A.

Transfer membrane to a hybridization tube (DNA side facing inward), add antibody solution (5 ml/100 cm^2), and incubate for 15 min at room temperature under constant rotation in a hybridization oven. Alternatively, spread membrane onto a glass plate (DNA side facing upward), pipette the antibody solution on the surface, and incubate for 15 min. Discard antibody solution. Fill hybridization tube with solution A, and transfer membrane to a tray. Remove excess antibody by three washes in solution A on a rotary shaker (15 min each).

Incubate membrane in solution B (2 × 15 min). Then spread membrane onto a glass or plexiglass plate (DNA side facing upward). Pipette AMPPD solution onto the membrane (5 to 10 ml/100 cm^2), and incubate for 5 min under gentle agitation. Decant the AMPPD solution (which can be reused several times), briefly drain the membrane on filter paper, cover it in plastic wrap, and incubate at 37°C for 15 min.

Document results by exposure to Kodak XAR (or equivalent) X-ray film at room temperature. Signal strength increases during the first 24 h, and slowly levels off afterwards. When X-ray films are developed within the first 48 h, 1 min to 1 h of exposure is usually sufficient depending on signal strength. This allows an indefinite number of different exposures.

Colorigenic detection

The first three steps are the same as above.

Prepare staining agarose, and pour it onto a glass plate. Allow the gel to solidify. Place the wet membrane onto the gel (DNA side facing downward). Overlay the membrane with molten 0.6% agarose cooled down to 50°C. Allow the gel to solidify. Incubate overnight in a dark, moist chamber at room temperature. Stop colour reaction by transferring the membrane into TE buffer.

Document results by photographing wet membranes in transmitting light.

Chemiluminescent detection of AFLP^T fingerprints

The PCR-based DNA fingerprinting technique, Amplified Restriction Fragment Polymorphism (AFLP), is a powerful technique to identify molecular markers for both plant and bacterial DNA. AFLP is performed by (1) restriction endonuclease digestion of genomic DNA and ligation of specific adaptors; (2) amplification of the subpopulation of genomic DNA by PCR, using primer pairs containing common sequences of the adaptor and one to three arbitrary nucleotides; and (3) gel electrophoresis analysis of the amplified fragments. The combination of different restriction endonucleases, the choice of selective nucleotides in the primers, and resolution of sequencing gels make the results highly reproducible and able to detect multiple polymorphic DNA markers. Therefore, the AFLP technique has become a well-accepted DNA fingerprinting technique for the construction of genetic linkage maps in plants and molecular typing for both eukaryotes and prokaryotes.

Detection of AFLP bands requires radioisotope-labelled primers. Non-radioisotopic detection of AFLP can use fluorescent-labelled primers and DNA sequencing instruments.

Chemiluminescence is also a sensitive nonradioisotopic detection technique. In general, chemiluminescent detection involves blotting nucleic acids onto membranes, followed by hybridization of a probe to add an enzyme to activate the chemiluminescent substrate.

Schedule
Genomic DNA of two soybean ecotypes was obtained (after Lin *et al.*, 1997).

For AFLP reaction, follow manufacturer's instructions. Briefly, digest 200 to 250 ng of genomic DNA with 5 units of *Eco*RI and Mse I at 37°C for 2 h and incubate at 70°C for 15 min. Then ligate the DNA fragments to *Eco*RI and Mse I adaptors at 20°C for 2 h. For preselective amplification, amplify 5 µl of a 10 fold diluted ligation mixture, using the premode pre-amplification mixture supplied in the AFLP core Reagent Kit for soybean samples. For the selective amplification of soybean samples, use *Eco*RI + AAA and different *Mse*I + 3 primers- + CCT or + CGA.

For AFLP gel electrophoresis and blotting, heat AFLP samples at 90°C for 3 min. Load on a 5% polyacrylamide sequencing gel. Electrophorese the soybean samples, after amplification by *Eco*RI + AAA and *Mse*I + CCT or +CGA for 3 h at 200 V on a 15 × 17 cm gel or 105 min at 2015 V on a 30 × 39 cm gel. After gel electrophoresis, transfer the DNA to a GIBCO BRL Biodyne B membrane overnight and bake at 80°C for 1 h.

Prehybridize the membranes at 42°C for 20 min in the ACES prehybridization buffer (500 mM $NaPO_4$, pH 7.2 + 0.1% SDS). Then hybridize with 6.5 µl of the AFLP non-radioactive probe (Cat. No. 10822) at 42°C for 30 min in the ACES hybridization buffer (500 mM Na_2PO_4, pH 7.2 + 1% Hammersten grade casein + 0.1% SDS). Wash the membranes twice with 50 mM Na_2PO_4 (pH 7.2) + 0.1% SDS at 42°C and twice in the final wash buffer (10 mM Tris-HCl, pH 8.6, 150 mM NaCl). Incubate the membranes with CDP star at room temp. for 5 min and expose to X-ray film at room temperature for 45 min.

PCR-based DNA fingerprinting

PCR with arbitrary primers
Techniques for generating fragments with RAPD analysis, arbitrarily primed PCR (AP-PCR), and DNA amplification fingerprinting (DAF) are given.

RAPD Technique Prepare a reaction mix for each primer ("master mix"), sufficient for all samples plus one negative control to which water is added instead of DNA. For the setup of the master mix, calculate 5 µl of buffer,

2.5 µl of dNTP stock, 0.2 µl of primer, 0.2 µl of polymerase, and 37.1 µl of water for a volume of 45 µl per sample. The final concentrations are 10 mM Tris-HCl, 50 mM KCl, 2 mM $MgCl_2$, 0.001% gelatin, 100 µM dNTPs, and 0.2 µM primer. Use a specially designated PCR pipette set. Master mixes for only a few samples are preferably made by using primers diluted to 5 instead of 50 pmol/µl, for easier handling. Divide the master mix into labelled reaction vials or into a microtiter plate. Add 5 µl of the template DNA solution or water (negative control). Mix the contents, and centrifuge briefly. Microtiter plates can be centrifuged in specially equipped centrifuges, but omitting this step does not seem to influence the reaction.

Overlay the reaction solution with two or three drops of mineral oil to prevent evaporation. Continue with ten high-stringency cycles: 1 min at 94°C; 1 min at 60°C; 2 min at 72°C.

Add to each reaction vial 90 µl of the following second master mix containing for each vial: 2.25 U Taq polymerase, 1× buffer, 0.2 mM dNTPs, and 5 µCi α [^{32}P] dCTP. Run the high-stringency cycles another 20 to 30 times.

Separate the amplified DNA on a 5% polyacrylamide and 50% urea gel in 1× TBE and visualise by autoradiography.

PCR with simple sequence and minisatellite primers

Dilute the genomic DNA to a concentration of 10 ng/µl. Mix the following components in a 0.5-ml microfuge tube on ice: 2.5 µl genomic DNA (10 ng/µl); 5.0 µl 10× Taq polymerase buffer; 5.0 µl 10× dNTP mix; 3.0 µl 50 mM magnesium acetate; 30 µl primer (10 ng/µl); 0.5 µl Taq polymerase (5 U/µl); 31.0 µl distilled water. The final volume is 50.0 µl.

For processing many samples, a master mix may be prepared consisting of water, 10× buffer, 10× dNTPs, magnesium acetate, primer, and Taq polymerase. This ensures the even distribution of all reaction components among the samples and reduces the number of pipetting steps.

Vortex briefly, and spin for 20 s in a microfuge. Overlay the samples with two or three drops of light mineral oil. Transfer to a thermocycler that is programmed with 40 cycles of: 20 s at 93°C (denaturation step); 60 s at the annealing temperature; 20 s at 72°C (elongation step), followed by a final extension period of 6 min at 72°C. Set the ramp time as short as possible. After the reaction is completed, mix the amplification products with 0.2 vol of gel loading buffer, and electrophorese in a 1.4% agarose gel. Stain with ethidium bromide, and photograph the gel using short-wavelength (302 nm) UV irradiation. For separation and detection of products, Agarose/PAG electrophoresis can be adopted and stained with ethidium bromide or silver nitrate.

SCOPE IN THE STUDY OF BIODIVERSITY

The study of biodiversity and the manifestation of biodiversity are exhibited at different levels. The first and the most obvious example of biodiversity is the visible difference in phenotypes at the species and genomic levels. Each group of plants is different from the other. The second is the ecological diversity where diversity is expressed in ecological preferences ranging from hydrophytes to xerophytes. The most important level of diversity is the genotypic diversity, especially expressed at the intraspecific level and between closely related species as well. This diversity, reaching upto the cryptic level including diversity in fingerprinting, is of special significance both from evolutionary standpoint and germplasm documentation.

The fingerprint of DNA sequences is a type of molecular documentation differentiating individuals. The term, originally coined by Jeffrey *et al.* (1985), in relation to the human system, permitted simultaneous detection of variable DNA loci by hybridization of specific multilocus probes with restriction fragments. It is now widely applied in plant systems for detection of genetic diversity and genotype identification, as relevant to population genetics, evolutionary studies, pathotype determination and molecular approach to the vast area of biodiversity.

The technique of fingerprinting, often termed as "DNA typing" or "DNA profiling" (Weising *et al.*, 1994), involves two approaches. The first and the initial approach was based on hybridization, involving restriction cut of the genomic DNA, electrophoresis and hybridization with labelled multilocus probes, yielding a distinct band pattern in the gel. The other approach takes advantage of polymerase chain reaction, amplifying specific DNA sequences with oligonucleotide primers, Taq polymerase, electrophoretic separation of amplified sequences and detection of polymorphism through staining the bands.

DNA fingerprinting, based on hybridization, is similar to RFLP technique, in the fact that two types of multilocus probes are used, namely the cloned DNA fragments complementary to minisatellites and microsatellites. Both are tandem repeat oligonucleotides, complementary to target DNA, the former of basic motif of 10–60 bp and the latter is a simple sequence of 4–5 base pairs with low degree of repeats. Both mini and microsatellites occur at several genomic loci, often interspersed between unique sequences. In view of the fact that the copy numbers of both micro- and minisatellites with related motifs are highly variable, these are grouped under the category of "Variable number of tandem repeats", otherwise termed as VNTRs, recorded in different groups of organisms, including plants.

With gradual refinements in methods using synthetic oligomers, synthesized repeated sequences complementary to microsatellites have been successfully applied as multilocus probes in DNA fingerprinting.

In addition to chromosomal data, DNA fingerprinting is ideal for documentation of germplasm at the genotypic level and is essential for any center dealing with plant genetic resources. This is of special significance in cereals as well as commercially important crops like tea, jute, mulberry, where a complete documentation of different clones is imperative. This molecular criterion for identification is utilized as markers or tags but it is more important for a check against biological piracy and a safeguard against infringement of intellectual property rights. It is true that biodiversity is a common heritage but even minute modification of that heritage brings it under the jurisdiction of property rights.

The importance of DNA fingerprinting in the characterization of endangered species is of signal value. It is well known that rapid industrialization and urbanization along with deforestation are leading to the rapid extinction of flora. It is estimated that at the present rate of extinction, several species could vanish before they are fully known or documented. The need for fingerprinting of the endangered genome can hardly be questioned.

It is necessary to improve the accuracy of DNA fingerprinting to the maximum possible extent. For the moment, it is a very powerful tool for establishing the truth.

Paradoxically the entire basis of fingerprinting is formed by the tandem repeats which may be located in the non-coding regions. Such non-coding sequences have yielded evidences of origin of land plants, origin of organelles from prokaryotes and even human ancestry. Such sequences, which form to a great extent the basis of fingerprinting, form also the ultimate technique of detection of biodiversity—the diversity which differentiates one individual from another.

CHAPTER III.6

ISOLATION AND MAPPING OF
GENES—PRESENT STATUS

The study of gene structure and DNA polymorphism has been greatly facilitated with the use of non-isotopic probes and polymerase chain reaction. The latter involving amplification, yields sufficient amount of DNA for electrophoresis. Polymorphism in genes is ultimately resolved through hybridization with probes. The preparation of genetic map with the use of markers has gone to the extent of mapping more than 1000 genes in wheat and a very high number in maize, cotton, rice, barley, tobacco and tomato (Law, 1995). The molecular markers are now exploited to tag genes for resistance to biotic factors. An extension of this method has led to a new approach combining a large number of genes for resistance to pathogen in one genotype of a crop. Such molecular markers are also utilized now for screening different lines of maize. The high yielding hybrids arising out of crossing inbred lines, can be more effective with inbreds strongly dissimilar in molecular markers. Lately, the discovery of homology of probe across wide taxonomic boundaries (Moore *et al.*, 1993) has further opened a new avenue of comparative plant genetics in different crops which has high potential for improvement.

In horizontal gene transfer, it is necessary to have a clear understanding of the localization, site of expression and the product of gene expression, before isolation. As the data are rather meagre in respect of multitude of genes of high crop value, three different approaches have been adopted to isolate and clone such genes and analyse their expression. These methods include the localization of a closely linked RFLP marker, cloning of the region, and isolation of the selected gene through chromosome walking. The other approaches are (i) transposon tagging, (ii) *Agrobacterium* mediated T-DNA tagging for direct insertion and (iii) creation or deletion of the region surrounding the desired gene followed by cloning through differential hybridization (Law, 1995).

Recognition and isolation of gene by transposon tagging have been successful in maize and *Antirrhinum*. However, in wheat and allied genera, the limitations are principally the absence of endogenous transposable element (Flavell *et al.*, 1993). Moreover, other disadvantages are the difficulty of achieving insertion through transposable element of maize, and poor response to *Agrobacterium* infection. Gene isolation through chromosome walking in crops is also limited by certain factors, such as, distance of the single copy gene from the marker, and repetitive elements. In cases, where the distance from the marker is too long and there are heavy repetitive sequences in between, the technique may not be successful. It is desirable to get an idea of the location of the single copy gene and the repeat elements before isolation.

As all these approaches, geared towards isolation of a single gene in crops, are rather difficult, the aid of a model system as the *Arabidopsis*, with small genome has been preferred. It is expected that the cloned genes from this model system acting as probe, may help in pulling out the desired gene from the crop species, and help in the identification of specific gene products through analysis.

A different strategy (Flavell *et al.*, 1993) for isolation and localization of genes in the wheat and its allies with large genomes has been suggested. This strategy is based on the premise that most cereals are derived from a common genome as evidenced in recent years. If the gene order is retained approximately on the same position in different cereals, it should be theoretically possible to prepare a tentative map of the gene order in the ancestral form, on the basis of analysis of present day cereals. The identification of different genes through mutation and molecular mapping in any species, and its integration into the common ancestral map, are expected to give a general idea of the gene order in the ancestral form. This would also give an idea of the position and function of homoeologous genes in different species. The isolation of genes is cereals with large genome through chromosome walking, utilizing YAC clones of small rice genomes, may thus be possible. The knowledge of synteny and colinearity is turning out to be a powerful tool in gene localization and study of evolution. It has further been demonstrated that the basic haploid genome exists as a real structural unit in the nuclear architecture and the spatial distribution of the heterologues in the natural karyotype can be predicted on the basis of a simple model (Bennett, 1996). A functional significance of the three dimensional genome structure has been visualized.

On the basis of a conserved marker synteny, and a set of homoeologous probes, the colinearity of the three wheat genomes has been established. The genome colinearity shows the conservation of the gene order in wheat, rice and maize, despite evolutionary diversification through 50 million years (Devos *et al.*, 1995). The application of this strategy based on synteny for gene identification and isolation is indeed full of promise, and further understanding may aid in securing results with predictable accuracy.

SECTION—IV

CHROMOSOME MANIPULATION AND ENGINEERING: GENOME AND GENE TRANSFER

The methods for chromosome manipulation have gone through various stages of development, ultimately culminating in horizontal gene transfer. Induction of gene mutations through physical agents provided the geneticist with a novel method of manipulation—the initial step for biotechnology and plant improvement. This discovery was followed by the demonstration of similar property by chemical agents—the mutagenic and radiomimetic chemicals. Desirable mutations could then be induced in various crop species through physical and chemical agents.

Coupled with gene mutations, alterations in structure and behaviour of chromosomes leading to polyploids, aneuploids and chromosomal variants, have been advantageously employed in crop improvement.

The knowledge, of the mutagenic and mutachromosomic properties of various agents, simultaneously created awareness of the genotoxic effects of environmental pollutants. The need for monitoring the effects on chromosomes, with plants as the test systems, was realized. Thus, the area of bio-indicators of genotoxicity came into prominence, as a sequel to the understanding of mutagenic property of physical and chemical agents.

The use of chromosome changes and mutations, coupled later with selective hybridization, resulted in methods for transfer of genes, chromosomes and genomes. The development of addition, deficient and substitution lines, along with mutations, not merely yielded elite varieties of cereals, but opened up new avenues of gene mapping and gene transfer.

These approaches, which may be categorized as conventional biotechnology, were supplemented later, with the use of totipotency of plant cells, permitting plant regeneration *in vitro*. Several genotypes and hybrids of different crops, could be raised through culturing of cells—during, prior to or after manipulation. The culture of cell, tissue and organ gradually turned out to be elegant methods, for manipulation and nurturing of aberrant cells.

A corollary to *in vitro* growth, is the culturing of naked protoplast. This method enabled the fusion of plant cells in culture and somatic hybridization. The aim of the somatic hybrids was to secure distant hybrids and genome substitution through nucleocytoplasmic exchanges. Such hybrids have been obtained in several species, including those of *Brassica*, *Lycopersicon* and *Solanum*.

The modern approach to genetic engineering and gene transfer is based on the application of certain refined biochemical techniques. The use of restriction enzymes for cutting of DNA sequences, suitable vectors as carriers of foreign genes, proper cloning medium, and *in vitro* system for transformation and regeneration have made transfer of genes across wide barriers—a reality. Recombinant DNA technology and tissue culture are the two methods which are crucial for the development of transgenics.

The transfer of identified sequences has further been aided through ingenious physical methods. These techniques principally involve the use of micromanipulator and laser beam application for isolation and transfer (Section V). The combination of different methods namely—multiple probes for simultaneous detection of several gene loci as discussed earlier, laser beam for dissection and isolation of target sequences, and PCR for amplification of dissected sequences have brought chromosome-mediated gene transfer at the visual level of the microscope.

CHAPTER IV.1

INDUCTION OF CHANGES IN CHROMOSOMES

The understanding of the effects of physical and chemical agents on chromosomes can be attributed principally to the pioneering works of Stadler (1928) in maize, Oehlkers (1943) on snapdragon and Blakeslee and Avery on *Datura* (1937). A systematic attempt to explore the properties of different chemical agents was first made by Levan (1949) and his collaborators. The technique devised, otherwise known as the Allium test, uses as experimental material the bulbs of *Allium cepa*, the common onion.

The chemicals to be tested are prepared in solution and kept in wide-mouthed jars. A bulb with roots intact is then placed over the mouth of the jar so that its roots dip in the solution and the jars are preferably covered with black paper to allow healthy growth of the roots. After treatment for a desired period, the roots can be excised, fixed, stained and observed directly or kept for recovery in water or nutrient solution under similar conditions, followed by processing. The above method is applicable only to root meristems, however, and for meiotic cells or pollen grains the entire inflorescence is generally treated by dipping the stalk in water. But the mode of treatment with these materials may vary and, if necessary, the anthers may be dipped in the fluid directly.

In order to study the effects of physical agents such as X-rays, ultraviolet rays or gamma rays from cobalt-60 source, plant materials are placed in front of the source and the required dosage is applied, the subsequent steps being the same as that with chemical agents. Tagged radioactive isotopes, such as tritium, P^{32} or C^{14} are administered in the growth medium.

CHEMICAL AGENTS

Induction of metaphase arrest and polyploidy

In the majority of plants, nuclear division within the different cells is not synchronous. Therefore the meristematic or the dividing zone represents a heterogeneous mass of cells, in which the nuclei are at different stages of division. The problems of using such materials are: (a) metaphase stages—the best nuclear phase for chromosome analysis—cannot be obtained in high frequency, and (b) with regard to the analysis of an effect, the exact stage affected cannot be ascertained. Synchronized division, is therefore necessary.

The most widely used chemical for securing a large number of metaphase plates is colchicine. It causes metaphase arrest by inhibiting the operation of the spindle mechanism. The organs can be treated directly with aqueous colchicine solution for a required period, being added to the culture or medium. The concentration, needed to secure this effect, may range from a very low 0.01%, to even 2% and the period of treatment from 10 min to 16 h. In *Allium cepa*, spindle in root tips can be arrested by just 1.5 h treatment in 0.2% colchicine solution. Before fixation and observation, a thorough washing in water for at least 15–20 min is necessary to remove any superficial deposits of this alkaloid. The characteristic appearance of metaphase stages, showing clear separate segments, is otherwise known as colchicine-mitosis or c-mitosis. In addition to colchicine, a number of other compounds such as gammexane, chloral hydrate, acenaphthene, actidione, etc., are all employed for metaphase arrest. Colchicine is significantly more effective than any of the others. If the generation time is known, treatment can be extended to that duration, leading to arrest of maximum number of cells at metaphase.

Arresting metaphases and the induction of polyploidy are inter-related phenomena. Polyploidizing chemicals like colchicine inhibit the formation of the spindle within the two poles and confine the chromosomes within one nucleus though their division remains unhampered. The narcotic action allows the tissue to recover as soon as the influence of the chemical is removed (Levan, 1949). Polyploid cells, which are formed by colchicine action, divide normally and give rise to polyploid shoots. The schedules for colchicine treatment on different organs are outlined as follows:

On seeds and seedlings
Immerse the seeds for 2–48 h in concentrations of colchicine solution varying from 0.02–0.1% before sowing. Just-germinating seeds can be treated with similar concentrations of colchicine solution for 12–48 h with the plumules dipped in the solution, or the entire germinating seedling can be immersed completely in solution.

On mature seedlings
Add colchicine in the form of soaked cotton plugs on the growing shoot, the period of treatment varying from 2–4 h, and the range of concentrations used being the same as the previous schedule. Moisten the cotton plugs, placed over the growing tip, at regular intervals by adding drops of colchicine solution with a brush. After the treatment, remove the plug and wash the tip by brushing with water. The same method can be followed for treating young inflorescences. When colchicine is used in the form of a paste mixed with lanolin or with glycerin, it has been found to be effective where the growing point lies within the plumules, as in monocotyledonous plants.

On pollen grain

Add colchicine in the agar medium meant for pollen tube growth (2 ml of 0.2% colchicine in 8 ml of agar medium or in culture medium for other tissues). Other chemicals such as chloral hydrate, gammexane and acenaphthene, which cause metaphase arrest, can also be applied for the induction of polyploidy. The success or failure to induce polyploidy can be determined as follows:

The chromosome number of young shoots, leaf tips and root tips can be counted, following acetic-orcein or Feulgen squash.

Polyploidy is often associated with an increase in stomatal size and decrease in stomatal frequency per unit area of the leaf. The lower epidermis of the mature leaf of the polyploid can be peeled off and mounted in 50% glycerin solution. Stomatal size and frequency per unit area can be noted and the result then compared with that of diploids obtained following a similar procedure adopted for a control diploid plant. Post-treatment with X-rays of colchicine-treated plants results in better survival of polyploids since X-rays induce greater damage in diploid cells, causing their elimination in selection.

Induction of chromosome fragmentation

Chromosome breakage can be induced by several chemical agents. Fragmentation, followed by translocation of some fragments, may bring about a new patterning of chromosome segments resulting in heritable phenotypic difference categorized as subnarcotic effect.

The study of chromosome fragmentation by chemicals has a special significance in bringing out the differential nature of chromosome segments. Several chemical agents such as 8-ethoxycaffeine induce chromosome breaks at certain specific loci. This differential break can be taken as an index of the distinct chemical nature of susceptible segments from the rest of the chromosome parts.

Different chemical agents, causing chromosome breaks, may act in different ways. Some of them affect sulphydryl groups of proteins whereas others act through their influence on hydrogen bonds of nucleic acids. Guanidine cross-linkages are held to be involved with mustard compounds. Some agents may affect the oxidation–reduction system within the nucleus. The final upset of the nucleic acid metabolism ultimately results in hazards in protein reduplication causing chromosomes to break at different loci. The effects of different chemical agents and their modes of action have been dealt with in detail by several workers including our group. The effects can be classified into: (i) *clastogenic*—showing chromosomal aberrations; (ii) *turbagenic*—affecting the spindle; (iii) *carcinogenic*—capable of inducing cancer; (iv) *mutagenic*—able to induce mutations; and (v) *mitogenic/mitostatic*—affecting the frequency of cell division (see Sharma, 1984).

These effects overlap and may be shown by the same chemicals under different conditions and using different concentrations and/or period of exposure. Several plant products, such as pigments, alkaloids, certain flavonoids, also induce chromosome breakage. Certain antibiotics and extracts of bacterial culture have been shown to affect DNA replication, transcription and/or protein synthesis and thus initiate clastogenic effects. On the other hand, several plant products have been shown to reduce the frequency of chromosomal aberrations induced by known clastogens. Well-known are chlorophyllin—a sodium potassium salt of chlorophyll and crude plant extracts containing high amounts of vitamin C, vitamin B or its precursors, sulphur compounds, anti-oxidants and fibres as shown by our group (see Sharma, A., 1995).

The mode of action of antibiotics like mitomycin D, streptomycin, puromycin and chloramphenicol has been thoroughly investigated. Most metallic compounds are mitotic poisons. When administered to higher organisms, most of them are clastogenic, the effects being S-dependent. In general, the clastogenicity is directly proportional to the increase in atomic weight, electropositivity and solubility of the cations in water and lipids within each vertical group of the Periodic table. The effects are modified by interactions between metals and the addition of external agents.

Protocols for inducing chromosome breaks

1. *Random breakage*
Place a healthy bulb of *Allium cepa*, with root tip intact, on top of a jar containing 0.005 M solution of pyrogallol. Keep the jar in a temperature of 25–30°C. After 6 h, cut a few roots, fix in acetic–ethanol mixture (1 : 2) for 30 min, and follow the usual method of orcein squashing or Feulgen staining for root tips. Mount in 1% acetic–orcein solution or 45% acetic acid and count the number of fragments in metaphase and anaphase. Continue the treatment of roots in pyrogallol solution up to 24 h and observe at regular intervals of 6 h to study the increase or decrease in the frequency of fragments.

2. *Localized breakage*
Place germinated seeds of *Vicia faba* on a sieve fixed over a jar containing 0.075 M 8-ethoxycaffeine solution, in such a way that the roots pass through the sieve in the solution. Continue the treatment for 6 h at 10°C. Allow the roots to recover in tap water at 20°C for 24 and 48 h. Cut the root tips and follow the same schedule for observation as for random breakage. The technique employed for observation must take into account the fact that

chromosome breaks may result during slightly prolonged heating with acetic–orcein–HCl mixture (Sharma and Roy, 1956), which is an essential step in the procedure for orcein squashing. As such, slight heating for a few seconds is recommended. Chromosome breaks have also been observed following prolonged water treatment in non-aquatic plants by Sharma and Sen (1954).

3. *Orcein breakage*

Treat excised root tips of onion in 0.002 M solution of 8-oxyquinoline for 2 h at 16–18°C. Fix in acetic–ethanol (1 : 2) for 30 min. Heat the root tips gently over a flame in a mixture of 2% acetic–orcein and normal hydrochloric acid mixed in the proportion of 9 : 1 for 30 s. After a few minutes, mount and squash in 1% acetic–orcein and observe. Fragments can be observed in metaphase and anaphase stages but if the heating is prolonged for a few seconds more, the frequency of fragments shows an increase.

Division in differentiated nuclei

Division in adult nuclei, when otherwise the nuclei have ceased to undergo apparent division, can also be induced with the aid of chemical agents. Huskins and Steinitz (1948) claimed that the adult nuclei, though apparently non-dividing, undergo endomitotic re-duplication of the chromonemata, thus exerting control over the process of differentiation. With the aid of a special technique using hormones they induced division in the adult nuclei, thus permitting the chromosomes to complete the nuclear cycle and reveal their polytenic and polyploid constitution.

Sharma and Mookerjea (1954) demonstrated that the sugar constituent alone of the nucleic acid can induce division, and claimed that the polytenic condition of the adult nuclei is due to deficiency in nucleic acid. Torrey (1960) used kinetin (6-furfuryl-aminopurine) and Sen (1970) 2,4,-dichlorophenoxy acetic acid for the induction of division in endomitotic plant cells and in vascular zone.

The occurrence of polyteny may possibly suggest that there is a limit for continued transcription of a strand, after which replication is necessary for the production of a fresh strand to be used in transcription in mature cells (Sharma, 1978).

Schedule for inducing division in differentiated nuclei by chemical treatment

Take cuttings or bulbs (*Rhoeo discolor*/*Allium cepa*) and remove all roots. Place in jars containing Knop's nutrient medium and allow to grow (Knop's medium: calcium nitrate 95 mg/ml in distilled water, ammonium nitrate 129; magnesium sulphate 180; potassium dihydrogen phosphate 133.5; ferric acetate 5.0; Mn 0.5; Cl 1.9; B 0.5; Cu 0.02; Zn 0.01; make upto pH 7.5). Keep the materials in the jar until freshly generated roots, 3–10 cm long, are obtained. Transfer to a similar jar containing 50–100 ppm indole 3-acetic acid and keep for 24–72 h. Re-transfer the materials to jar containing fresh nutrient medium and keep for a maximum period of 72 h. Take out the materials after every 24 h. Remove about 4 mm from the root tip and then cut out 1–2 mm to obtain differentiated zone. Follow the usual procedure for fixation in acetic–ethanol and orcein squash technique. Control preparations, allowing the roots to grow in culture solution, do not show any division in differentiated zone.

Induction of somatic reduction

The occurrence of reductional separation of chromosomes in tissues, other than the gonadal ones, was demonstrated on slides prepared by Huskins and his associates after treatment with sodium nucleate, in root-tip cells.

Somatic reduction, induced through treatment with sodium salts of nucleic acid, evidently suggests that the balance of the nucleic acid within the cell is at least one of the principal controlling factors of mitosis and meiosis.

Schedule for the induction of somatic reduction

Take a healthy young bulb of *Allium cepa* and let it grow in a jar containing tap water for root growth. When the roots are just 2–3 mm long, fit the bulb with roots on top of a jar containing 0.1–0.2 % sodium nucleate solution in water; keep for 6–12 h. Cut the tip portion of the root (meristematic region) and follow the usual method of fixation and Feulgen or orcein squashing.

Reductional separation of chromosomes can be observed in a number of cells, showing eight chromosomes at each pole, instead of normal 16. Abnormal separation of chromosomes is also seen.

PHYSICAL AGENTS

In addition to X-rays, other types of radiations, such as γ-rays, β-rays, fast neutrons, ultraviolet and infrared rays have all been used to induce mutations and chromosomal aberrations. The biological action of these agents depends on their ionizing and non-ionizing properties. Ionizing radiations dissipate energy during their passage through matter, by ejecting electrons from the atom through which they pass. The ionized atom, losing the negative balance, becomes positively charged and is known as an ion. The result of ionization is a chemical change of the molecule concerned. Whenever a binding electron between the two molecules is affected, serious after-effects ensue. Non-ionizing radiations, such as ultraviolet rays, infrared rays, etc., cause dissipation of energy within the tissue by molecular excitation. The principal biological action is attributed to the absorption of energy by particular cellular constituents, the most important one being nucleic acid (Table 1 from Sharma and Sharma, 1960).

Table 1 Chromosome breakage induced by radiation in some plants

Radiation	Material	Dosage	Radiation	Material	Dosage
Alpha rays	*Tradescantia*, pollen tube	Different dosages	X-ray	*Hyacinthus*, root tip	150–1000 R
	pollen grain	Inflorescence treated in radon solution		*Lilium* p.m.c.	150 R
	Vicia faba, root tips	Treated in radon solution 7–8 units		*Scilla*, endosperm	50 R
Beta rays	*Tradescantia*, pollen grain	^{32}P plaque, externally (a) from ammonium carbonate at 0.9–8.2 µCi/ml for 4–8 days		*Tradescantia*, p.m.c.	18 R and 360 R
				anther	150 R
	Tradescantia, anthers	(b) ^{32}P from sodium hydrogen phosphate at 1–10 µCi/ml for 1–9 days		pollen grain early	360 R 30–300 R
		(c) ^{3}H thymidine 1 µCi/ml for 8–56 h		pollen grain late	180 R, 250 R
Gamma rays	*Tradescantia*, pollen grain	(a) 5/min for 100–2000 min (b) ^{40}Co 100–400 R (c) ^{40}Co: 1.1–1.3 MeV in air and nitrogen	X-ray	*Trillium*, root tip p.m.c	5–45 R 50 R
Near infrared	*Tradescantia*, pollen grain	(a) 1000 nm approx. for 3 h before and after 100 R X-rays		pollen grain	45–375 R
		(b) 1000 nm approx. for 3 h before and after 90 and 350 R X-rays		*Uvularia* p.m.c.	90 R
	Tradescantia, meiosis	Different dosages		*Vicia*, root tip	50–200 R
Neutrons	*Tradescantia*, pollen grains			*Zea mays*, pollen grain	800–1500 R
Ultraviolet	*Tradescantia*, pollen tube	(a) 254 nm at 2×10^{-3} ergs/mm^2 for 60 s (b) 253.7 nm before and after X-rays			
	Zea mays, pollen grain	253.7 nm at 546.000 ergs/cm^2			

In addition to gene mutation, chromosome breakage may involve both the chromatids of the chromosome or a single chromatid depending on the stage in which the nucleus has received the radiation. The subsequent effects of chromosome breakage are translocation, deletion, inversio, rejoining as well as stickiness, pyknosis and polyploidy, and can all be studied in treated materials (see Table 2).

Table 2   Types of chromosomal alterations

Term	Definition
Chromatid gap	An achromatic region in one chromatid, the size of which is equal to or smaller than the width of the chromatid
Chromatid break	An achromatic region in one chromatid, larger than the width of the chromatid. It may either be aligned or unaligned
Chromosome gap	Same as (i) only in both chromatids
Chromosome break	Same as (i) only in both chromatids
Chromatid deletion	Deleted material at the end of one chromatid
Fragment	Separated part of single chromatid or part of chromosome
Acentric fragment	Two aligned (parallel) chromatids without an evident centromere
Translocation	Obvious transfer of material between two or more chromosomes
Triradial	An abnormal arrangement of paired chromatids resulting in a triradial configuration.
Quadriradial	An abnormal arrangement of paired chromatids resulting in a four-armed configuration
Pulverised chromosome	A spread containing one fragmented or pulverised chromosome
Pulverised chromosomes	A spread containing two or more fragmented or pulverised chromosomes, but with some intact chromosomes still remaining
Pulverised cell	A cell in which all the chromosomes are totally fragmented
Complex rearragement	An abnormal translocation figure which involves many chromosomes and is the result of several breaks and mispaired chromatids
Ring	A chromosome which is a result of telomeric deletions at both ends of the chromosome and the subsequent joining of the ends of the two chromosome arms
Minute (min)	A small chromosome which contains centromere and does not belong to the karyotype
Polyploid or endoreduplication	A cell in which the chromosome number is an even multiple of the haploid number (n) and is greater than $2n$
Hyperdiploid	A cell in which the chromosome number is greater than $2n$ but is not an even multiple of n
Dicentric	A chromosome containing two centromeres

Different methods have been devised for the application of different physical agents but before undertaking any work on the effects of physical agents on chromosomes certain factors which control their sensitivity should be taken into account.

The factors affecting the expression of radiosensitivity may be internal and/or external. The internal factors include the nature of genotypes; chromosome morphology and number; stage of cell division, level of ploidy and age of the tissue. External ones include moisture, temperature, presence of oxygen and other atmospheric gases, dosage, conditions and durations of storage, presence of other chemicals and ionization density of the rays.

The comparative action of different rays and chemicals has also been dealt with by different authors (see Ghosh and Sharma, 1968). The importance of nucleic acid and protein in the manifestation of chromosome breaks has also been studied (Sharma and Sharma, 1960).

Schedules for irradiation with different agents

X-rays
Seeds Place dry seeds, in a single row, in a petri dish and expose to different doses of X-rays. Germinate in moist sawdust. Remove root tips when about 1 cm long, treat in acetic–ethanol mixture (1:3) for 30 min and squash following orcein or Feulgen schedule. If soaked in water before irradiation, the effect is usually more and may become drastic.

Seedlings and bulbs with root tips Place young seedlings on a petri dish with their radicles pointing in the same direction. For bulbs, take healthy ones with a tuft of healthy roots about 2 cm long and place them with the root tips facing the source. Expose to X-rays for the desired period. Transfer them to sawdust or to nutrient medium. For immediate effect, remove root tips at regular intervals of an hour, fix for 30 min in acetic–ethanol (1:3), followed by staining by Feulgen or orcein squash methods. For prolonged effects, study the root tips at intervals of 24 h.

Inflorescences Grow the plants in flower-pots. Expose the young inflorescences to X-rays by bending them to face the source. For immediate effect, select a flower bud of suitable size and observe the pollen mother cells after smearing in acetic-carmine solution. Allow the inflorescence to develop and observe meiotic stages at intervals of 24 h. Endosperm and pollen grain mitosis can be studied in a similar manner.

Ultraviolet rays
The apparatus consists, in simpler forms, of a mercury lamp or a quartz mercury arc in conjunction with a monochromator. The method of treating the material is similar to that followed with X-rays.

Infrared rays

The apparatus generally consists of special types of tungsten lamps used for rapid drying. In some later models, arcs, for the production of both ultra-violet and infrared rays, have been built within one source with separate controls. The unwanted rays are screened off with suitable filters. The method of application is similar to that of X-rays.

Treatment of tissue with both physical and chemical agents

Radiation effects and the consequent cytological irregularities have often been found to be considerably reduced in the absence of oxygen or in the presence of certain chemicals. Such chemicals are referred to as 'protective chemicals', against radiation damage.

Protective chemicals can be applied before, after, as well as during, the time of radiation, the procedure varying according to the type of chemical and the nature of the radiation. From a cytological aspect, protection is afforded against chromosome breakage and inhibition of cell division.

Schedule

Place young denuded bulbs of *Allium cepa* in jars containing tap water and allow the roots to germinate till they are 1–3 cm long. Place the bulbs horizontally in large glass petri dishes filled with a solution of the chemical whose protective action is to be studied. For this purpose 4×10^{-3} M sodium thiosulphate may be used. Treat for 30 min in this solution. Irradiate the bulbs in the same solution with X-rays. After irradiation, keep the bulbs in the fluid for another 10 min. Transfer the bulbs to beakers containing tap water where they can be kept for 4–5 days for immediate observation and for observation of the root tips at regular intervals. Cut healthy young root tips, treat in 0.5% colchicine solution for 1 h to secure a large number of metaphase stages necessary for the study of chromosome interchanges and deletions. Fix in acetic–ethanol (1:3) and follow the usual procedure for orcein or Feulgen staining of root tips and observe the interchanges and deletions at the metaphase stages. To measure the protection afforded by the chemical, set up a control experiment in which all the above steps are followed, substituting distilled water in place of the chemical.

SYNCHRONIZATION OF CELL DIVISION

Synchronization is a unique strategy for securing cell mass in an identical phase. The principle of cell synchronization involves blockage of cells at a specific phase of the cell cycle, achieved through different means, including the use of cytotoxic agents and depriving the medium of some of the essential constituents needed for division (after Sharma, 1996). Synchronized populations are essential for study of the cell cycle-related events, specially since metabolic processes are quite different at different phases in the cell cycle including the transitional phase. The cell cycle follows the normal course, provided the supply of metabolites and cellular environment remain unhampered. Conditions restrictive for progression at one stage lead to accumulation of cells at a particular phase, and further resumption of growth is through input of the metabolite. Such arrest may occur at any of the phases, though the arrest of the synthetic phase may often lead to cell death. In general, the methodology for cell synchronization should have as prerequisites, the action on a target metabolic phase without any damaging effect, and potential of substantial harvesting of synchronized cells. The success and extent of synchronization depend to a great extent on the method adopted for securing synchrony. Any technique should, however, take into account the cyclin-dependent kinase and the checkpoints necessary for the identification of the desired phase prior to induction of synchrony.

The enzyme kinase which controls the behaviour of other proteins by attaching phosphate groups, and the protein cyclin which triggers the kinase at a critical period, bringing about changes in cell cycle, are both essential (vide Murray and Hunt, 1993). Cyclins and kinases control several points in the cell cycle. Cyclin activates Cdc_2 kinase at mitosis at the time of dissolution of nuclear membrane and chromosomes moving to poles. It is also active in late G_1 to start replication. The cyclins generally belong to two distinct classes, cyclin A and cyclin B, the mitotic cyclins being of the latter type. The cyclin A may be active in activating kinase at the beginning of "S", being different from the "Start" which is active in late G_1 phase.

Events in the cell cycle require to be completed in a definite sequence (Hartwell and Weiner, 1989). Chromosome replication should precede chromosome segregation and as such the progress of mitosis requires completion of "S" phase. Checkpoints permit monitoring of progression through the succession of essential steps and thus coordinate cell cycle for replication and segregation of components.

The complex of cyclins and cyclin-dependent kinases is deeply involved in the transitional pathways of DNA replication and mitosis. Two mechanisms operate in the interdependent pathways. The events acting on G_1 and needed for S phase and events on G_2 necessary for mitosis, constitute one pathway. The other process exerts controls on G_1/S to check premature entry into mitosis, and on G_2 to check re-entry into G_1 and triggering of new S phase. The failure of any of these mechanisms would lead to alteration in genome dosage. Successive divisions without S phase may lead to reduction in number and often lethality, whereas successive S phases without mitosis cause endoreplication.

Two important events are clearly involved in eukaryotic gene replication (1) at the end of mitosis and (2) at the beginning of S phase (Wang and Joachim, 1995). The initiation competence at G_1 is capable of triggering S phase and initiation competence at G_2 phase. The initiation competence is lost in S phase with the onset of replication. Further reversal of competence starts from G_2 to G_1 through the M phase. The competence evidently is established much before S phase, though not adequate to initiate replication. At G_1/S boundary, the process reaches its culmination, triggering S phase and DNA replication. With replication, there is again loss of competence.

Several authors have shown that entry into mitosis proper is caused by a promoting factor along with complex of Cdc_2 kinase and cyclin B. Assembly of spindle is a landmark point. The activation involves breakdown of nuclear membrane, condensation of chromatin and formation of spindle. The termination involves a reverse cycle-cycle B degradation and inactivation of maturation promoter (Nishimoto et al., 1992). The monitoring point in the spindle may be attachment of microtubules to kinetochore. The treatment of spindle with microtubular poisons like benonyl or mocodazole blocks progression and mitosis.

Methodology for synchronization

For synchronization, several techniques are in vogue, both physical and chemical. In the plant system, specially in higher plants, mostly chemical method is applied.Amongst the physical methods, the *detachment* method is based on the principle of detaching the cells, initially done through enzymes and microscopic detection of cells at the maximum frequency of division. It is followed by separation of mitotic cells by shaking and later plating. Synchrony can be evaluated by continuous tritiated thymidine followed in successive cycles. Such systems undoubtedly permit the study of G_1 events but post S phase events often require further treatment with certain chemicals.

Of the chemical methods applied in plant systems, the target for inducing synchrony, is the M phase—checking spindle assembly or blocking the cells in S phase. Of the different spindle poisons, colchicine is considered to be the most potent one, inhibiting spindle formation through its effect on spindle proteins (vide Sharma and Sharma, 1980, 1994). The advantage of using colchicine is that it does not affect the biochemical pathways in G_1, S and in other phases of mitosis. The effect being reversible, the cells recover from a narcotized stupor and show a high degree of synchrony on recovery.

The cells accumulated in metaphase after a specified period of treatment in colchicine can be induced to divide synchronously following the application of medium without colchicine. Similarly, okadaic acid and topoisomerase inhibitors allow premature condensation and derangement in spindle formation. A number of other compounds has also been worked out inhibiting the normal cell cycle, including gammexane, chloral hydrate and some coumarin derivatives.

In the laboratory, synchrony in cells, involving metaphase arrest, can be induced by treating the organs with dividing cells in dilute aqueous solution of colchicine followed by washing, recovery in nutrient solution and observation through fixation and staining as usual. However, to secure optimum effect, colchicine treatment should be carried out for the entire period needed for full cell cycle, which varies from species to species. With hydroxyurea for blocking in S phase, the treatment with 1.25 mM in aerated Hoagland solution for a long period (18 h in *Vicia faba*) followed by recovery has proved to be successful (Schubert *et al.*, 1993).

CHAPTER IV.2

STUDY OF THE EFFECTS OF ENVIRONMENTAL
TOXICANTS USING PLANT SYSTEMS

The exposure of the living organisms, including man, to a wide variety of physical, chemical and biological agents, both environmental and industrial, has enhanced the importance of evolving techniques for biomonitoring. The techniques can be used to monitor damage caused to the cell, including chromosomes and different organelles, without necessarily indicating the specific source. The changes can be monitored at the phenotypic, cellular, chromosomal and genetic levels. The use of multiple protocols and test systems, both *in vivo* and *in vitro*, has been responsible for identifying the toxicity of different agents, which affect either the genetic architecture, for example the DNA skeleton, or act as promoters or inhibitors of genotoxic effects. Such genotoxic effects reflect the cytotoxic, mutagenic, clastogenic and turbagenic activity of the toxicants. The endpoints for monitoring are generally mutational events affecting phenotype, metabolism or chromosomal architecture (Sharma, 1994; Sharma and Sharma, 1989; Sorsa *et al.*, 1982).

Both single test systems with multiple endpoints and multiple batteries of tests have been advocated for correct assessment of genotoxicity of individual toxicants and complex mixtures.

The principal endpoints for monitoring can be tested both *in vivo* and *in vitro*. These are either (i) mutational events, involving DNA, or (ii) chromosomal aberrations, involving breaks and interchanges at different levels. These changes may occur in the somatic cells or in the germ cells.

A large number of compounds also have been tested to detect their potential for inducing aneuploidy or alterations in chromosome number, initially developed as an ancillary to clastogenesis.

The test systems range from microbial systems to plants and animals and finally human leucocyte in culture. In this chapter, the more commonly used plant systems *in vivo* have been described.

SOME COMMON TEST SYSTEMS USED

Yeast and fungi

Strains of yeast have been utilized *in vitro* as indicators of reverse mutations, gene conversion, mitotic recombination, aneuploidy and other forms of chromosome damage. Growing cells of different species of *Saccharomyces* produce the enzymes necessary to activate many promutagens which would not be able to induce mutations *in vitro* without S9 mix. Mutation studies have also been carried out with species of *Neurospora* and *Aspergillus*. These systems, however, have a relatively limited scope.

Yeast-agar overlay protocol
Suspend stationary-phase cells of an appropriate tester strain in saline. Take the test extract up to 4% and mix with yeast minimal medium kept at 45°C with requisite number of yeast cells. 35×10^{-2} or 35×10^{-3} per ml are adequate for the detection of histidine-independent and tryptophan-independent prototrophs respectively. If required, add an exogenous S9 mix. Supplement the minimal media with 20 µg/ml of tryptophan and 0.1 µg/ml histidine for detection of respective prototrophs used. This allows at least 3 cell divisions during incubation along with the tissue extract. Keep the plates in the dark for 9 days at 28°C and count the frequency of prototrophic colonies (after Parry, 1985).

Higher plants

Higher plants have been used extensively for monitoring genotoxicity of chemicals due to the complexity of their genomes, their capacity in some cases to activate promutagens to mutagens and the ease of their use. The more commonly used ones are different species of *Vicia* and *Allium*, *Tradescantia*, *Hordeum* and maize. For the localization of chromosomal alterations, karyotypes with marker chromosomes are required. Pretreatment of root tips with other chemicals prior to exposure alters the susceptibility in some cases. Gene mutation studies have been carried out on *Arabidopsis*, *Glycine*, *Hordeum*, *Tradescantia* and *Zea*. Of these, *Zea mays* has been found to be a very good bioindicator because of the convenience of detecting mutation in the specific gene locus in identified chromosomes.

Tests for gene loci

Maize
The two tests which are widely employed involve chromosome 9.

Waxy locus pollen test: The gene is located on the short arm of chromosome 9 and is responsible for the synthesis of the carbohydrate amylose. The starch of individuals carrying the dominant allele *Wx* is a mixture of amylose and amylopectin, whereas amylose is absent in the double recessive individuals. Iodine test colours amylose blue-black. Its absence gives a red or tan colour. Since the pollen grain is haploid, containing either *Wx* or *wx*, the colour can be used as monitor.

Take an early-synthetic type of maize and allow it to grow, till anthesis, in an area containing the mutagenic pollutants. Harvest each tassel from double recessive wx-C/wx-C and label. Dehydrate to 70% ethanol for 48 h. Shake the selected tassels in pure ethanol and remove a few unopened flowers. Dissect the anthers from the floret and place in a stainless steel cup of microhomogenizer. Prepare: Stock A-Iodine solution with 25 ml of water, 100 ml of KI and 95 mg of iodine. Stock B-Gelatin solution with 25 ml of water, and 15 g of gelatin. Mix A and B in equal proportion. Take 0.6 ml of mixture and add in the cup. Add a drop of Tween-80. Homogenize the minced anthers for 30 s. Filter the homogenate through cheesecloth on a microscope slide and put the coverglass on the suspension. When the suspension solidifies, examine the slide under the microscope. The black grains would indicate the revertants and the percentage of mutation can be worked out.

Yg-2 locus on short arm of chromosome 9 in maize induces pale yellow green and slowly maturing leaves in homozygous recessive plants (Yg-2/yg-2). The dominant Yg-2 causes dark green coloration. In the heterozygous state (Yg-2/yg-2) leaves are a normal green in colour. In a heterozygous individual, mutation of the dominant allele leads to deficiency in the production of chlorophyll in the leaf sector. If a leaf primordial cell is involved, yellow green sectors develop in the green leaf.

Choose heterozygous Yg-2/yg-2 seeds from Early-Early synthetic type. Expose the seeds under aeration to specific environmental pollutants by immersion. Germinate the seeds and grow the plant in growth chambers at 28°C till the emergence of 4 to 5 leaves. Dissect the mature leaves with ligule and observe under ultraviolet light.

Tradescantia

Staminal hair method: A change in colour in high mutagen-sensitive staminal hairs has been utilized to monitor mutagens in *Tradescantia*. The mutational events change the colour from blue to pink. Observe in the stamen under the microscope after dissection.

Microspore method has been used to measure the extent of radiation as absorbed by the plant when exposed. In general, the microspores of *Tradescantia* are excellent materials for scoring the effects of pollutants as shown by the formation of chromosomal aberrations during mitosis of pollen grains.

Pick up pollen with a brush from anthers of plants growing in polluted environments. Heat 12 g of lactone and 1.5 g of agar with 100 ml distilled water till agar dissolves. Add 0.01 g of colchicine when the temperature cools down to 80–60°C. Keep at 60–70°C. Coat slides with a thin layer of egg white. Dip in the medium at 60°C until it warms up. Withdraw the slide, drain off the medium, and wipe the back with a piece of clean cloth. Dust a thin layer of pollen on the medium after it has set. Immediately place the slide in a moist growing box which is a horizontal glass staining dish lined with damp blotting paper on the two sides and top, at 20–25°C. Observe at intervals till the optimum period for maximum division is found. Add a drop of 1% acetic–carmine solution to the pollen tube, squash under a coverglass and observe.

Somatic meristems

Barley, *Vicia faba* and species of *Arabidopsis* offer good systems for environmental monitoring. In the last species, due to the short life cycle, the effect of the pollutant can be measured in all phases of growth.

Allium cepa, Allium sativum

The *Allium* test, initially worked out by Levan is one of the most convenient methods of monitoring genotoxicity. The chromosomes of *Allium*, being quite large, allow a detailed analysis. The bulbs can be easily handled and roots with meristems grow profusely at regular intervals. Such rapid successive growth enables an analysis of both long and short term effects. The method involves the culturing of healthy bulbs on the top of small jars containing the solution to be tested for genotoxicity, with the growing roots immersed in the solution. The roots are cut out and processed at regular intervals to study the effects of different periods of treatment at different concentrations.

In order to study the recovery from the toxic effects, if any, the bulbs with roots after treatment are transferred to Knop's nutrient medium for observation at specific intervals. The effects can be classified into subnarcotic, narcotic and lethal. The subnarcotic effects include chromosomal alterations and consequent lethality. Certain vital processes are switched off but enough of the basic processes are maintained so that the tissue can recover when the external influence is removed. On the other hand, certain compounds are highly toxic, leading to pycnosity, liquefaction of chromatin and subsequent lethality.

Place healthy bulb of *Allium cepa* at the mouth of jar containing the chemical to be tested. Keep the jar at room temperature. Excise young healthy root tips and fix in acetic acid–ethanol mixture (1:2) for 30 min. Transfer to 45% acetic acid for 15 min. Prepare a mixture of 2% acetic–orcein in 45% acetic acid and N HCl in the proportion of 9:1. Slightly warm the root tips on flame for 4 to 5 s for staining, separation of middle lamella and clearing. Keep in the stain mixture for 1 h. Mount in 1% acetic–orcein or 45% acetic acid. Observe the number of chromosome fragments in metaphase and anaphase stages. The frequency of breaks gives an indication of the extent of chromosome damages caused.

Micronucleus test

The root tip micronucleus test, initially established in *Vicia faba* by Ma (1982), has been used extensively for the detection of environmental mutagens

(Grant, 1994; Ji and Chen, 1996; Sandhu *et al.*, 1994) and clastogens with *Allium cepa* (Ma *et al.*, 1995). In Early-Early synthetic type of maize, the presence of micronuclei in root tips has been taken to indicate chromosome aberrations produced as a result of exposure to environmental toxicant.

Micronucleus test has also been utilized in different plants, following exposure to radiation. Micronuclei are taken to indicate non-disjunction, resulting in whole chromosomes or acentric fragments being left out of the spindle.

Collect young tips from germinating seeds of plants grown in a specific environment which is to be monitored. Fix for 24 h in ethanol–acetic acid mixture 3 : 1. Rinse in demineralized water. Hydrolyse in N HCl at 60°C for 7.5 min. Rinse the root tips again and place in Feulgen solution for 1 h and squash in 45% acetic acid.

Alternatively, Treat the root tips with pectinase solution for softening and clearing. Squash the root tips in 45% acetic acid. Screen the interphase cells under the microscope. At least 1000 interphases are scored in each slide. Five slides are used per treatment making a total of 5000 cells. It is a convenient and rapid assay method for monitoring acute or chronic exposure and shows direct correlation with the degree of pollution.

For short term exposure, treat seedlings in the toxicant for 6 h at room temperature. Transfer to distilled water and keep for 22 h to recover. Cut out and fix root tips in acetic–ethanol (1 : 3), hydrolyse with 1 N HCl at 70°C for 7–8 min, stain with Feulgen solution for 2 h and squash as usual. Only the micronuclei localized inside the cell wall and in the cytoplasm surrounding the main nucleus, are counted.

The comet assay

The comet assay is a sensitive and rapid method for DNA strand break detection in individual cells. Its use has increased significantly in the past few years. The process can be summarized as given in Fig. 1 (taken from Fairburn *et al.*, 1995)

The comet assay, also called the single cell gel assay (SCG) and microgel electrophoresis (MGE) was first introduced by Ostling and Johanson in 1984 as a microelectrophoretic technique for the direct visualization of DNA damage in individual cells. A small number of irradiated cells suspended in a thin agarose gel on a microscope slide were lysed, electrophoresed, and stained with a fluorescent DNA binding dye. The electric current pulled the charged DNA from the nucleus so that relaxed and broken DNA fragments migrated further. The resulting images, which were subsequently named for their appearance as 'comets', were measured to determine the extent of DNA damage.

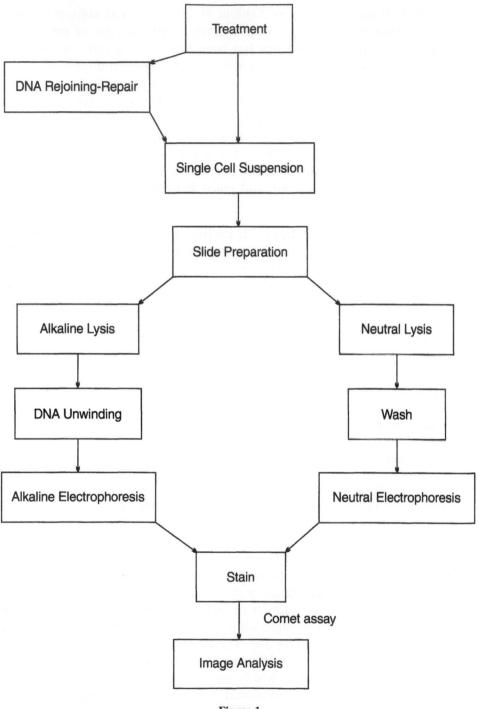

Figure 1

CHAPTER IV.3

MANIPULATION AND TRANSFER OF CHROMOSOMES AND GENOMES THROUGH HYBRIDIZATION

Chromosome engineering had its inception in the pioneering contribution of Sears (1956), who utilized irradiation as a means to translocate a gene for leaf rust resistance from *Aegilops squarrosus* to wheat. The establishment of aneuploid lines permitted manipulation of individual chromosomes and genetic analysis of wheat and its relatives. The production of a large number of interspecific hybrids and amphiploids aided the generation of several addition and substitution lines in this group. Such lines proved to be of much help in transfer of foreign genes in this group of plants (Feldman, 1988; Sybenga, 1995).

In the *primary* pool of wheat, the donors of all the three genomes, AA, BB, and DD, are included. Transfer of genes from the primary pool could be achieved through hybridization which permitted crossing-over of homologous genes and through back-crossing.

Closely related species of amphiploid *Triticum* and *Aegilops*, having genome affinity with wheat, are included in the *secondary* pool. The latter also covers the Sitopsis section of *Aegilops* contributing possibly to B genome. The transfer from secondary pool is possible by hybridization, back-crossing and selection of homologous genome. In wheat and its allies, homoeologous pairing and crossing-over can be induced genetically in interspecific and intergeneric hybrids. Transfer of alien segment through induced homoeologous pairing can be achieved by hybridizing alien species, synthetic amphiploids, or addition and substitution lines to certain genotypes, which allow homoeologous pairing. Several genes of common wheat, such as Phl, the pairing gene located in 5B which restricts pairing between homologues, can be deleted to achieve homoeologous pairing. The latter objective can also be fulfilled by counteracting the effect of Phl by *High pairing* gene of *Aegilops*, which allows several Triticineae chromosomes to pair with their wheat homologues.

Finally, both diploid and polyploid species, containing genomes non-homologous to wheat but related through possible homoeologous segments, are included in the *tertiary pool*. Here no gene or chromosome transfer is possible by normal hybridization and recombination. For transfer in such cases, induction of chromosome aberration through irradiation, *in vitro* technique, as well as special genetic methods is necessary.

Of all the special methods applied, wide hybridization coupled with embryo rescue, has been very successful. The technique involves hybridization between the recipient and the donor species, which are genomically widely different from one another, such as, wheat, maize, *Elymus, Hordeum, Secale* and others, in addition to the members of the secondary pool with homoeologous and non-homoeologous segments. In such cases, limited number of embryos can be formed. In wheat and maize hybrids, the embryo initially showed clear haploid complements of wheat and maize. But in successive cell cycles, all maize chromosomes were eliminated, leaving only one haploid set of wheat. The haploid embryo, on culturing, resulted in the production of haploid plant. As such, wide hybridization is now considered to be very effective for chromosome manipulation, utilizing the production of haploids (Laurie *et al.*, 1990; Matz and Mahn, 1994).

The technique (Matz and Mahn, 1994), as adopted for wheat with pollen parent of *Pennisetum* sp., involves: emasculation at least 1–2 days prior to anthesis and pollination between 1–7 days after anthesis; treatment of pollinated spikes with synthetic auxins, namely 10–1000 ppm of p-chlorophenoxyacetic acid, 2,4D; 2,4,5T or dicamba (3,6-dichloro-2-methoxy benzoic acid) by spraying, dropping and dipping for stimulation. Embryo formation for non-imasculate crosses was tested at different temperature regimes. Further study involved observation of the frequency of stimulated grain formation and differentiation of embryos after 14–18 days. If necessary, excised embryos are cultured in Kruse's medium. Chromosome doubling is carried out, if needed, in plantlets following exposure to low colchicine (0.02%) for 48 h in the medium.

Elimination of alien chromosomes from some crosses has been taken advantage of in gene transfer. In crosses involving *Hordeum bulbosum* and wheat, partial elimination of barley chromosomes was noted, resulting in wheat × *bulbosum* addition lines. Similarly, wheat–maize addition lines as well as oat and maize crosses were also obtained. In order to secure such addition lines for gene transfer, it is desirable to use a wide range of genotypes, so that some of the genotypes may permit the retention of alien chromosomes. For alien gene transfer, addition and substitution lines are often necessary.

ADDITION LINES

In gene transfer experiments, utilizing F_1 hybrid, it is necessary to produce an amphidiploid, particularly if the F_1 hybrid is sterile. Such amphidiploids permit the expression of alien genes, avoid loss of chromosome integrity in addition lines, and serve as a source material for gene transfer. However, because of difficulty in securing addition lines from a basic alloplasmic source (Jiang *et al*., 1994; Sharma and Gill, 1983) and to have it in a euplasmic background, the first backcross involved an amphiploid as the male parent. In the production of amphiploids, colchicine treatment is always favoured for chromosome doubling. Success in the production of amphiploid addition lines through colchicine doubling, however, would greatly depend on the response of the genotype. For the accuracy of the method, the concentration and duration of treatment, which differ from species to species, as well as the target organ, are influencing factors. Genotypes of wheat, an allied species, often differ in their response at an intraspecific level.

There are certain limitations in securing a full set of addition lines of one alien genome. The principal problems are—genetic sterility associated with certain chromosomes, genocidal effects under certain combinations, and even poor pollen transmission in certain genotypes (Jiang *et al*., 1994). Moreover, though it is possible to add alien chromosomes to wheat-producing alien addition lines, yet because of the imbalance, the genotype may ultimately lose the alien chromosome and return to euploid state. In order to avoid this difficulty, Sears (1981) worked out the strategy of substituting the alien chromosome pair for a related homoeologous pair of wheat. It was achieved by crossing the alien with a proper wheat monosomic, and recovering F_2 plants deficient for wheat pair and disomic for alien pair.

SUBSTITUTION LINE

Substitution lines are those in which a chromosome or a complete genome of an alien species has been substituted in the target recipient. Such chromosome substitution with foreign gene can also be obtained as a special strategy without using addition lines. The method involves (Kota and Dvorak, 1985) a cross between a monotelosomic and a different diploid species to secure a nullisome amphidiploid, followed by backcross to a monotelosomic with male nullisomic amphidiploid (Zhang *et al*., 1992). The homoeologous partner of the diploid species will act as substitute for the missing chromosome in the nullisomic amphidiploid. Following this method, several substitution lines can be prepared, each line utilizing an amphiploid.

The monotelosomic amphidiploid should be checked for the fertility or sterility of the male. Several substitution lines can be prepared utilizing an amphidiploid for each line. The availability of amphiploid and fertile nullisomic lines is obligatory for this approach.

The substitutions including intervarietal transfer require the availability of monosomics and nullisomics. Such monosomics and nullisomics for all chromosomes are available in wheat, maize, *Nicotiana* as well as cotton

(Endrizzi *et al*., 1985; Proc. Wheat Genet. Symp., Beijing 1993; Kuspira and Unrau, 1957; Sears, 1954). Their importance in gene mapping has been fully realized, as in rice.

ORIGIN AND USE OF MONOSOMES

The monosomics can also be achieved through a variety of procedures. Of all the methods, the mediation of haploids as done in wheat by Sears involved the crossing of *Triticum aestivum* ($n = 21$) with *Secale cereale* ($n = 7$) resulting into a few haploids. The latter, when crossed as the female parent with *T. aestivum*, led to the production of a few monosomics. These monosomics, on selfing, further gave a series of aneuploids including nullisomics and triploids.

Another method is to secure an interspecific hybrid involving a polyploid and a diploid progenitor, followed by back-crossing with the polyploid parent, as done in case of *Nicotiana*. Several monosomics can be obtained through this method. Partial asynapsis is often noted in interspecific hybrids, as in asynaptic mutants. Such asynapsis often leads to the formation of univalents, and later monosomics following union, as in wheat and tobacco. Irradiation with ionizing rays also causes disbalance in chromosome movement and separation, leading to the presence of unequal number of chromosomes in gametes. If viable, such gametes result in the production of monosomics and trisomics.

In addition to such induced methods, spontaneous production of monosomics is on record in several crops including wheat, oat, and tobacco. The most convenient method of the identification of monosomics, is meiotic analysis showing univalents.

In addition to the use in alien chromosome substitution, monosomics and multisomes are utilized in gene mapping as well. Gene mapping in chromosomes, with the aid of mono and multisomics, involves a comparative analysis of different traits in normal and monosomic individuals. The absence of certain traits in monosomics can be attributed to the location of these genes in chromosomes in which the monosomic plant is deficient.

GENE AND CHROMOSOME TRANSLOCATIONS

In order to secure translocations of alien gene, substitution or addition lines, as used in wheat, are intermediate steps. Several strategies have so far been adopted for securing translocations.

The first strategy is to use homoeologous pairing in substitution or addition lines. In wheat, this approach requires the elimination of 5B chromosomes, controlling pairing, using pH 1B mutant (Sears, 1981); or inhibiting the expression of Ph1 gene. Ph1 gene is obtained from *Aegilops speltoides*; the high pairing gene (Chen *et al.*, 1994), suppresses the effect of Ph1 and induces homoeologous pairing. Being dominant in single dosage, it can promote alien translocation, without removing Ph1. But it is not as effective as the removal of 5B in inducing homoeologous pairing (Jiang *et al.*, 1994).

An ingenious method was worked out by Sears (1972, 1981), substituting one arm of one alien chromosome for a corresponding wheat arm. The method takes advantage of the fact that univalents in monosomic wheat often misdivide at meiosis leading to telocentric chromosomes. Similarly univalent alien chromosome too may become telocentric by misdivision. The fusion of the two telocentrics may result in a heterobrachial chromosome, one arm being of wheat and the other of the alien chromosome. In view of the probability and complicated procedure, this method is less used as compared to other methods of substitution. Such plants have stable constitution as observed in East European cultivar of wheat with 1R of rye substituted for 5B of wheat (vide Sears, 1981; Zeller, 1973).

However, as mentioned earlier, the discovery of chromosome 5B of wheat controlling pairing within homologues (Riley *et al.*, 1968) and pairing suppressor Ph1, controlling homoeologous pairing (Sears, 1977), has been of much use in inducing stability of alien chromosome transfer. Several genes of wheat act as promoters or inhibitors of chromosome pairing.

The presence of Ph1 on the long arm of chromosome 5B restricts pairing only to homologues. If it is absent, homoelogous chromosomes also pair but less than the homologues. The deletion of Ph1 or counter-acting its effect by high pairing types of *A. speltoides* or its mutant as mentioned earlier, can permit several chromosomes of Triticinae to pair with wheat homoeologues. Such induced pairing is necessary to transfer genes from chromosomes to wheat.

The pollination of monosomic 5B with the desired alien species leads to a large percentage of offspring deficient in 5B chromosome. This method could lead to transfer of genes of *Aegilops* and rye to wheat (Joshi and Singh, 1979). However, even then, the most efficient method is the use of multisomic for 5B or a mutant line Ph1b, Ph1b, deficient for Ph1, the pairing suppressor for homoeologous pairing (Sears, 1977).

Regarding survival, breakage and translocations may or may not involve centromeres (Jiang *et al.*, 1994). If such translocated chromosomes contain fertility restorer S (*Rf*) genes, there is selective survival of such plants. In order to succeed in alien translocation, several factors are to be kept in view. It is always desirable to utilise an unaltered chromosome as the recipient and not the structurally modified ones such as 5A, 5B, which are already involved in translocation (Liu *et al.*, 1992). Similarly, the essential sequences needed for vital functions such as diploidization and fertility, are to be kept intact for normal functioning of translocated sequences without impairing fertility. The location of the target genes is one of the crucial factors for survival. The genes located near the telomeres have more chances of recombination than those located near the centromeres (Jiang *et al.*, 1994; Werner *et al.*, 1992).

IRRADIATION AND GENE TRANSFER

The other methods involve irradiation, causing random breakage, somatic instability in tissue culture or use of gametocidal chromosomes (Endo, 1990). The latter often leads to random breakage and translocation.

The advantage of irradiation is that its effect is independent of the location of the gene, and does not involve any loss of chromatin. However, it suffers from the limitations of imbalance associated with translocation (Jiang *et al.*, 1994); association with undesirable genes and long gestation period for stabilization of genotypes with translocations. Further researches are necessary for the separation of desirable and undesirable genes and the removal of unnecessary alien chromatin in translocations through radiation treatment involving small segments and little imbalance. In view of these disadvantages, very few cases of irradiation translocations have been utilized in wheat. Tissue culture has its use in inducing translocations as well as regenerant lines with additions and substitution in hybrids. But the exact mechanisms of causing such predictable aberration and viability of regenerants are yet to be worked out.

TECHNIQUES FOR IDENTIFICATION OF ALIEN SEGMENTS

The identification of alien segments involves three aspects namely, identification of the chromosome with translocation, break points and amount of foreign chromatin. The analysis of karyotype and meiotic behaviour may to a great extent permit detection of alien sequences. However, the banding technique, with its capacity to detect dispersed repeats, is a valuable tool. Moreover, in recent years, the use of molecular markers and *in situ* hybridization with different fluorescent probes for different genomes has become powerful tools for identifying different genomes or gene sequences in a complex crop. These methods have been outlined in the chapters on *Banding* and *In situ hybridization*. Interchange breakpoints can however be located through analysis of karyotypes; pachytene chromosomes: chromosome pairing, recombination and chiasma frequency and finally linkage studies utilizing genetic markers. Crossing with monosomic or trisomics, if available, may give a trivalent or tetravalent configuration in case two chromosomes are involved in translocation (Burnham, 1954; Menzel *et al*., 1991; Tsuchiya, 1961).

TRANSLOCATION TESTER SETS

For a long time, genetic manipulation, involving induced interchanges has been in practice in genetics, the breaks and translocations being achieved through ionizing radiations. Several crops including rice, barley, bajra, cotton, maize and others thus contain translocated lines. In order to identify and locate translocations, tester sets are often used as in rye, tomato and cotton.

The technique for the preparation of translocation tester sets involving all chromosomes in the set, requires initially seed irradiation of the material and induction of translocation and selection of heterozygote in M_1 through the detection of translocation ring. The seed selfing of the selected M_1 heterozygotes leads to the formation of normal and translocation heterozygotes showing bivalents, in addition to heterozygotes with rings. The selected homozygotes are then crossed with normal individuals leading to production of individuals in M_2 showing normal meiosis. Each individual represents a translocation line. Translocation lines involving each chromosome are then prepared and can be used as translocation tester set. Such translocation sets are ideal materials for testing induced translocation in any chromosome. This technique, if coupled with chromosome banding methods, can delineate exactly the location of breakage points, and translocations with even the amount of chromatin involved.

CHAPTER IV.4

GENOME MANIPULATION IN CULTURED CELLS

TISSUE CULTURE IN ARTIFICIAL MEDIUM

The cultivation of cells, tissues and organs *in vitro* forms the theme of tissue culture, where the objective is to simulate *in vivo* conditions in artificial medium. The technique has immense application in chromosome research, including their manipulation and engineering. In genetics, cell, tissue, and embryo culture has acquired the status of a routine technique in the laboratory. The importance of tissue culture lies principally in its capacity to secure rapid propagation of disease-free clones and of embryo culture in the production of interspecific hybrids and in overcoming the incompatibility barrier. The potential of protoplast culture in foreign gene incorporation and of pollen culture in the production of haploids is immense. The totipotency of plant cells in regeneration of whole plants from a single cell, is the basic premise underlying all strategies in tissue culture.

Principal steps involved in culture

Due to the restricted nature of growth in certain specialized and local regions, not all the tissues and organs form convenient material for cultural studies. The highly active growing points, such as apices of stem, leaf and root as well as buds, provide excellent materials for tissue culture. Pollen, embryos and endosperm of higher plants, as well as spores and prothalli of lower groups, can be cultured in a suitable medium. Protoplasts isolated from cells are cultured and form an ideal substrate for genetic manipulation.

All nutrient media for culture contain essential macro- and microelements, sucrose as carbon source, supplemented by vitamins, auxins, amino acids, cytokinins, etc. along with in some cases, antibiotics. The nature of the medium is also an important factor. A mixture which is effective for callus formation in a liquid medium may not be so in a solid medium. Some commonly used media are listed at the end of this section. *Callus culture* can be done either in solid or liquid media. For solid media, the most convenient and simple ingredient used for solidification, is agar. The liquid media may be either stationary or agitated, the latter being of wide use as adequate aeration is ensured. Callus can be utilized for organogenesis, formation of somatic embryos and regeneration. Chromosomes can be studied at all stages.

For suspension cultures, cells or colonies are grown in a liquid medium through dispersion and movement. Materials for suspension culture may be obtained from pieces of friable callus or grinding the tissue or embryo in a homogenizer. The cell suspension after a period of growth may be plated on agar when colonies from single cell can be obtained. The suspension culture filtrate may be subjected to cellular count after pectinase treatment for separation by low speed centrifugation and removal of the supernatant. The suspension of the required density is mixed with a sterile medium in agar, pre-cooled and then finally plated in sterile petri dishes.

Strictly aseptic conditions are to be maintained at all phases of tissue culture, including sterilization of the object, tools and medium. Several chemical agents are used for tissue sterilization, the most common ones being calcium hypochlorite (9–10%), sodium hypochlorite (2%), mercuric chloride (0.1–1%), hydrogen peroxide (10–12%), bromine water (1–2%), silver nitrate (1%) and antibiotics (4–5 mg/l).

Cell cycle

The term 'cell cycle' implies the sequential occurrence of different phases of the cell, initiated from a division till the completion of the next cell division. In the actual method of analysis, it involves the time span between one point in a cycle to the same point in the next cycle. A recent important finding has been the necessity of cyclin-dependent kinase in cell cycle. The kinase which controls the behaviour of other proteins by attaching phosphate groups and the protein, cyclin, which is responsible for triggering kinase at a critical period, are both necessary (Murray and Hunt, 1993).

At the time of counting in culture, cells are in the logarithmic phases of growth, in which every cell is active. In this type of culture, the cell count at each stage is nearly proportional to the time taken by the cells to complete this phase. For example, if the generation time is 15 h and the frequency of dividing cells is 5%, the time required for mitosis is 15×0.05 h. This method of calculating the generation time also depends on the nature of the medium and the type of the cell. Mitotic indices can also be utilized in working out the duration of the cell cycle. The technique is based on the fact that the generation time (T) is inversely proportional to the division frequency of cells per unit time. If M is the mitotic index and d the period or duration of mitosis, then the number of cells entering into mitosis per hour is M/d. The following formula for working out the generation time fits well with the observational data:

$$T = \log 2 \frac{d}{M} = 0.693 \frac{d}{M}$$

Mitotic index at any given time is,

$$M(t) = \frac{n(t+d) - n(t)}{n(t)}$$

where n is the number of cells at the time t, and d is the mean duration of mitosis. In the logarithmic phase of cell multiplication, as in cultures, the number of cells at any time t is also $n(t) = n_0 2t/T$ where n_0 represents the number of cells counted at the beginning of this phase.

Determination of the duration of mitotic cycle: Treat germinating seeds or bulbs with 5–10 mm long roots with 0.1% caffeine solution for 1 h. Wash and then transfer to Knop's solution. At successive hourly intervals, fix root-tips to observe the development of binucleate cells induced by caffeine. The criterion by which the duration of mitotic cycle is calculated, is based on the first observation of telophase in a binucleate cell, since binucleate cells are obtained through inhibition of cytokinesis in telophase of the previous mitotic division. Confirmation of the mitotic cycle time can be made with a second treatment of caffeine when quadrinucleate cells form. The entire mitotic cycle is divided into four phases namely, G_1—growth phase, post telophase; S—DNA synthetic phase; G_2—post-synthetic growth phase, and M—rest of the mitotic phase. Checkpoints permit monitoring of progression through succession of steps and coordinate cell cycle for replication and seggregation of components. Species differ with respect to the durations of these different phases, G_1 phase in general being variable and taking the maximum and G_2 the minimum periods of interphase. The method of calculating the duration of the different phases in culture is usually through autoradiographic procedure. After pulse labelling or short treatment with tritiated thymidine and by varying the time between labelling and fixation, the duration of the three different phases can be worked out. If the cells are fixed after a long interval, only those which were at the S phase at the time of treatment show labelling.

Chromosome analysis

For chromosome analysis from liquid medium if necessary, the medium should be centrifuged and pellet of cells taken. From callus, it is desirable to take (50 mg or so) samples from different parts and place in fresh medium for growth for a week or more. The period of maximum frequency of division should be chosen for fixation. Usual schedules may be followed for fixing, hydrolysing and staining of the cells (Sharma and Sharma, 1994), the most common fixative used being acetic acid–ethanol (1 : 3) . For Feulgen staining, hydrolysis is preferably carried out in 5 N HCl for 1 h at 24–28°C. Dry ice technique may be adopted for making permanent preparations of squashes followed by mounting in euparal.

For softening of the tissue in case of hard callus, developing organs or somatic embryo, treatment with mixture of pectinase and cellulase (0.5% of each in 0.1 M sodium acetate buffer, pH 4.5) for 1–2 h has been recommended. For securing a high frequency of division and large number of metaphases, placing the culture for 12–14 h in darkness at 15°C followed by 11 h at 27–29°C and application of colchicine 5 h before harvesting have also been suggested.

For ultrastructure analysis of chromosomes of callus the most adequate method for fixation is for overnight with 6% glutaraldehyde in 0.1 M phosphate buffer (pH 6.9) at 5°C (initial 2 h being at 25–28°C) followed by washing in phosphate buffer for 3 h and then post-fixation for 1 h in 1–2% buffered osmium tetroxide. Dehydration in ethanol, embedding through propylenearaldite to araldite and staining in uranyl acetate are recommended.

Microculture for chromosome study in living cells

Prepare hybrid tobacco (*Nicotiana tabacum* x *N. glutinosa*) single cell clones, isolated from stem callus and grown in liquid 'tobacco' supplemented by coconut milk (150 ml/l), calcium pantothenate (2.5 mg/l), naphthalene acetic acid (0.1 mg/l) and 2,4-dichlorophenoxyacetic acid (6.0 mg/l), in tubes within a shaker. Place a drop of paraffin oil near each end of a standard microscope slide. Lower a No. 1, 22 mm square coverglass onto each droplet to form risers for a shallow central chamber on the slide. Put a drop of mineral oil in a rectangle on the slide, connecting the two coverglass risers and covering the inner end of each. Place a droplet of liquid medium at the centre of a third square coverglass. Isolate a single cell or a small cluster of cells from a culture-tube with a pair of flattened teasing needles under a dissecting microscope and transplant it in the droplet of liquid medium on the third coverglass. Invert the coverglass over the rectangle of mineral oil on the slide in such a manner that the mineral oil surrounds the liquid medium with its enclosed cells and the ends of the top coverglass lie upon the inner ends of the coverglass risers. The culture thus lies in a liquid medium in a tiny micro-culture chamber filled with liquid paraffin. Observe the microcultures directly under the Phase contrast microscope. Keep them in sterile petri dishes in the dark at 26°C at controlled humidity. By this method, the different cytological changes during the growth of the culture in living cells can be observed under both ordinary and phase contrast microscopes (after Jones and colleagues, 1960).

Pre-requisites for culture of cells and tissues

Preparation of media
For liquid nutrient medium (a) Add 100 ml of 0.005% aqueous ferric sulphate solution to 500 ml of 8% aqueous sucrose solution, 200 ml of standard salt solution, 2 ml of organic accessory solution and add distilled water to prepare

1000 ml of *stock* solution. (b) Further add 100 ml of distilled water to the stock solution. Distribute in 50 ml portions in flasks and in 1 ml portions in test-tube. Plug with sterile cotton wool and autoclave at 18 lb pressure for 20 min.

For agar nutrient medium (a) Prepare stock liquid nutrient medium as described in (a) above. Add an equal quantity of 1% hot agar solution in water to the liquid nutrient, mix, keep half of the mixture thus prepared in portions of 15 ml in test-tubes. (b) To the remaining half, for every 100 ml of agar nutrient, add 1 ml each of calcium pantothenate, biotin and naphthalene acetic acid solution, mix, divide into test-tubes, plug and autoclave.

Preparation of explant

For root, shoot or other organs, sterile water for washing and sterile scalpel for cutting before transfer to medium are essential. Roots are to be taken from germinating seeds after washing and soaking in sterile water. Temperature required is between 26–30°C. Specific photoperiods are needed for shoots and dark period for roots.

Callusing and suspension

The requirements vary according to the genotype and the explant. For leaf, hypocotyl and stem explant, MSI or II medium (1.0 mg/l NAA, 0.1 mg/l kinetin, 3% sucrose, and 0.8% agar) gives good results. Inoculation may need continuous light (3000 lux) at 25–27°C for 4–6 weeks. After subculturing in same medium for 3–4 weeks, the callus can be transferred to liquid S-2 medium (2.0 mg/l NAA, 0.5 mg/l BAP, 3% sucrose) and incubated at 25–27°C on a rotary shaker at 100 rpm to secure suspensions.

ANTHER AND POLLEN CULTURE

The anther and pollen culture technique is adopted to induce embryoids as well as haploids. It is always desirable to choose anthers, where the pollen is undergoing first mitotic division, which may be checked before selecting the anther.

The schedule for anther culture involves surface sterilization of unopened flower buds at proper stage, removal of sepals and petals, and plating the anther immediately in basal agar medium of Murashige and Skoog (1962) or floating on liquid medium. Several species may require hormones and cyto-kinins in conjunction with sucrose for their growth.

The schedule for pollen culture involves initially the gentle grinding of anthers of suitable size in a glass homogenizer containing liquid culture medium as in anther culture. The suspension is filtered through nylon sieve to hold the large particles allowing the pollen suspension to be separated as filtrate. The filtrate is centrifuged at 100× g for 4 min and the pellet is washed repeatedly in culture medium. After the addition of adequate medium, the suspension is transferred to culture vessel and may be gently agitated for aeration, if necessary. The suspension can also be cultured in agar medium.

Tissues grown *in vitro* have been utilized for chromosome manipulation, including induction of polyploidy. Treatment with chemicals like colchicine, oryzalin and amiprophosmethyl, which cause microtubule inhibition and spindle arrest, can be applied in the medium during growth or directly to the tissue, callus or embryos. For induction of haploidy, pollen culture or embryo rescue method is adopted. The chromosomal variants, grown *in vivo*, often die because of loss of balance with the environment, consequent to genetic change. The artificial setup *in vitro*, on the other hand, through providing adequate nutrients with necessary growth promoters, enables such aberrant types to thrive and enter into organogenesis. The growth of regenerants and finally their establishment in the field require hardening under controlled conditions in the intervening period.

Representative schedules

1. Induction of polyploidy through shoot-tip culture in diploid banana (Musa acuminata)

Grow multiple shoot-tip cultures of *M. acuminata* clone SH-3362 in 250 ml Erlenmeyer flasks, containing 15 ml of liquid MS minimal organic medium (Murashige and Skoog, 1962), supplemented with $40 \, gl^{-1}$ sucrose, $40 \, mgl^{-1}$ cystein and $20 \, \mu M$ benzylaminopurine (BAP). Keep the flasks on a gyratory shaker (60 rpm) in a growth chamber at 28°C and 16 h light per day (3500 lux).

Isolate single root-tip meristems, 3 to 5 mm long, from actively growing cultures after 14 days of growth and dissect longitudinally. Filter-sterilise (0.22 μm) a 1 mM solution of oryzalin (3,5-dinitrodipropyl sulphate, Analar, Riedel-de-haen, GmbH, Germany) and a 0.1 M solution of colchicine (Plant Culture grade, Sigma). Both are anti-microtubule agents. Add each to separate sets of the liquid medium to reach final concentrations of 15, 30 and 60 μm oryzalin and 2.5, 5.0 and 10.0 mM colchicine. Keep for 48 h in media containing colchicine and for 7 days in those containing oryzalin (after van Duren *et al.*, 1996). After treatment, wash meristem tips three times with distilled water. Transfer to single shoot proliferation medium (MS), supplemented with $40 \, gl^{-1}$ sucrose, $40 \, mgl^{-1}$ cystein, Gamborg's BS vitamins, $5 \, \mu M$ 2iP (6-[-dimethylallylamino] purine and $0.1 \, \mu M$ IBA (indolyl butyric acid).

Measure weight and number of shoots per explant after 14 days of post-treatment growth. Propagate each shoot-tip for three vegetative cycles in liquid medium. Then place the plants on rooting medium (MS minimal organics supplemented with $40 \, gl^{-1}$ sucrose, $40 \, mgl^{-1}$ cystein, $12 \, \mu M$ BAP and $3 \, \mu M$ indole-3-acetic acid (IAA), solidified with 0.75% agar). Transfer rooted plants to Jiffy pots with a peat : soil mixture (1 : 1) and then to the greenhouse after an acclimatization period.

Prepare suspension of nuclei from leaf tissue as described elsewhere in this book. Stain nuclei with $2 \, mg \, l^{-1}$ DAPI (4,6-diamidino-2-phenylindole, Sigma) and analyse fluorescence using a Partec CA II flow cytometer (Partec GmbH, Germany). Routinely screen at least 10,000 nuclei per sample to determine ploidy level. The level can be confirmed by chromosome analysis and/or evaluation of the DNA content.

Propagate each selected tetraploid plant for two vegetative cycles.

After 14 days in culture, the survival rates for colchicine treated explants were 0–73% and for oryzalin treated ones 0–80%.

2. Induction of tetraploidy through ovule culture
In *Beta vulgaris* (after Hansen *et al.*, 1994).

Clone selected diploid genotypes *in vitro* by the micropropagation techniques described earlier. Grow under glasshouse conditions for 6 weeks before 3 months of vernalization at 4°C and a 16 h daylength. Transfer the plants to two-liter pots with a mixture of 1% leca, 10% perlite and 89% peat and store in a growth chamber at 10°C/5°C day/night. Raise for flowering in a growth chamber. Culture 10 ovules in 5 cm plastic petri dishes containing 1 ml of induction mixture.

For colchicine treatment, transfer ovules after culturing for 7 days to multidishes (35 mm diameter) containing 2 ml liquid induction medium supplemented with 0.05 to 0.4% colchicine and 1.5% DMSO. Gently spin (50 rpm in dark) during treatment. Remove colchicine solution after treatment, rinse the ovules twice with 5 ml liquid medium and transfer to a solid induction medium for continued *in vitro* culture. Determine the ploidy of regenerants when the plant reaches the 3–6 leaf stage *in vitro*. A short treatment of 5 h with a high level of colchicine (0.4%) gave the best results of 5.0 diploid plants per 100 ovules. 62.1% of the regenerated plants were diploids.

3. Induction of double-haploidy through anther culture
Two wheat (*Triticum aestivum* L.) genotypes (after Navarro-Alvarez *et al.*, 1994).

Evaluate the effects of genotype, media, sugar and colchicine concentration using a split–split-plot design with the whole plots arranged in a randomized complete block. Place anthers from two genotypes in nine embryoid initiation media with three sugar sources (0.26 M maltose, 0.26 M sucrose and 0.13 M maltose + 0.13 M glucose; Sigma, St. Louis, MO) and three colchicine concentrations (0, 0.1 and 0.2 g/l, Sigma). Prepare 17 petri dishes (60 × 15 mm) of 30 anthers per dish using modified Liang's embryoid initiation media with wheat starch as gelling agent for each genotype-sugar-colchicine concentration combination.

Keep anthers on the colchicine treatment media in dark at 25°C for 72 h. Remove, wash three times with liquid initiation media (prepared without wheat starch or colchicine) and then transfer to the corresponding colchicine-free medium including wheat starch for embryoid initiation. Keep in dark at 25°C for 90 days, observing periodically after the first 14 days.

Transfer 1 mm long embryoids to a regeneration medium having the same sugar as the initiation media for regeneration into green plants. Before transferring the regenerants to vermiculite, remove root-tips from a sample of each combination and count chromosome number following acetic–carmine squash technique. The low levels of colchicine added to the initiation media effectively increase the frequency of doubled haploid plants.

4. Initiation of double haploidy through microspore culture

In *Brassica napens* (after Chen *et al.*, 1994), direct colchicine treatment of isolated microspores resulted in a chromosome doubling efficiency of 70% of the whole plants. The results were much higher than after colchicine treatment of microspore-derived plants or embryos.

Culture isolated microspores in NLN-13 medium containing 0.05 or 0.1% colchicine at a density of 0.5–2 buds/ml. Incubate for 8–22 h at 32°C. Wash the microspores once with NLN-13 medium and reculture in fresh NLN-13 medium at 32°C at a density of 0.25–1 bud/ml for 3–5 days. Then transfer to 25°C. Keep in dark for 14 to 20 days and then transfer to light for embryo formation. The protocols for embryo culture and plant recovery were similar to those described earlier. Colchicine treatment also increased rate of embryogenesis. The doubled haploid plants were morphologically indistinguishable from diploid plants grown from seeds but could be easily distinguished from the haploid plants. This method of producing doubled haploid plants from microspores reduces the time needed for doubling at embryonic or whole plant stage.

5. Induction of diploidy on haploid shoots by oryzalin

In haploid apple shoots (after Bouvier *et al.*, 1994).

Take *in vitro* shoots of haploid variety of apple, produced by *in vitro* culture of immature seeds obtained by pollination with gamma-irradiated pollen. Remove the leaves along the stem, without damaging axillary meristems. Retain the apex. Cultivate the shoots *in vitro* at 30°C with a photoperiod of 16 h, using a medium with MS salts and the phytohormones, benzylaminopurine (BHP 1 mg/l), indolacetic acid (0.5 mg/l) and gibberellin (0.5 mg/l). Prepare 5,15 and 30 μM solutions of oryzalin in 5% DMSO and sterilise by filtration. Add to a low melting point agar and allow to settle.

Embed stem shoots with intact axillary meristems in the agar and keep in contact with oryzalin for 20 to 30 days. Transfer the shoots to fresh multiplication media and subculture.

Count chromosome number in cells from the newly formed axillary bud apex for determining ploidy level according to protocol for studying chromosomes from shoot apex as described elsewhere. A relatively high number of shoots with doubled chromosome number was induced with the lower concentrations of oryzalin.

6. *Induction of haploidy through embryo rescue technique*

In crosses between wheat × maize and wheat × pearl millet (Matz and Mahn, 1994).

Use varieties of wheat (*Triticum aestivum*) as female parents and varieties, both diploid and tetraploid, of *Pennisetum americanum* and *Zea mays* as pollen parents.

Emasculate the spikes one or two days before anthesis, hand-pollinate with freshly harvested pollen on the day of anthesis. In order to stimulate embryo development, treat the pollinated spikes with the synthetic exogenous auxins, CPA (*p*-chlorophenoxy acetic acid), 2,4-D (2,4-dichlorophenoxy acetic acid), 2,4,5-T (2,4,5-trichlorophenoxy acetic acid), or dicamba (3,6-dichloro-2-methoxy benzoic acid). Compare the different methods of application—spraying, dropping and dipping. Use concentrations between 10 to 1000 ppm and apply after pollination between 1 and 7 days.

Record the frequency of stimulated grain, number of embryos and degree of embryo differentiation 14 to 18 days after pollination. For chromosome doubling, culture plantlets *in vitro* with one passage on a medium containing 0.02% colchicine for 24–48 h. Count chromosome number in male p.m.c.s following staining with acetic–carmine.

The successful steps recommended are: (i) Single treatment with an aqueous solution of *Dicamba* for embryo stimulation *in vivo*; (ii) application by spraying or dipping the spikes at 2 to 4 days after pollination; (iii) embryo rescue 15 to 18 days after pollination. This method is successful in producing haploids in wheat × maize and wheat × pearl millet crosses.

CULTURE OF PROTOPLASTS

The naked protoplast in culture has become a convenient substrate for regeneration, securing cell fusion, and incorporation of chromosomes and foreign genes in the regenerants.

The aim is to isolate protoplast without causing any irreversible damage to its structure. The maintenance of a correct osmotic level without excessive plasmolysis is an important factor in protoplast isolation. The actual concentration of osmotic stabilizer varies from tissue to tissue. In order to liberate protoplasts from the cell, digestion of cellulose and pectin and breakdown of xylans are essential. Therefore, for isolation, enzyme preparations, such as pectinase (macerozyme) derived from the fungus *Rhizopus*, cellulase (Onazuka) derived from *Trichoderma viride*, driselase—a cellulose enzyme complex derived from a basidiomycete as well as hemicellulase to break xylans, are widely used. Helicase from snail has been used in some cases. In view of the fact that enzyme preparations often contain some low molecular weight chemicals, which may cause toxicity by affecting osmotic concentration, it is often necessary to desalt the enzymes through centrifuging in a salt solution, and gel filtration. Antibiotics are added to eliminate contamination.

Protoplasts can be isolated in two ways: (i) isolation of single cells, followed by isolation of protoplast or (ii) by direct treatment in an enzyme mixture, which serves both objectives.

For protoplast isolation, it is always desirable to grow plants under controlled growth conditions such as 1000–10,000 lux light intensity for 16 h at 22–25°C. For a good yield of protoplasts, growth under low light intensity (2.52–10.8 MJ/m^2/day for 15 h a day) with balanced fertilizer containing calcium nitrate may ensure continued supply of mesophyll cells. The entire isolation technique should be carried out before a laminar air flow cabinet.

Protoplast isolation from mesophyll cells

Technique 1
Sterilise the leaf surface, immersing in 70% ethanol for 30 s followed by 2.5% sodium hypochlorite for 30 min and thorough washing with sterile water. Peel off the lower epidermis of the leaf in strips (4–6 g) and place in 20 ml of maceration medium (0.5% macerozyme, 13% mannitol and 1% potassium dextran sulphate). Adjust the pH to 5.8 with 2 N HCl, before filter sterilization. Shake the materials in the maceration medium at 25°C in a reciprocal shaker (100–120 cycles/min, 4.5 cm stroke). Isolate cell fractions after 30 min, 1 h, 1–15 min, 2 h and 3 h, replacing the macerating medium each time. Centrifuge the last two fractions which are almost pure (100–200× g, 3 min) and resuspend twice in fresh 13% mannitol and centrifuge again. Put the isolated cells in 40 ml, 4% Onazuka cellulase in 13% mannitol, pH 5.2 (adjusted with 2 N HCl before filter sterilization). Incubate the suspension at 36°C for 3–3.5 h with gentle swirling. Harvest the protoplasts from the medium by slow centrifugation (100× g) for 1 min. Resuspend twice for washing in 13% mannitol with 0.1 mM calcium chloride and centrifuge again. Suspend the protoplasts in 5 ml, 13% mannitol; count a sample in counting

chamber and sediment again by centrifugation. Resuspend in fresh 13% mannitol to get a concentration of $1-4 \times 10^6$ protoplasts/ml.

Technique 2
Surface sterilise the leaves and wash in sterile water. Take slightly flaccid leaves, remove the lower epidermis. Cut pieces of peeled areas and float exposed surface downwards in mixture of 13% mannitol and CPW salts (KH_2PO_4 , 27.2 mg/l; KNO_3, 101.0 mg/l; $CaCl_2 \cdot 2H_2O$, 1480 mg/l; $MgSO_4 \cdot 7H_2O$, 246.0 mg/l; Kcl, 0.16 mg/l; $CuSO_4 \cdot 5H_2O$, 0.025 mg/l pH 5.8) in a petri dish (14 cm in diameter) for 1–2 h. Remove the mixture and replace with 20 ml (approx) filter sterilized enzyme mixture (4% w/v meicelase, 0.4% w/v macerozyme, 13% mannitol, CPW salts, pH 5.8, adjusted with 5 N HCl) and incubate for 18 h in the dark. With forceps, agitate slowly to release protoplasts, tilt the dish and allow the protoplasts to settle for 30 min. Pipette out the enzyme mixture. Transfer the protoplasts to screw-capped tubes, suspend in the mannitol/CPW salts and centrifuge at $35\times g$ for 10 min. Remove the supernatant, suspend the protoplasts again in 20% sucrose solution containing CPW salts and centrifuge at $50\times g$ for 10 min. Remove the protoplast from the supernatant with a pipette and resuspend in 1 ml mannitol/salts. Count the sample. Sediment the protoplasts for 5 min at $35\times g$ and resuspend in nutrient medium to get a final concentration of 1×10^5/ml. The technique may be modified specially in relation to enzyme level and plasmolyticum to suit different species.

Protoplast isolation from callus and cell suspensions

Protoplasts can also be isolated from cultured cells and cell suspensions. The friable nature of the cells allows rapid penetration of the enzymes.

Induction of callus and initiation of cell suspensions
Plant: Leaf/hypocotyl/stem explants from *in vitro* raised shoot cultures from seedlings.

Media/Solutions: MS-1 (1.0 mg/l NAA, 0.1 mg/l kinetin, 3% sucrose and 0.8% agar). MS-2 medium (2.0 mg/l NAA, 0.5 mg/l BAP, 3% sucrose).

Place sterile or *in vitro* raised leaf/hypocotyl/stem explants on MS-1 medium for callus induction. Incubate cultures under continuous light (3000 lux) at $25 \pm 2°C$ for 4–6 weeks. Subculture the callus once at 4–6 week stage on the same medium. Transfer uniformly growing callus at 3–4 week stage after sub-culture to liquid MS-2 medium in 1 : 3 ratio of callus : medium. Incubate liquid cultures at $25 \pm 2°C$ on continuous rotary shaker at 100 rpm. Draw growth curves from above cultures and determine optimal time (exponential phase of growth) for protoplast isolation.

Isolation of protoplasts from cell suspension
Plant: 4 day old cell suspension culture as above.
Media/Solutions: Enzyme solution-II, CPW13M, CPW21S PCM.

Allow the cell clusters to settle down in the flask. Decant the medium slowly and replace with CPW13M. Leave for 30 min to 1 h. This step facilitates maintaining osmotic stability of protoplasts that are released. Add 20–25 ml of enzyme solution per 250 ml flask and incubate overnight (15 h) at 28°C preferably on a slow rotary shaker (30 rpm). Pour the digested contents of the flask onto a 75 μ sieve. Agitate and wash with CPW13M. Collect the filtrate in screw-capped centrifuge tubes. The remaining steps are the same as those described for mesophyll.

Isolation of protoplasts from other plant parts
1. *From cotyledons* Sterilise seeds and germinate in half strength MS medium with 1% sucrose. Keep seedlings at 3–4 day stage for 6–48 h at 6–10°C. Chop the cotyledons into thin slices with a scalpel blade and plasmolyse in CPW salts with 13% mannitol for 1 h. Use a smaller petri dish. Replace CPW13M with enzyme solution (same strength). Incubate overnight (15 h) at 28°C. Pour the incubation mix onto a 50 μ sieve and release protoplasts by squeezing with a pasteur pipette.

2. *From roots* The procedure for the isolation of protoplasts remains essentially the same as that for cotyledons except that seedlings are chosen at a relatively earlier stage, i.e., 2–3 days prior to emergence and opening of cotyledons. Depending upon the plant system that one is working with, protoplasts can be isolated from hypocotyl stem and petiole explants.

3. *From microspores or pollen* Microspore or pollen tetrad derived protoplasts do not divide. However, they can be used to develop a selection method for the production of 3n plants by gameto-somatic fusions and to study the inheritance of organelles in gametic:somatic hybrids.

Sterilise flower buds in 0.1% mercuric chloride for 2 min and rinse with sterile distilled water 5–6 times.

Excise anthers and squeeze the pollen tetrads directly to a 5 ml enzyme solution in a petri dish. Maintain an aseptic state throughout. Incubate for 1 h at 27°C in dark.

Pass through 40 μ sieve and wash with PCM solution (MS salts + 3% sucrose, + 9% mannitol, 1.0 mg/l NAA, 0.2 mg/l 2,4D, 0.5 mg/l BAP, pH 5.8). Purify by centrifugation (60g, 10 min) 2 times and suspend.

Methods of protoplast culture

Protoplasts can be cultured in different combinations of various media, mostly based on Murashige and Skoog or Gamborg's formulations. They can be plated on liquid, agar, liquid–agar combination or on filter paper or even as a microdrop in a petridish. Cultures are initially maintained for 10 to 15 days in dark at 25–30°C.

For liquid layers: Plate 10 ml of the protoplast suspension at requisite density into 90 mm petridish. Seal with parafilm and incubate in moist environment.

For embedding in agar: Melt double strength agar (1.4%) medium with 9% mannitol and bring to 45°C; pour 5 ml of protoplast suspension into a 90 mm petri dish at double the required density and then pour 5 ml of molten double strength agar medium into 5 ml of protoplast suspension in a 90 mm petri dish. Mix thoroughly and allow agar to set. Seal dishes with parafilm.

For liquid agar combination: Allow measured volume of agar solidified PCM to set as a thin layer—10 ml/90 mm petri dish. Pour 10 ml of the protoplast suspension in culture medium and at twice the optimum density over the thin layer of agar. Seal with parafilm.

For filter paper placed on agar, place a sterile filter paper over the agar substratum as given before and then pour gently 1 ml of a highly concentrated protoplast suspension on the filter paper, taking care that the liquid does not overflow.

Chromosome analysis
For chromosome analysis, centrifuge the protoplasts and take the sediment with nuclei. Pre-treat with saturated solution of aesculin in water for 30 min at 12°C before fixing in ethanol:acetic acid (3:1) for 2–3 h. Warm in a mixture of acetic–orcein/N HCl (9:1) for a few seconds and after 1 h, squash in 1% acetic–orcein or 45% acetic acid. Carbol fuchsin stain can also be employed.

Purification of enzyme mixtures
Dissolve required concentration of enzyme mixture (for 100 ml) in 40 ml water. Centrifuge at 8000g for 15 min and collect the supernatant.

Load 40 ml of above supernatant on to a Biogel P6 column (2.5 × 35 cm; bead size 80–150 μ) equilibrated with distilled water. Add enzyme to run into column, followed by continuous dripping of distilled water to elute the enzyme. Collect the first peak of approximately 50 ml using a fraction collector and add to extract.

Alternatively, as soon as the light brown fractions are collected, begin protein test as follows:

Mix 1 part of eluted enzyme with 2 parts of 10% TCA. Foaming indicates presence of protein. Collect first 50 ml of the protein fraction. Supplement with CPW salts for 100 ml volume; 13% mannitol and 2-(morpholino) ethanesulphonic acid (MES) as buffer. Make volume to 100 ml and use.

SOMATIC HYBRIDIZATION

Protoplasts, once isolated and devoid of pectin and cellulose, can be utilized for regeneration or *somatic fusion*. The fusion technique brings together the cells of two widely differing genotypes in culture, followed by regeneration. Initially applied to the members of Family Solanaceae, the fusion experiments have been successful in other plant genera as well. Fusion can be achieved through both physical and chemical means, including electroporation on one hand and the use of polyethylene glycol, high pH and calcium or sodium salts on the other. The strategy is to bring membranes of the two fusing partners in close proximity (vide Wilson *et al.*, 1971; Power *et al.*, 1976). Of the different methods worked out, polyethylene glycol and high pH and calcium techniques are found to be most successful, along with electrofusion.

Selection systems in plants for the detection of hybridity or heterocaryon nature are rather specific. Refined chromosome techniques are essential for checking fusion. Carlson, Smith and Dearing (1972) utilized nutritional requirements as selection criteria of *Nicotiana glauca* and *N. langsdorfii*. Chlorophyll deficient mutants have been used as markers in varieties and mutants of *Nicotiana*. In *Sphaerocarpos*, Schieder's method (1975) of utilizing auxotrophic mutants is analogous to the method adopted in animal system. In maize, mutant strains have been used. Sensitivity to drugs, such as Actinomycin D, has also proved to be an effective marker. The advancement in the method of banding pattern analysis of chromosomes is an effective method of detection of somatic hybridity. Application of refined techniques for chromosome analysis using differential fluorescence patterns has immense potential in the identification of hybrids.

Somatic hybridization also permits the production of *cybrids* by fusing an enucleated protoplast with a normal recipient. The protoplast can be enucleated through irradiation or high speed centrifugation for at least 45 min. Cybrids often show male sterility as well as resistance to antibiotics and herbicides, as noted in cybrids of *Brassica* species.

1. Protoplast fusion with sodium nitrate

Suspend isolated protoplasts in a mixture of 5.5% $NaNO_2$ in 10% sucrose solution for 5 min at 35°C. Centrifuge at 200× g for 5 min. Transfer the pellet to a waterbath and keep for 30 min. Slowly replace the supernatant (mixture) with Murashige and Skoog's medium with an addition of 0.1% NaNO$_2$. After some interval, wash twice with the medium and plate (after Power, Cummins and Cocking, 1970).

2. Protoplast fusion using high pH and Ca^{2+} solution

Adjust the density of protoplast suspension to 1×10^6 protoplasts/ml and mix 1 ml suspension of each of cell suspension and mesophyll protoplasts in a centrifuge tube. Centrifuge at 30g for 5 min so as to settle the protoplast mixture and remove the supernatant. Resuspend the protoplasts in 5 ml solution of high pH Ca^{2+} and centrifuge at very slow speed so as to settle the protoplasts (0.74% w/v CaCl$_2$·6H$_2$O; 0.375% w/v glycine; 9% w/v mannitol—filtered, sterilized, pH 10.6). Incubate in a water bath at 30°C for 20 min. Gently remove supernatant without disturbing the pellet of protoplasts. Gently add 8–10 ml of stabilizer solution (sterilized CaCl$_2$·2H$_2$O— 3.5% w/v aq. soln.) without disturbing the pellet and incubate at room temperature for 30–45 min.

Remove stabilizer solution without disturbing the protoplasts and replace with PCM (8–10 ml). Incubate at room temperatute for 5–10 min. Remove PCM and repeat washing as above. Finally suspend the fusion mixture at a final plating density of 1×10^5 protoplasts/ml and plate 8–10 ml of the suspension in a 90 mm petri dish.

3. Protoplast fusion using single step polyethylene-glycol—high pH calcium

Mix 0.5 ml protoplast suspension at a density of 1×10^6 ml from each partner in a centrifuge tube. Centrifuge at 50g for 5 min. Remove most supernatant leaving 0.5 ml of protoplast suspension.

Take out the suspension and place it in 90 mm sterile petri dish in the form of 0.1 ml droplets (5–6 droplets/dish). Agitate the dish gently to accumulate the protoplasts in the centre of the drops and allow to settle for 5 min.

Add PEG–high pH Ca^{2+} solution (~ 0.2 ml/drop—20% PEG 8000, 35 g/l CaCl$_2$·2H$_2$O; 4 g/l glycine, pH 10.5 adjusted with KOH) around the drops containing protoplasts. Incubate at room temperature for 10–12 min. Add stabilizer solution (~ 0.5 ml/drop CaCl$_2$·2H$_2$O—3.5% w/v aq. soln.) slowly and elute the same with PEG–high pH Ca^{2+} which was initially added. Add fresh stabilizer. Incubate for 5 min, remove the stabilizer solution along with some fusogen. Repeat this process 2–3 times without further incubation. (This whole process should take 15–20 min).

After 2–3 washes with stabilizer, wash again 4–5 times with either CPW 13 M or PCM in the same manner as above. Repeat to remove the traces of PEG–high pH Ca^{2+} and stabilizer. The final volume of the PCM should be such that the overall plating density remains at 1×10^5 protoplasts/ml. Seal the petri dish with parafilm. Incubate the culture in humid chamber at 25°C in dark.

4. PEG + Ca method adopted for *Petunia* by Cocking and his associates

Isolate leaf protoplast of *P. parodii* and *P. hybrida*, as outlined under tissue culture. Suspend the protoplasts in 9% (w/v) mannitol solution containing inorganic salts. Keep the samples in screw-capped 8 ml tubes and use equal volume of each species for fusion. Centrifuge the tubes at $80 \times g$ for 10 min and remove the supernatant. To induce fusion, add 2 ml 15% (w/v) polyethylene glycol (mol. wt. 6000), 4% (w/v) sucrose and 0.01 M $CaCl_2$ in the tubes. Re-suspend protoplasts and keep at 25°C for 10 min. Add after every 5 min, M/S medium 0.5, 1.0, 2.0, 3.0 and 4.0 ml, continually re-suspending the protoplasts after each addition. Centrifuge at $60 \times g$ for 15 min. Remove the supernatant and add 8.0 ml M/S medium in each tube. Keep the tubes for 1 h before plating. (A sample count may be taken at this stage which shows nearly 4% of the nuclei in a fused state.)

Culture the protoplasts on liquid agar medium. Take 9 cm plastic petri dishes and add 8 ml M/S medium with actinomycin D (1.0 µg/ml) solidified with agar (0.5% w/v). Add 4 ml M/S medium with actinomycin D (2.0 µg/ml) and 4 ml protoplast suspension (2.0×10^5 ml) in M/S medium without actinomycin D on the surface of the agar. After this dilution the concentration of actinomycin D on the liquid becomes 1.0 µg/ml at a protoplast density 1×10^5/ml. Keep the culture at 27°C using daylight fluorescent tubes for 28 days. Transfer the cultures to M/S medium with 3% mannitol solidified with 1% agar, without actinomycin.

After 60 days, transfer the cultures to M/S medium without mannitol for the formation of callus. After 10 weeks, transfer the fused hybrid callus to M/S medium with IAA (2.0 µg/ml) and 6-benzylaminopurine (1.0 ml) for shoot regeneration. After shoot formation, transfer to M/S medium with NAA (0.1 µg/ml) and 0.3% agar. After the formation of plantlets of suitable size, transfer to pots to grow till maturity. Following this method, hybrids of *P. parodii* and *P. hybrida* were obtained with purple flowers and chromosome number ranging between $2n = 28$ and $2n = 24$ showing tetra and hypotetraploid constitution (after Power *et al.*, 1976).

Electrofusion to induce cell to cell fusion

The principle involves the generation of high voltage electric field pulses to increase the permeability to cells. Electrically induced protoplast fusion has been used to produce somatic hybrids. The fusion is achieved in a multi-electrode fusion chamber containing a mixture (1 : 1) of mesophyll protoplasts of both species. Following an alignment of protoplasts induced by an AC field of 125 V/cm and 1 MHz, fusion was initiated by an exposure of proto- plast samples to a chain of 3–4 DC pulses of 1.2 kV/cm each 20 ns. The fusion rate was estimated as 20–40%, 30% being binary fusions. A large number of somatic hybrid plants were recorded following fusion experiments in *Nicotiana* and other crops.

A simple method of electrofusion was devised by Watts and King (1984), in which 1.0 ml of protoplast suspension of suitable density can be fused within 20 s. It involves application of a radio frequency field from an oscillator (Sinusoidal 10 V RMS 0.5 W and range of 0.1 to 0.9 MHz) and causes mutual attraction of induced dipoles on the protoplasts. This resulted in movement and aggregation.

Fusion through irradiation/uv/laser treatment

Irradiation with gamma or X-rays of protoplasts has been used in the transfer of foreign genes through cell fusion. Similarly, UV laser light of nanosecond pulse length has been used as apical needle to prick cells. It causes a self-healing perforation in the membrane, through which foreign DNA kept in the medium can pass. Irradiated nucleus often undergoes elimination in the fused cell. As such, the donor cell can be irradiated and the recipient can be treated with iodoacetate for inactivation of protoplast. This method of fusion leads to production of *Cybrid*. The cybrids are particularly useful while dealing with cytoplasmically inherited traits, such as cytoplasmic male sterility (Ichikawa *et al.*, 1988). The gameto-somatic hybrids can be obtained by fusing protoplasts of pollen culture with those of leaf protoplasts through PEG technique (Lee and Power, 1988). The ultimate result is the substitution of a male nucleus in a female cytoplasm.

CHAPTER IV.5

CHROMOSOME TRANSFER USING CULTURED CELLS

Somatic fusion is gradually becoming a promising approach in transferring desirable traits, specially those controlled by polygenic sequences. This technique has considerable potential in crop breeding (see Waara and Glimelius, 1995). However, a limitation of somatic fusion is the involvement of the entire genome in the process, thus carrying with it, the undesirable genes as well. Repeated backcrossing and spontaneous chromosome elimination may minimise the problem but the entire procedure is time-consuming, and to some extent depends on chance.

Identification of hybrid cells can be done through the use of albino or azotrophic mutants and drug sensitivity as markers as well as fluorescence activated cell sorting and chromosome analysis. Chromosome analysis can be carried out of callus and regenerants utilizing p-dichlorobenzene/aesculine as pretreatment agent, fixing in Carnoy's fluid (6:3:1) overnight, mild heating in 2% acetic–orcein: N HCl (9:1) mixture, keeping for 3 to 4 h and then squashing in 45% acetic acid.

In somatic cell fusion another objective is to achieve limited gene transfer through the production of asymmetric hybrids. Through the accelerated elimination of non-viable chromosomes and judicious application of selective pressure, it may be possible to develop asymmetric hybrids of importance (Glimelius, 1988). The scope of controlling chromosome elimination in somatic hybrids offers great possibility in gene transfer.

Of the different methods, irradiation of donor protoplasts before hybridization may cause damage to chromosomes and elimination is not predictable (Puite, 1992). For partial gene transfer through asymmetric hybridization, two methods, namely, metaphase chromosome transfer and microprotoplast fusion, have been found to be promising. The former involves chromosome sorting and flow cytometry, which have been dealt with elsewhere. The latter initially results in the formation of micronuclei. It is followed by isolation of microprotoplast with one or a few chromosomes as a prerequisite for fusion with the recipient, having undisturbed genome.

In the method involving microprotoplast, the principal steps are:

(i) Induction of micronuclei in cell suspensions with the use of compounds which affect microtubule organization, cause metaphase arrest and chromosome decondensation and form several nuclei within a single cell. Normally, APM (amiprophosmethyl), oryzalin or crematin are used. High mitotic activity and synchronization of division are essential for this purpose.

(ii) The next step is the isolation of micronuclei with the aid of cell wall digesting enzymes such as macerozyme and pectinase in presence of the above compounds. These compounds increase the number of micronuclei through their action on decondensation and microtubule inhibition. The enriched microprotoplasts are then repeatedly washed and filtered through a nylon mesh of proper porosity.

(iii) Subsequently, the isolation of microprotoplasts is carried out by sucrose loading on iso-osmotic gradient of percoll and subjected to high speed centrifugation. The ultracentrifugation leads to bands containing mixture of protoplasts and microprotoplasts. A large number of microprotoplasts with a few chromosomes has to be selected. In order to secure a large number, microprotoplasts can be collected at a lower density or they can be filtered sequentially through filters of different pore size which ultimately may lead to protoplasts with one or a few chromosomes.

FUSION WITH MICRONUCLEI

For the use of micronuclei as donor, the cells should be highly active, with selectable markers such as Kanamycin resistance Kn^R or Glucuronidase sequences (S Gus). The media should be properly chosen, and isolation procedure should not affect the physiological activity, *vis-à-vis* growth.

The most important criterion for the choice of recipient system is that the system must have well established protoplast regeneration protocol. Moreover, it must have a selectable marker such hygromycin resistance ($Hygr^R$) or Nitrate Reductase deficiency (NR^-). This is necessary for double selection after fusion. The fusing agent may be polyethylene glycol or other compounds. The mixture of two protoplasts at the initial phase may be in equal ratio, but gradually there will be an increase in the ratio of microprotoplasts.

The next step is the immobilization of the fusion products in a thin layer of agarose and identification under adequate UV excitation. The identified portions are then selected and growth monitored.

The hybrid products are confirmed through different parameters. Confirmation of hybridity is obtained through chromosome analysis, biochemical criteria, fluorescence and other parameters. Selection of clones for regeneration and study of chromosome characteristics are then followed through the usual schedule.

METHOD TO INTRODUCE CHROMOSOMES IN PROTOPLAST

Stain isolated chromosomes in 0.1% 4,6-diamidino-2–phenylindole (DAPI) for 1 h in dark. Wash off the unbound stain through centrifugation in protoplast buffer (1 mM $CaCl_2$, 5 mM 2 (n-morpholino) ethanesulfonic acid (MES) and 10% mannitol, pH 6.5). Prepare mesophyll protoplasts of recipient species by overnight incubation in enzyme mixture (2% cellulysin Calbiochem, 1% macerase and 10% mannitol, pH 5.5). Wash off the enzymes through centrifugation and suspension. Incubate 1×10^6/ml of protoplast for 20 min and 1×10^7/ml of isolated chromosomes in 35% PEG 4000, 2% mannitol, and 12 mM $CaCl_2$, pH 6.0. Add 4 vol of 50 mM $CaCl_2$ and 10% mannitol, pH 8.5 to stop the reaction. Collect protoplasts by centrifugation and wash in protoplast buffer. Use fluorescence microscope for visualization of stained chromosome in protoplast. DAPI stained chromosomes fluoresce yellow green against red background.

MICROPLAST-MEDIATED TRANSFER OF SINGLE CHROMOSOME (after Ramulu *et al.*, 1995, 1996)

A transformed triploid cell line of potato, *Solanum tuberosum* (cell line 413 $2n = 3x = 36$), carrying a selectable marker Kan^R and the reporter gene *gus*, was used as the donor source. The sources of recipient protoplast were shoot cultures of tobacco *Nicotiana tabacum* ($2n = 4x = 48$) cv. Petit Havana SR 1 and the wild tomato *Lycopersicon peruvianum* PI 128650 ($2n = 2x = 24$), a highly regenerable hygromycin-resistant (Hyg^R) line.

Cremart treatment of donor cells

One day after subculturing, treat early logphase suspension cells of the potato line with Cremart (butamiphos, *O*-ethyl-*O*-(3-methyl-6-nitrophenyl) *N*-*Sec*-butylphosphorothioamidate, Sumitomo, Osaka), which inhibits microtubule formation, 7.5 μm for 48 h. Incubate the treated cell suspension for 18 h in a cell wall-digesting enzyme mixture (1% cellulase R-10, Onozuka and 0.2% macerozyme R-10, Yakult Honsa, Tokyo) in half-strength V-KM medium containing 0.2 M glucose and 0.2 M mannitol but no hormones. Add cytochalasin B (CB) 20 μM and Cremart 7.5 μM to enzyme mixture. Incubate in 9-cm petri dishes containing 1.5 ml packed cell volume and 15 ml enzyme solution on a gyratory shaker (30 rpm) at 28°C. Adjust osmolality to 500 mosmol kg^{-1} and pH to 5.6. After incubation, filter protoplast suspension through 297 μm and 88 μm nylon meshes and wash with halfstrength VKM medium with macro- and micronutrients and 0.24 M NaCl (pH 5.6).

Isolation of microprotoplasts

Load the purified dense suspension of protoplasts onto a continuous iso-osmotic gradient on Percoll (Sigma). Centrifuge at 100,000× *g* for 2 h in ultracentrifuge. Prepare the Percoll gradient by adding 7.2% (w/v) mannitol to Percoll solution. After centrifugation, one large and two small bands are formed from the top of the centrifuge tube, containing successively cytoplasts, evaculated protoplasts and microprotoplasts.

Sequentially filter the bands through nylon sieves of decreasing pore size (48, 20, 15, 10 and 5 μm). Collect the fractions as pellets in mannitol after centrifugation at 80× g for 10 min. Recentrifuge the supernatant at 160g for 10 min. Examine by microdensitometry, flow cytometry and microscopy to determine nuclear DNA content and viability of microprotoplasts.

Recipient protoplast culture and regeneration

Isolate protoplasts from leaf pieces of shoot cultures of the recipient lines after incubating overnight (18 h) in 1% (w/v) cellulose R-10 (Onozuka) and 0.2% (w/v) macerozyme R-10 (Yakult Honsha) dissolved in halfstrength V-KM medium with 0.2 M each glucose and mannitol but no hormone. Culture protoplasts of *Lycopersicum peruvianum* (Hyg^R) from shoot cultures in liquid medium (TM2 + 25 mg/L^{-1} hygromycin) for callus induction. Transfer to callus growth medium (TM3 + 50 mg/L^{-1} hygromycin) on day 18; and then to regeneration media (TM4 + 0.5% sucrose for shoot bud initiation and TM4 + 2.0% sucrose for root growth–elongation). 70–80% of the calli usually regenerate (about 30 shoots per callus) within 90 days. Higher sucrose is detrimental to regeneration.

Microprotoplast fusion and selection of hybrid cell

Carryout fusion between microprotoplasts obtained from the donor and the protoplasts of the recipient, using polyethylene glycol (PEG) based mass fusion protocol. Mix microprotoplasts and protoplasts in a ratio of 2:1 (3.0×10^5 microprotoplasts : 1.5×10^5 protoplasts) in 6-cm petri dishes in W5 medium and plate at a density of 1×10^6 mL^{-1}. After 7 min, carefully remove PEG and W5 medium using a pasteur pipette and slowly add a high pH buffer solution. Keep for 20 min. After the mixture has settled down, rinse with TM2 for fusion samples of potato + tomato. Culture the mixture at a density of 0.2×10^6 mL^{-1}. Transfer the hybrid cells on day 5 to callus inducing liquid medium TM 2. Select hybrid calli on day 12/19 in media supplemented with kanamycin and/or hygromycin at 60 mg/L^{-1} each, depending on type of fusion. When the resistant calli turn green on solid callus growth medium, transfer to the regeneration medium without adding kanamycin or hygromycin. Root the regenerated shoots on MS medium, supplemented with 2% sucrose.

Kanamycin resistance and GUS assays

Determine kanamycin resistance of shoot regenerated from Kan^R calli on the basis of root induction from shoots grown on MS media supplemented with 3% sucrose and kanamycin at 50 mg L^{-1}. Perform GUS assay as described by Jefferson *et al.* (1987). Incubate in an extraction buffer (50 mM sodium phosphate buffer, pH 7.5, 10 mM disodium—EDTA and 1% v/v Triton x—100), to which iron cyanide solution (0.5 mM potassium ferricyanide in water) and x-Gluc solution (1 mM 5-bromo 4-chloro 3-indolyl β-glucuronide in dimethylformamide) had been added, for 16 h at 37°C. A blue colour appears, indicating GUS activity. Remove chlorophyll from the stained tissue by extraction with ethanol and observe GUS staining under a dissection microscope.

Karyotype and genomic *in situ* hybridization

For karyotype study of plants regenerated for Kan^R calli, excise root-tips, fix and stain following Feulgen schedule and squash in 45% acetic acid. Observe as described under general methods for chromosome study (Chapter II.4).

For *genomic in situ* (GISH) hybridization, pretreat actively growing young root tips of regenerant plants in aqueous 2 mM 8-hydroxyquinoline solution for 2.5 h at 17°C. Fix in acetic acid–ethanol (1:3) for 24 h. Wash in water. Incubate in an enzyme mixture containing 0.1% pectolyase Y23, 0.1% cellulase RS and 0.1% cytohelicase in 10 mM citrate buffer, pH 4.5 for 1 h at 37°C. Transfer root-tips to a grease-free slide and macerate as described under chapter for study of chromosomes. Perform DNA denaturation, *in situ* hybridization and detection amplification as described under the respective chapters in this book (see also Isaac, 1994).

Use total genomic DNA isolated from the donor leaf material as probe and leaf DNA from the recipient plants as blocking DNA.

Labelling of donor DNA can be done by an indirect or a direct method. In the indirect method, shear the DNA by passing through a syringe until the fragments attain a size of 1–10 kb. In the direct method, sonicate the DNA so as to obtain 1.0–2.0 kb fragments. In the indirect method, label with digoxigenin-11-dUTP (Boehringer-Mannheim) and detect with anti-digoxigenin Fluos (fluorescein isothiocyanate), raised in sheep and amplified with anti-sheep-FTC raised in rabbit according to standard random primer labelling protocol. In the direct method, label using Fluorescein-high prime kit fluorescein-12–dUTP (Boehringer-Mannheim).

Sonicate the DNA from recipient plant for 10 s (12 micron amplitude) to obtain fragments of about 700 bp. Prepare hybridization mix (100 μl per slide) with deionized formamide, sodium dextran sulphate (Sigma); 2× SSC;

sodium dodecyl sulphate (Sigma); $200\,ng\,\mu l^{-1}$ of donor probe DNA and $10\,\mu g\,\mu l^{-1}$ of recipient blocking DNA. Denature the mix for 10 min at 70°C and then place on ice for 5 min. Hybridise for 18 h at 37°C.

Wash the slides in 2× SSC buffer for 30 min at 20°, in 0.1× SSC for 3×10 min at 42°C and again in 2× SSC for 15 min at 20°C. Counterstain chromosomes with DAPI $(2\,\mu g\,ml^{-1})$ and propidium iodide $(1\,\mu g\,ml^{-1})$.

CHAPTER IV.6

GENE TRANSFER THROUGH
TRANSFORMATION—TRANSGENICS

METHODS FOR GENE TRANSFER

Several methods are available for transformation of plant cell through transfer of foreign DNA, chromosome fragments or chromosomes, involving cell cultures. The insertion of chromosome or chromosome fragments has been dealt with elsewhere in this section.

GENE TRANSFER—ASPECTS

The plant system, because of its flexibility and totipotency, has been subjected to different techniques of gene manipulation. These techniques, which are all aimed at horizontal transfer of genes directly from one organism to another, utilize in general naked protoplast system as the recipient.

In horizontal transfer, the earlier methods principally concentrated on *Agrobacterium tumefaciens*, the crown gall inducing bacterium in plants. It is needed for introducing the foreign DNA into the host genome with *Agrobacterium* plasmid DNA serving as the vector. But one of the serious limitations of *Agrobacterium* mediated transfer is that the monocotyledonous plants, which cover most of the crop species, do not respond well to *Agrobacterium* infection. As such, several other supplementary methods had to be developed for gene transfer, including irradiation, electroporation, microinjection, direct uptake and finally the biolistic method of microprojectile bombardment (Potrykus, 1991; Vasil, 1994). In all these methods, the basic objective is to incorporate the foreign gene either as a gene construct with or without the vector, or as the sequence directly, into the recipient protoplast. The incorporation of a marker and a reporter permits monitoring of entry and expression of the vector in the system. Once transformed, expression of foreign gene in the recipient background is tested through analysis of gene expression and the plants are regenerated.

So far, the incorporation of alien genes in plants to secure transgenics has been done principally to solve some of the standing problems in agriculture and horticulture. They include induction of resistance to biotic factors, such as pathogens and insects; increase in content of seed protein; prolongation of the shelf life of fruits and vegetables, change of flower colour; acceleration of photosynthetic capacity; control of flowering and most important of all, nitrogen fixation. As compared to species of food crops, data on medicinal and other commercial species are rather meagre. Medicinal and aromatic species possess an immense potential for direct incorporation of foreign genes.

Of all the methods so far tried for gene transfer, the most widely used method is *Agrobacterium* mediated transformation (Maliga *et al.*, 1995).

The strains of *Agrobacterium tumefaciens* cause a disease called *crown gall*. It is the result of transfer of a defined piece of DNA (T-DNA), from the bacterium's tumour-inducing plasmid (Tip), 200–250 kb long, into the chromosome of a range of plants. The T-DNA contains a number of genes that encode enzymes involved in the synthesis of hormones and specific metabolites called opines. As a result, the transformed plant cells form tumours capable of growing on hormone-free medium under *in vitro* conditions. In binary vector system of *Agrobacterium*, the host strain contains a disarmed Ti-plasmid, where the tumour-inducing genes excepting the borders are deleted and that carrying the VIR function replaced by pBR 32 or some other plasmid sequences. The T-DNA inverted border repeats are located on an alternative compatible replicon (plasmid) which is capable of functioning in *Agrobacterium*. The desired sequences of DNA are inserted between the T-DNA borders. These are transferred to and maintained stably within the plant genome upon infection. It is necessary to include: (i) a marker gene for detection of transformants and (ii) a reporter gene for qualitative as well as quantitative measure of gene expression. These genes are inserted between the inverted repeat sequences of the T-DNA. However, for the expression of the prokaryotic genes in a plant genome, promoter sequences of eukaryotic origin are incorporated, making it a chimaeric gene such as 35S promoter of cauliflower mosaic virus.

Two species, namely *Agrobacterium tumefaciens*, the tumour inducing type and *A. rhizogenes*, a hairy root-inducing wild type, are often used. The commonly used marker genes are hygromycin, streptomycin and kanamycin resistance genes whereas reporters are Nopaline and Octopine synthase as well as Beta glucuronidase and finally Luciferase (Ow *et al.*, 1986; Jefferson *et al.*, 1987).

GENE TRANSFER BY CO-CULTIVATION OF TOBACCO LEAF DISCS WITH *AGROBACTERIUM TUMEFACIENS*

Bacterial strains grown in YMB or AB medium

Plant Material: Leaves from 6–8 weeks old *in vitro* raised shoot cultures of *Nicotiana tabacum* (var. Samsun) grown at $25 \pm 2°C$ at 3,000 lux light intensity in the culture room. Alternatively, healthy glasshouse grown leaves at 6 weeks stage are sterilized and used.

Media needed

For raising aseptic shoot cultures, *MT-1 1/2* strength MSO.

For initially inoculating infected leaf discs, *MT-2 B5* pH 5.5 medium + 250 mg/l NH_4NO_3.

For growth of callus and shoots on leaf discs, *MT-3 B5 medium* + 250 mg/l NH_4NO_3 + 2% glucose, 0.5 g/l MES (pH 5.7); 400 mg/l adenine + 0.8% agar; 0.5 mg/l BAP and 0.1 mg/l IAA; 5 mg/l cefotaxime and 50–100 mg/l kanamycin.

MT-4 for root induction and growth of plantlets 1/2 strength MS salts (pH 5.7 + 2% sucrose + 200 mg/l cefotaxime. *4 MT-5* for growth of *Agrobacterium tumefaciens* YMB + 50 mg/ml of kanamycin.)

Method

Take young leaves from healthy plants at 6–8 week stage, preferably grown under *in vitro* conditions. Remove midrib and cut out discs of about 0.3 cm and avoid desiccation.

Place leaf discs upside down in a 9 cm petridish containing 10 ml of bacterial suspension (density of about 10/ml). Blot dry the leaves on sterile filter paper and place the discs on a filter paper overlaid on Medium MT-2 for co-cultivation for 24–48 h. Remove the discs, wash with liquid medium MT-3 and place them on MT-3 medium for 4 weeks.

After 4–5 weeks separate regenerating callus from the leaves and subculture on the same MT-3 medium. Excise developing shoots and place on MT-4 medium. Plantlets develop within 2 weeks. Assay shoots for transformation. Propagate the plants either as sterile shoot culture or transfer them to soil for further growth and analysis.

YMB medium g/l: Magnesium sulphate, 2.0; dipotassium hydrogen phosphate 0.5; NaCl, 0.1, mannitol 10.0; yeast extract, 0.4; agar 15.0, pH 7.0.

AB medium: 10 ml of AB minimal stock (K_2HPO_4, 3 g; NaH_2PO_4, 1 g; NH_4Cl, 1 g; KCl 0.15 g) + 1 ml $FeSO_4$, 0.15 M $CaCl_2$, 2.5 ml glucose, 0.13 ml $MgSO_4$ + 86.33 distilled water.

Gene transfer by co-cultivating protoplasts with *Agrobacterium tumefaciens*

This method of transformation depends on the availability of regenerating systems via protoplasts.

Required: (i) Protoplast isolated and cultured as described earlier. (ii) *Agrobacterium tumefaciens* with p BI 121 vector. Stock solution of cefotaxime and kanamycin (preserved at −20°C in PCM).

Culture isolated protoplasts at a final density of 1×10^6 protoplasts/ml in a 9 cm sterile petri dish. Monitor cell wall formation and cell division using inverted microscope at 4–5 days stage, when the cells are changing shape and beginning to divide.

Add 1 ml of the *Agrobacterium tumefaciens* with pBI 21 vector (800 bp 35 S promoter cauliflower mosaic virus) suspension grown up to a density of $10^7 - 10^8$ cells/ml. The *Agrobacterium* is suspended in the protoplast culture medium (PCM) prior to addition.

Incubate/co-cultivate plant cells and *Agrobacterium* for 24–48 h (depending on the species) in dark at 25°C. Wash cells preferably with pre-conditioned medium and add cefotaxime to a final concentration of 500 µg/ml. (Cefotaxime kills agrobacteria). At the 50 cell stage of colony development, add suitable antibiotics for selection of transformants. Follow similar methods to regenerate and grow plants through protoplast cultures as stated earlier.

GENE TRANSFER BY ELECTROPORATION

The method of electroporation, involving electric field-mediated membrane permeability, is based on the observations that electric pulses can open cell membrane and allow the penetration of alien DNA into the plant cell. The principle involves the generation of high voltage electric field pulses in a multi electrode fusion chamber. Heat shock, in combination with electroporation, has been seen to result in high efficiency (Asano and Ugaki, 1994). An essential factor for the success of electroporation is the choice of suitable protoplasts, with an established capacity for regeneration. Amongst plant food crops, callus cells of sugarcane (Arencibia *et al.*, 1995) and intact seed embryos of rice have also been transformed through electroporation (Chaudhury *et al.*, 1995).

Introduction of gene and regeneration (after Kaufman *et al.*, 1995)

Prepare protoplast suspension as described earlier.

Measure its resistance by adding 0.35 ml of the suspension to the chamber of the electroporator. Add an appropriate amount of $MgCl_2$ solution to adjust the resistance from an initial 1.4 to 1–1.1 kV. Heat shock the protoplasts at 45°C for 5 min and cool to room temperature, then place on ice.

Aliquot 250 µl of the protoplast suspension into three to five sterile or disposable tubes. Add 20 µl of DNA solution and 125 µl of PEG solution. Mix well and let stand for 10–15 min. Transfer the samples to the chamber of the electroporator and pulse three times at 10 s intervals with an initial pulse field strength of 1.4 kV/cm.

Transfer each sample to a 6 cm diameter petri dish in a laminar flow hood and let stand for 10 min. Add 3 ml of a 1:1 mixture of K3 and H media containing 0.6% (w/v) SeaPlaque agarose to the dish. Gently mix the protoplasts in the agarose medium and allow to harden without any disturbance.

Seal the dishes containing the electroporated protoplasts with parafilm to prevent contamination. Incubate at 24°C for 24 h in the dark followed by 4 6 h of continuous dim light (500 lux) depending on the plant species.

Cut the agarose-containing protoplasts into small quadrants using a clean razor blade and place the agarose blocks from one dish into a culture vessel (10 cm in diameter and 5 cm in depth) containing 50 ml of medium A with 50 µg/ml of kanamycin sulfate for selection of transformants. Culture at 24–28°C in continuous dim light (500 lux) with shaking at 80 rpm. Resistant clones will be visible 3–4 weeks later, depending on the plant species.

When the clones are 2–3 mm in diameter (5–6 weeks after culture), transfer to medium A containing 0.8% (w/v) agar with 30 g/ml mannitol and 50 µg/ml of kanamycin sulfate. Allow the colonies to grow for 2–5 weeks, depending on the plant species. Transfer the colonies to medium A without mannitol and allow to grow for 2–3 weeks.

Induce root formation by culturing the colonies in medium A without mannitol but with 20 µg/ml sucrose and 0.25 µg/ml BAP (6-benzylaminopurine) hormone. Incubate the dishes in dark for 1 week followed by illumination at 3000–5000 lux until shoots are generated from the callus. Cut off the shoots (1–2 cm long) from the callus and place on medium B lacking hormones in order to produce roots. When the shoots are 3–5 cm long, gently wash away the agar once the root system is established and transfer the plantlets to pots of soil. Transfer gradually to a regular greenhouse.

GENE TRANSFER THROUGH MICROINJECTION

Microinjection is an effective method for injecting chromosome fragment or DNA in the recipient cell (Delaat et al., 1989; Blaas, 1989). This technique involves injection of a small amount of DNA solution into the recipient cell through pipette, capillary tube or injection under pressure. Since the procedure is manual, only a few cells can be treated. It also requires expensive instruments like micromanipulator and inverted microscope. Injection of chromosome fragment into recipient cell is also included in this procedure. The advantage is that the material can be directly injected into the cell. Initially used in *Brassica*, linear and supercoiled DNA was injected into the cytoplasm of embryoids derived from pollen. Several chimaeras were obtained. Other successful examples are introduction of chromosomes of

Petunia alpicola into protoplasts of *P. hybrida* (Griesbach, 1987). One method is to separate nuclei and cytoplasm before microinjection and then to grow nuclei individually or after rejoining with cytoplasm by electrofusion. In injecting the chromosome with microneedle of micromanipulator, the first step is to obtain evacuolated plant protoplasts.

For securing evacuolated protoplasts, the following method may be adopted: Isolate mesophyll protoplasts from leaf tissue (1×10 mm) by overnight treatment with 0.8% cellulysin, 0.4% macerase, 10% mannitol, 5 mM-MES (2N morpholin-o-ethanesulfonic acid) and 10 mM $CaCl_2$, pH 5.7. Clean the protoplast suspension by layering on top of 30% sucrose solution and centrifuging at $200 \times g$ for 15 min. Wash the cleaned protoplast band at interface twice in 5 mM-MES, 10 mol $CaCl_2$ and 10% mannitol, pH 6.0 by centrifugation at $100 \times g$ for 5 min. Mix 0.5 ml of 2×10^6 protoplasts/ml with 45 ml Percoll containing 100 mM $CaCl_2$, 5 mM N-2 hydroxyethyl piperazine–N2 ethanesulphonic acid (HEPES), pH 7.0 and 8% mannitol. Introduce Percoll/protoplast suspension through pipette into a 5 ml polycarbonate centrifuge tube and centrifuge at 40,000 rpm for 60 min at 23°C (Beckman SW 56 rotor). Aseptically remove the band of evacuolated protoplasts with a pasteur pipette. Wash three times with MS regeneration medium with 2 g/l bactopeptone, 20 g/l sucrose, 100 g/l mannitol, 1 mg/l NAA and 2 mg/l BAP supplement at pH 6.0.

For uptake of chromosomes in the protoplast, the schedule has been described earlier in this chapter.

GENE TRANSFER THROUGH BIOLISTIC METHOD

In this method, the DNA or the gene construct is coated on the surface of tungsten or gold particles and introduced into the cells through a particle acceleration gun. To the plant tissue the tungsten particles are more toxic than particles of gold (Kjellsson *et al.*, 1997). The gun can be driven by air pressure or gunpowder. At present helium gun is available (Biorad Laboratories) and widely used for coating of DNA on the tungsten particles (Maliga *et al.*, 1995). Normally GUS glucuronidase or Kanamycin (Kan gene-encoding neomycin phosphotransferase) genes are added for checking expression after transformation. These two genes are normally carried by the plasmids PFF 19 G and PFF 19 K present in the cassettes, utilizing 35 S promoter of cauliflower mosaic virus. The blue colour indicating GUS expression and resistance to Kanamycin of the cells, in the medium, serves as indicator for transformation.

Despite the limitation of high cost of the biolistic gun and random nature of the result which is a disadvantage, the method has numerous advantages for which it is widely applied with success. The technique is extremely simple and convenient and transformation can be carried out in many cells simultaneously. All types of cells ranging from protoplasts to germ cells and meristems can be subjected to bombardment. The fast and random hits of the microprojectiles add to the frequency of transformation as well. Biolistic transformation has been proved to be useful in the transformation of embryogenic maize culture of wheat (Vasil, 1994) as well as rice and cotton. The method and its application and advances have been extensively reviewed (Yang and Christon, 1994).

The method described below covers the transformation protocol, confirmation through GUS assay and method for finding out the number of copies of the integrated gene in the transgenic plant (taken from Kaufman *et al.*, 1995).

Remove two to six fully expanded, young leaves from plants grown under sterile conditions or in a greenhouse. Surface-sterilise the leaf in 5 volumes of 10% Clorox solution for 10–15 min, in a laminar flow hood, followed by thoroughly rinsing four to five times in 5 volumes of sterile distilled water. Peel off the lower epidermis of the leaf using a pair of jeweler's forceps, if needed.

Slice the leaves into approximately 1×0.5 cm pieces using a clean razor blade. Transfer the excised leaf pieces onto Grade 617 Whatman filter paper in disposable petri dishes (60×20 mm) containing 15 ml callus medium with 100 µg/ml kanacycin (medium A). Orient the leaf pieces in the centre of each dish for maximal exposure to bombardment.

Carry out macroprojectile bombardment as follows: Sterilize 100 mg tungsten (1.3 µm microprojectiles) in 1.5 ml of 95% ethanol in a sterile 15 ml centrifuge tube for 5 min. Sonicate on ice for 10 min with a continuous pulse using a 20% duty cycle at level 2 output. Transfer the sonicated microprojectiles into a microcentrifuge tube and centrifuge at $12,000 \times g$ for 2 min. Decant the ethanol supernatant and gently suspend the pelletted microprojectiles in 1.5 ml double distilled (dd) water. Centrifuge again. Decant the supernatant, add 1.5 ml double distilled (dd) water and recentrifuge. Remove the supernatant, resuspend in 1.5 ml double distilled water, and sonicate the vial. Aliquot 5 µl of the samples into microcentrifuge tubes. Resonicate after every two aliquots to maintain uniform bead concentration for each aliquot.

Add 10 µl of gene constructs (1 µg/µl) to each aliquot of microprojectiles and mix well. Add 25 µl of 2.5 M $CaCl_2$ to each DNA/microprojectile mixture and mix well. Add 10 µl of 100 mM spermine to the mixture, mix by vortexing, and let it set for 20–30 min. Centrifuge at $12,000 \times g$ for 2 min and carefully remove the supernatant to a final volume of 30 µl.

Sonicate the DNA/microprojectile mixture and pipette 1.5–2.0 µl onto a sterile macroprojectile. Resonicate after every two aliquots. Place the macroprojectiles in the gun barrel and the power level 1 blank in the chamber. After inserting the stopping plate and tissue sample in place, attach the detonator and draw vacuum. Fire the gun when the vacuum reaches

68–71 cm Hg. (See the manufacturer's instructions for details, DuPont Company or GTE Products Corporation.)

After bombardment, transfer the dishes to a growth room at 28°C with 12–16 h day length at $100 \, \mu E \, m^{-2} S^{-1}$, depending on the plant species, and maintain the bombarded leaf strips on the callus medium (medium A) for 2–3 weeks.

Transfer the leaf strips to regeneration medium (medium B) and allow to grow for 2–4 weeks with transfer to fresh medium every 2 weeks. Plantlets will develop from the callus (after Kaufman *et al.*, 1995).

Confirmation of transformation by assay of transient GUS activity

A gene that encodes for β-glucuronidase (GUS) is usually used as a reporter gene in the chimeric gene constructs for transformation. The enzyme hydrolyses 4-methyl-umbelliferyl-β-D-glucuronide (MUG) and produces 4-methyl-umbelli-ferone (4 MeU), generating a blue fluorescence and can be assayed by quantitative measurement or by tissue staining.

Homogenise or grind 5–10 mg of tissue from transgenic and nontransgenic plants in 0.1–0.2 ml of lysis buffer. The lysis buffer is made up of 50 mM sodium phosphate (pH 7.0), 10 mM EDTA, 0.1% (v/v) Triton X-100, 10 mM 2-mercaptoethanol, and 0.1% (w/v) sarcosyl. Transfer the homogenate to a microcentrifuge tube and centrifuge at $12,000 \times g$ for 15 min at 4°C. Transfer the supernatant to a fresh tube and store at 4°C until use.

Carry out protein measurement as described elsewhere. Determine the background fluorescence from nontransgenic tissue. Determine the maximum amount of protein from the protein extracts, which can be used for the GUS assay using pure GUS as a positive control. For the fluorescence assay of the sample, add protein extract to two microcentrifuge wells containing 2 and 4 μg total proteins, respectively. Add lysis buffer to final volume of 45 μl in each well. Add 5 μl of MUG to start the reaction. Cover the plate and incubate at 37°C for 30 min. Add 150 μl of 0.2 M Na_2CO_3 to terminate the reaction. Read the fluorescence and subtract the blank value.

Calculate the specific activity (units/ng protein/min) of GUS in the tissue, based on the total proteins and incubation time.

Assay through light microscopy

Fix whole seedlings or sliced tissues in 25 mM sodium phosphate buffer (pH 7.5) containing 0.1–1% (v/v) glutaraldehyde for 30 min at 4°C or at room temperature. Wash the tissue for 3 min five times in 25 mM sodium phosphate buffer (pH 7.0).

Quickly and completely cover the tissue with the substrate mixture and vacuum filter. Incubate in dark at 37°C overnight or for 1 day. The substrate mixture can be made from stock solution to a working concentration of 10 mM sodium phosphate buffer (pH 7.0), 0.5 mM potassium ferricyanide, 0.5 mM potassium ferrocyanide, and 1 mM X-glucuronide (5-bromo-4-chloro-3-indolyl glucuronide). Rinse the tissue for 5 min twice in sodium phosphate buffer until the tissue shows an intense blue colour.

Directly observe and photograph the stained tissue under a microscope without further processing. To improve the image remove the pigments (chlorophyll) by passing the tissue through 25, 50, 70, 95, 100, 100, and 100% ethanol with 15 min per step. Rehydrate through ethanol series and progressively infiltrate with glycerol. Remove any air bubbles by final vacuum filtration (after Kaufman *et al.*, 1995).

Determination of the copy number of the integrated gene in the regenerants

Genomic DNA is isolated from transgenic plants and digested with appropriate restriction enzymes. The fragments are separated by agarose gel electrophoresis followed by transfer to a nitrocellulose or a nylon membrane. The membrane is then hybridised with a labelled DNA probe that is the introduced gene or gene fragment. Stable transgenic plants show a positive signal. The positive fragment(s) of DNA may be sequenced. If the sequence shows both host and foreign DNA sequences, the introduced gene has been integrated into the chromosomes of the host cell showing strong evidence of stable transformation.

Isolate genomic DNA from transgenic plants and measure the concentration of DNA in the sample. Calculate the quantity of DNA that corresponds to a specific copy number. One copy of the gene per haploid genome is equivalent to the size (bp) of the transferred gene (g) divided by the size (bp) of the genome (G), or g/G. Therefore, if 1 µg of the genomic DNA is used for the experiment, the quantity of the transferred gene will be equal to $g/G \times 1\,\mu g$.

Set up copy number standards of known gene equivalents of the query sequence. If the query sequence (bp), the gene, or cDNA insert (I) is purified from a recombinant plasmid, the amount (N, µg) that is equivalent to one copy per genome of the target sequence can be calculated as: $1 \times N = g/G \times 1\,\mu g$. If the I (bp) is in the cloning vector such as plasmid (V, bp), the N (µg) can be calculated as: $I/V \times N = g/G \times 1\,\mu g$. Prepare a set of copy number standards that is equivalent to 1, 2, 3, 4, 5, 10 and 15 copies of the target or transferred gene.

Carry out restriction enzyme digestion of 1 µg of genomic DNA using an enzyme that cannot cut within the target sequence. Load the digested genomic DNA and a set of copy number standards into separate wells of an agarose gel. Carry out electrophoresis and Southern blot hybridization using purified, labelled cDNA, the monomeric sequence of a repetitive DNA family, or a linearised plasmid DNA containing the insert as a probe. Analyse the detected band(s) by scanning with an integrating densitometer. The copy number of the transferred gene in the genome can be estimated by measuring the extent of hybridization signals of genomic bands and comparing the intensity of the signal with those of known gene standards. With a family of tandem repeats, the sum of the extent of all hybridised bands is the total number of gene copies, as determined from the copy number standards (after Kaufman *et al.*, 1995).

Chromosome analysis

For detection of integrated sequences in the host chromosome, the procedure involves the use of leaf-tip smear technique for chromosomes, and FISH technique as adopted at the chromosome level. The sequence desired for integration is then biotinylated and used as the probe. This method identifies the exact location of sequences in the chromosome after integration.

TRANSGENIC TRAITS

The entire process of transgenesis in plants can be subdivided into several stages:

1. Cloning, isolation, characterization, and subcloning of the gene of interest to be transferred.
2. Selection and/or purification of promoter, enhancer, poly(A) signal, reporter gene, selectable marker gene, and the gene of interest from recombinant plasmids.
3. Construction of chimeric or fusion genes.
4. Transformation of plant cells or tissue with the chimeric genes.
5. Selection and regeneration of transgenic plants.
6. Analysis of the expression of the introduced gene in transgenic plants.

So far a large number of traits has been incorporated both in relation to resistance to biotic and abiotic stresses as well as desired morphological traits (Kjellsson *et al.*, 1997). The tolerance to herbicides, insect pest, virus, fungal and bacterial pathogen deserves special mention. Against the non-selective herbicides, such as glucosinate, ammonia, bromoxynil, sulfonyl urease as well as 2,4,D, tolerant plants have been raised. Similarly introduction of toxin gene of *Bacillus thuringinensis* against lepidopteran insects; protease inhibitors; amylase inhibitors, nematode resistance as well as other defence genes, mediated by lectin, have opened up new possibilities. The viral attack repelled through coat protein mediated resistance, fungal attack through phytoalexins and chitinases and bacterial attack through lysozymes, toxin genes as well as glucose oxidases, are also involved in transgenic traits.

Of the abiotic stresses, genes for proline, betaine and fructans against drought as well as superoxide dismutase and glutathione reductase against oxidative stress are utilized in experiments on transgenesis in crops. Cold and salinity tolerance is introduced through genes coding for glycerol-3-phosphate and mannitol phosphate dehydrogenase respectively.

The research on transgenics suffers from the serious constraint of expression of integrated gene in the foreign recipient. Despite confirmatory evidence of the expression of markers and reporters, the transformants often fail to yield optimum gene expression. The genotypic control on transgenic expression plays a crucial role. Such control becomes evident even after the use of organ-specific promoters, specially in crop species. The codon usage, introduction of introns and methodology of introduction, are rather specific for genotypes as noted in cotton, rice, potato and other crops. As such, in order to have the optimum benefit of transgenesis, the integration, transformation, regeneration and hardening protocols are to be critically worked out for each genotype.

HORIZONTAL GENE TRANSFER—SCOPE AND ADVANCES

Gene transfer through transgenesis has been employed either for rapid transfer of desirable character to secure the desired crop or to gain an understanding of the mechanism of gene expression in biological systems. Genes for several desirable traits have been manipulated including resistance to insects, pathogens, and herbicides, tolerance to stress, increased shelf life of fruits, desired proteins in the seeds, alteration in flower colour and male sterility. The transformation in dicotyledonous species could be achieved comparatively easily through Agrobacterium-mediated system whereas in monocotyledonous crops such as rice, maize, rye and others (Greenberg and Glick, 1993), the biolistic method of transformation prior to or during embryogenesis has proved to be effective.

The genes for traits so far inserted in transgenic plants through horizontal transfer, as stated before, cover to a great extent resistance against undesirable biotic factors. Of the two different strategies so far adopted to introduce the trait of insect tolerance namely, introduction of a toxin gene of *Bacillus thuringinensis* or natural protease inhibitor, such as cowpea trypsin inhibitor, the former has been widely used because of its wide applicability, nonhazardous nature and efficacy at low concentration. The latter however has the advantage of being a natural inhibitor and of eukaryotic origin. The *Bt* toxin gene confers resistance through the production of endotoxin, which undergoes dissolution in the midgut of the insect and causes lethality. The endotoxins coded by different sequences are toxic either to lepidoptera (*Cry* I), lepidoptera and diptera (*Cry* II), coleoptera (*Cry* III), diptera (*Cry* IV) and also lepidoptera and coleoptera (*Cry* V). The continued application of *Bt* toxins may however develop resistance against insects. The protease inhibitors, on the other hand, are antimetabolic proteins and the control is manifested through increased mortality and decrease in growth rate. It constitutes a natural defense system in plants against insect attack. The protease inhibitors have been found to be effective against pests of several crops including tobacco, pea, and common bean, *Phaseolus vulgaris*. The action is mostly tissue specific and may not cover plant parts or organs meant for human consumption.

The resistance against pathogenic organisms involves different strategies, depending on the type of pathogen to be dealt with. In several crop species such as tobacco, potato and tomato, the resistance against viral attack has been induced through the introduction of coat protein genes of viruses which in excess, cause disruption in virus assembly or inhibition of the expression of viral genome. In addition to gene for coat protein, introduction of antisense sequences of viral genome (Greenberg and Glick, 1993), interfering with viral replication or translation of viral proteins, has proved to be of promise. But high specificity restricted only to its parental sequences, is a barrier to its wide application. The resistance against fungal pathogens is achieved through the introduction of phytoalexin gene in solanaceous crops, ribosome inactivator in tobacco against *Rhizoctonia solani*, and chitinases and gluconases against the same fungus in tobacco and tomato. The action of these two enzymes is synergistic (Kjellsson *et al.*, 1997).

The bacterial pathogenesis has been contained through lysozyme as in transgenic potato. Similarly, introduction of cholera toxin gene sequences into tobacco led to a decrease in susceptibility against *Pseudomonas tabaci*. The gene glucose oxidase generating hydrogen peroxide conferred tolerance to the bacteria *Erwinia corotophora* and the fungus *Phytopthora infestans* in transgenic potato (vide Kjellson *et al.*, 1997).

In transgenics, the trait of herbicide tolerance is of special significance in strategies for weed control in agriculture. The natural herbicide tolerance in plants is achieved through three mechanisms namely, (i) reduced sensitivity at the site of action by alteration of action site or gene amplification, (ii) herbicide degradation, inactivation of the herbicide after absorption or (iii) herbicide avoidance. In a series of solanaceous crops as well as cotton and *Arabidopsis*, transgenics have been raised, resistant to specific chemical herbicides. Transgenic tobacco transformed by the introduction of bacterial nitrylase gene has shown resistance to the toxic effects of the herbicide.

Of the abiotic factors, the different aspects of stress tolerance such as drought, salinity, oxidative stress, and cold temperature have led to promising results in a few cases. The prevention of osmotic stress and increase of osmotic potential could be recorded in transgenic tobacco, rice and *Vigna aconitifolia*—the moth bean. The strategy was to introduce genes for glutamic acid kinase–Proβ and D-pyrroline-5-carboxylate synthase, to increase the proline level. In tobacco, increased root yield and flower development under drought stress have been observed.

For achieving tolerance to salinity, the bacterial gene coding for mannitol phosphate-dehydrogenase (mtlD), which leads to the production of acrylic poly mannitol, has been successfully introduced into *Arabidopsis* and tomato. However, for salt tolerance, extensive programme has been undertaken in rice, through introduction of porterasia gene to rice in India.

As far as cold tolerance or frost resistance is concerned, genes from *Arabidopsis* which have a high proportion of unsaturated fatty acids leading to cold tolerance, have been introduced into tobacco. Similarly, to cope with oxidative stress, superoxide dismutase genes from different sources including pea and tomato, have been introduced into potato, tobacco and cotton, to increase the activity of antioxidant enzymes.

Of the other traits, the alteration of fruit ripening pattern in tomato has been successful in transgenics. The antisense gene for polygalactouronase introduced in tomato permits ripening but retards the softening associated with it. The approach to interfere with ethylene production led to slowing of the process of ripening and softening. It was achieved through the use of antisense for amino cyclopropane carbolic acid, i.e. ACC synthase or ethylene forming enzyme (Hamilton *et al.*, 1990). This strategy may prove to be fruitful for increasing the shelf life of other fruits as well.

The genes for flavonoid synthesis have been manipulated and novel colours in petals have been obtained (Vander Krol *et al.*, 1990). Attempts are being made to check early floral senescence, through acceleration of scavenging capacity of pre-radicals by Mn-SOD gene. The control of flowering is also proving to be a promising approach.

In order to control compatibility and production of hybrid seed, cytoplasmic male sterility and self incompatibility have been induced in species of *Brassica* (Mariani *et al.*, 1990). The genes RNAse Ti and Barnase from *Bacillus amyloquefaciens* under the control of another specific promotor from tobacco have been used in these transgenics.

Transfer of desirable character like seed protein has been attempted. Several genes have been cloned, such as glutenin and glindin from wheat, hordein from barley and zein from maize and wheat glutenin has been transferred into tobacco. It is necessary to see the extent to which these genes on incorporation into crops like *Sorghum* can increase the protein content, or increase the content in its parent species.

The research on transgenics through horizontal gene transfer is undoubtedly paying dividends at present. However, the pace of success with biotic stress is more as compared to that with abiotic factors. It is true that promising results have been obtained in a few cases with transgenics against drought or salinity. But in such abiotic stresses, the control is often complex and exerted through a large number of genes. The absolute resistance to such stresses can only be achieved through a proper understanding of triggering multigenic and coordinating control, to overcome the stress factors. The research on transgenics is gaining momentum but ultimate success would depend on its capacity for sustainable development without causing any ecological hazard or disbalance in the biosphere.

One of the recent discoveries in the area of gene transfer is the development of oral vaccines utilising the plant system. The principle involves the development of transgenic plants, containing subunits of toxic virus or enterotoxin genes. The oral administration of potato or tobacco transgenic tissues lead to the development of potato or tobacco carrying Immunoglobin G and A antibodies. Such oral administration of plant tissues, is in effect oral vaccination, which is indeed inexpensive. Plant antibody technology is still in its infancy and requires much more perfection.

APPENDIX 1

Table 1 Advantages and disadvantages of Gene Transfer techniques

Method	Advantages	Disadvantages
Agrobacterium (Ti and Ri)-mediated gene transfer Transfer to leaf discs Transfer to protoplasts Transfer to roots	Well established and high efficiency for dicots	Not suitable for monocots
Polyethylene (PEG) Promotes protoplast fusion and the uptake of DNA	Relatively simple	PEG is toxic to humans
Electroporation The use of short electrical impulses of high field strength that increases the permeability of protoplast membranes	High efficiency and used for both dicots and monocots	Up to 50% cells die due to highvoltage damage to plasma membranes of protoplasts
Microinjection Precisely inject recombinant DNA into specific compartments of protoplasts	Specific, highly efficient, and used for all plants	Requires special skills, special equipment, and high cost
Particle bombardment Particle gun shoots coated recombinant DNA into protoplasts or cells	High efficiency, and DNA is delivered simultaneously into many cells	Requires special skills, expensive equipment, and high cost

*After Kaufman *et al*. (1995).

APPENDIX 2

Table 1 Plant tissue culture media (Nitsch and Nitsch, 1956)

Constituent	mg/l	Constituent	mg/l
KCl	1500	$CaCl_2$	25
$MgSO_4 \cdot 7H_2O$	250	$MnSO_4 \cdot 4H_2O$	3
$NaH_2PO_4 \cdot H_2O$	250	$CuSO_4 \cdot 5H_2O$	0.025
KNO_3	2000	H_3BO_3	0.5
IAA	0.18–1.8	$Na_2MoO_4 \cdot 2H_2O$	0.025
Sucrose	34,000	$ZnSO_4 \cdot 7H_2O$	0.5

Table 2 Plant tissue culture medium (White, 1954)

Constituent	mg/l	Constituent	mg/l
KCl	65	H_3BO_3	1.5
$MgSO_4 \cdot 7H_2O$	720	$Fe(SO_4)_3$	2.5
$NaH_2PO_4 \cdot H_2O$	16.5	Sucrose	20,000
KNO_3	80	Glycine	3
Na_2SO_4	200	Cysteine	1.0
$Ca(NO_3)_2 \cdot 4H_2O$	300	Vit B_1	0.1
$MnSO_4 \cdot 4H_2O$	7	Vit B_6	0.1
KI	0.75	Nicotinic acid	0.5
$ZnSO_4 \cdot 7H_2O$	3	Ca D-pantothenic acid	1.0

Table 3 Plant tissue culture media (Murashige and Skoog, 1962)

Constituent	mg/l	Constituent	mg/l
$MgSO_4 \cdot 7H_2O$	370	$CuSO_4 \cdot 5H_2O$	0.025
$CaCl_2 \cdot 2H_2O$	440	H_3BO_3	6.0
NH_4NO_3	1650	$Na_2MoO_4 \cdot 2H_2O$	0.25
KH_2PO_4	170	Sucrose	30,000
KNO_3	1900	Glycine	2
$FeSO_4 \cdot 7H_2O$	27.8	Myo-inositol	100
$MnSO_4 \cdot 4H_2O$	22.3	IAA	1–30
KI	0.83	Vit B_1	0.1
$CoCl_2 \cdot 6H_2O$	0.025	Vit B_6	0.5
$ZnSO_4 \cdot 7H_2O$	8.6	Nicotinic acid	0.5
EDTA (disodium salt)	37.3	Kinetine	0.04–10

SECTION—V

MICRODISSECTION AND CHROMOSOME ENGINEERING

Advances in Chromosome Engineering have been attributed principally to the refinements in methods for microdissection, microcloning and amplification of chromosome and chromosome segments. The method basically involves physical excision of chromosome fragments and cloning of the purified DNA sequences. The isolation of chromosomes and dissection of fragments at the light microscopic level, and direct cloning for analysis of purified DNA from the target sequence, are of crucial advantage to genetic manipulation.

The chromosomal DNA, following excision, can be subjected to direct analysis and manipulation, but the low yield is a limiting factor in this approach. In order to overcome this barrier and increase the amount of target sequences, microcloning through lambda vectors has been resorted to. This method provides with adequate amount of DNA after isolation from the vector.

In recent years, the invention of the method for DNA amplification through polymerase chain reaction and laser dissection of chromosomes has added new dimensions in procedures in microsurgery. Refined techniques are available to ligate the purified DNA into vectors suitable for PCR amplification. Such amplification through PCR can be coupled with non-isotopic labelling, for detection of target sequences through specific fluorescence. The method has now been further developed to include dissection of stained chromosomes as well.

Simultaneously, advances in laser optics and chromosome banding have been advantageously employed in different techniques of cell and molecular biology (vide Fukui *et al.*, 1995). In the field of chromosome research, its application covers laser poration of protoplast aiding fusion through protoplast, microdissection of chromosomes and use as optical tweezers in chromosome-mediated gene transfer.

The entire process of microdissection, cloning and amplification, thus requires a critical control of dissection—manual, or through laser collection of chromosome fragments, vector ligation and amplification.

In microdissection and microcloning, proper chromosome fixation is essential. Normal acetic acid–methanol fixation for prolonged period causes acid deprivation and heavy hydrolysis which may lead to low yield of microclones. Moreover it is difficult to clone small restriction fragments.

Too much hydrolysis and depurination make the target DNA a poor substrate for PCR amplification. Fixation for a short period, on the other hand, yields large number of microclones of suitable sized fragments. In order to overcome this limitation, fixation for a few seconds and fresh preparation of metaphase spreads from single cell are preferred (Brown and Carey, 1994).

In several plant systems, such as barley, rice and other crops, computerized argon laser dissection has proved to be very useful for microdissection. In the technique, the chromosome preparation on a plastic coverslip is covered with a polyester membrane, irradiated with microlaser beam for dissection, followed by recovery of fragments in Eppendorf tubes. If tinted membrane is used, the dissected fragment can be conveniently recovered (Fukui et al., 1995).

REPRESENTATIVE SCHEDULES

1. Microdissection through micromanipulator

Incubate seeds or seedlings on filter paper subsequently soaked with tap water at (i) 4°C for 3 days; (ii) 22°C for 5 h; (iii) 1.25 mM hydroxyurea at room temperature, 18 h (after Schondelmaire et al., 1993); (iv) distilled water at room temperature, 5 h; (v) 4 µM APM (O-methyl-O-(2-nitro-p-tolyl) N-isopropyl-phosphoroamido-thioate; amiprophosmethyl) (Bayer-Leverkusen) at room temperature for 3 h.

Cut root-tips, rinse in distilled water, incubate in ice water overnight. Store in 70% ethanol for one day.

Wash in distilled water and incubate in enzyme solution (2.5% pectolyase Y23, 2.5% cellulase R10, 75 mM KCl, 7.5 mM EDTA, pH 4) for 45 min. Treat in 75 mM KCl for 15 min.

Wash protoplast suspension three times with 70% ethanol and centrifuge for 5 min at 75 g. Resuspend cell sediment, in fresh fixative (acetic acid–ethanol 1 : 3) and centrifuge for 2 min. Remove supernatant, drop suspension on ice cold slides and use for microdissection after drying.

For microdissection, use an inverted microscope (Zeiss 1 M 35) with programmable stage, micromanipulator and phase contrast optics for selecting and manipulating suitable metaphase spreads at a maximum magnification of 640×.

Deposit a collection drop (2 ml) of 10 mM Tris-HCl, pH 7.5; 10 mM NaCl, 0.1% SDS, 1% glycerol, 500 µg/ml proteinase K on a depression slide overlaid with liquid paraffin (Merck spectroscopic).

Collect the chromosomes or fragments in that drop. For cloning after lysis in the collection drop (2 ml), purify the DNA, followed by restriction digestion, ligation with vector containing universal sequencing primer, amplification in PCR and cloning in a standard plasmid vector.

When all the dissections have been completed, take up the collection drop into a micropipette and transfer to a small, siliconized watchdish (1.5 cm diameter) that lies within a petri dish filled with oil equilibrated with *Rsa*I reaction buffer, and cover the drop. Carry out all subsequent micromanipulations under oil.

For Proteinase K and SDS treatment, add an equal vol (1 nl) of collection solution. Incubate the drop for 90 min at 37°C in a large humidity chamber to prevent evaporation.

Perform four phenol extractions in the following way. Add an equal vol (2 nl) of phenol equilibrated in *Rsa*I buffer to the collection drop. The phenol surrounds the collection drop and is left for 5 min before being removed by micropipetting. Remove the residual phenol by diffusion by changing the oil in the petri dish and incubating overnight at 4°C in a humidity chamber. Change the oil again, this time incubating for 1.5 h.

Digest the dissected DNA with *Rsa*I by fusion with a drop of equal vol (2 nl) of *Rsa*I in a 2X reaction buffer and incubation in a humidity chamber for 2.5 h at 37°C. Repeat with a second aliquot (4 nl) of *Rsa*I in IX reaction buffer. Extract the digested DNA with phenol four times as previously described, to inactivate the enzyme.

The drop containing the DNA is now 8 nl in vol. Add to this 8 nl (8.5 ng) of *Sma*I-cut pUC vector. The pUC vector used has been modified to incorporate a single *Sma*I site flanked by two *Eco*RI sites. Leave the drop for 5 min. Add polyethylene glycol (40%) (relative molecular mass 8000) twice (16 and 32 nl) to give a final concentration of 15% in the ligation mixture.

Add an equal vol (64 nl) T4 DNA ligase in ligation buffer (at 6 U/µl) and incubate overnight at 12°C in a humidity chamber. Take up the microdrop in 2 µl water and transfer to an Eppendorf tube for the amplification steps. Inactivate the DNA ligase by heating for 5 min at 65°C.

Perform PCR on this sample using standard methods, the M13/pUC sequencing, and reverse-sequencing primers. Release the amplified DNA inserts by *Eco*RI digestion, and remove the primers. Spin the mixture through two successive Sepharcyl S-200 columns, equilibrated with ligation buffer, at 500g for 2 min to separate the inserts from the primers and unincorporated nucleotides. Collect the eluate.

Seventeen microliters of a total of 150 µl column eluate are used in a standard ligation reaction with *Eco*RI-cut pUC 13 DNA. An aliquot of the ligation mixture is used to transform commercial competent cells (DH5α) using standard methods (after Brown and Carey, 1994).

2. Laser microdissection

Pretreat and fix root-tips in methanol: acetic acid (3:1) following the usual schedule (after Fukui *et al.*, 1992). Wash the tips thoroughly and subject them to enzymatic maceration (2% cellulase Onozuka RS, 0.3% pectolyase Y-23, Sheishin Pharma, Tokyo and 1.5% macerozyme R 200, adjusted to pH 4.2) on a glass slide as well as on a polyester membrane fixed at the bottom of a 35 mm plastic petridish at 37°C for 30 to 60 min. Wash off the enzymic mixture and macerate root tips with fine forceps into almost invisible fragments in a drop of the fixative. Air dry. Stain with 2% Giemsa solution covering the surface of the membrane and dipping in case of glass slides. Wash and air dry. Carry out microdissection in ACA = 470 (Meridien Instruments, Okemos, Mich, USA) which consists of an argon-ion laser tube, an acausto optic modulator (AOM) and a controlling microcomputer (CPU 80286). The single laser beam of 488 nm is used for microdissection of chromosomes. The intensity of the laser beam is controlled by power supply and modulator. The laser beam, which is introduced into the center of the axis of the inverted microscope, is focussed to 1 μ by a 40 × objective and the target region is irradiated. In order to have a beam, less than 1 μ, focussing on the chromosome, a 100× objective is used. In general, with the increase in intensity of beam, the band cut will be wider.

For very fine chromosome micromanipulation, a band width of 0.5 emu obtained through regulating power supply and AOM is optimal. A beam intensity of 1 μ can be focussed in unnecessary regions—intra or extra-chromosomal. Such laser treatment of unnecessary parts or segments can be monitored by negative result with DNA fluorescence of the remaining segments. The targetted region is dissected out after removal. The entire operation can be completed in 10 min. Transfer of chromosome fragments or chromosome to Eppendorf tube is carried out by the dissection of the tinted polyester membrane carrying the chromosomes. For this purpose, a strong laser beam of 10 μm to 2 mm is used to divide the membrane in octagonal pieces. These octagonal pieces are picked up with fine forceps under stereo-microscope and placed in Eppendorf tube for storage at −20°C before use.

3. Microamplification of specific chromosome sequences
(after Ponelies *et al.*, 1997)

Prepare chromosome spread and utilize short arm of chromosome 6 of maize for manipulation under microscopical control. Both glass needles and laser light give similar results.

Scrape off segments and transfer to a collection drop (100 nl) of GP-buffer (87% glycerol 4 vol, 0.05 M sodium/potassium phosphate, 1 vol, pH 6.8) on a coverglass. Adjust the volume of the drop to 5 μl with 2 mg/ml proteinase K (Boehringer); 0.25% SDS; 10 mM tris-HCl pH 7.5; 1 mM EDTA. Incubate coverglass in a moist chamber at 37°C for 1.5 h. Transfer the reaction mixture to a 0.5 ml test tube and extract three times with 5 μl phenol–chloroform (1 : 1) saturated with TE (10 mM Tris-HCl pH 7.5; 1 mM EDTA). Microdialyse the solution (0.025 μm membrane pore size), according to manufacturer's protocol (Millipore) against 10 ml TE for 45 min. Then digest DNA with the restriction enzyme *Mbo*I (10U) in a volume of 10 μl of reaction buffer (50 mM Tris-HCl, pH 8.0; 10 mM MgCl$_2$; 50 mM NaCl) at 37°C for 1 h. Stop the reaction by heat inactivating at 65°C for 20 min. Ligate to the adaptor molecules in 15 μl 0.5 × buffer (1 × buffer = 50 mM Tris-HCl, pH 7.6; 10 mM MgCl$_2$; 1 mM ATP; 1 mM DTT; 5% poly-ethylene glycol 8000 with 1 Weiss unit T-4 ligase (Life Technology) and 0.45 μM *Mbo* adaptor at 15°C for 12–16 h.

After ligation, amplify fragments in a volume of 100 μl, containing 15 μl ligation product; reaction buffer (50 mM KCl, 1.5 mM MgCl$_2$, 10 mM Tris-HCl, pH 9.0); 250 μm of each αNTP and 1 μM primer *Mbo* 20.

Denature at 95°C for 10 min. Add 2.5 U of *Taq* polymerase (Pharmacia). Amplication conditions for 40 cycles are 94°C for 1 min; 45°C for 1.5 min; 72°C for 2 min and at 72°C for 5 min. Perform the reaction under mineral oil in a thermal cycler (Gene ATAQ, Pharmacia). Use the same protocol to test the sensitivity of the PCR reaction and controls.

This method does not need oil chamber or micropipetting. Only microliter volumes are required for all purification and biochemical reactions. The compensation of the very small quantities of dissected chromosomal DNA in enlarged volumes is driven to completion by adding a large excess of synthetic linker-adaptors. Libraries generated from adaptor PCR products yield a 10–100 times higher number of recombinants as compared to using vector PCR. There is no limitation to a target region on the genome. Coding or non-coding regions may be selected by choosing suitable sites for restriction endonucleases. Quantities between 2 and 3 femtogram (fg) of DNA can even be detected and amplified by this method.

4. Microdissection and FISH for localization of gene sequences
(after Kamisugi and Fukui, 1992)

Prepare chromosome samples by the ordinary cytological method described previously (Fukui *et al.*, 1992) with minor modifications. Sow seeds in petri dishes for germination at 28°C. Excise root tips 1–2 cm long. Pretreat with distilled water at 0°C for 18 h. Fix in methanol : acetic acid (3 : 1). After fixation, wash thoroughly and macerate with an enzymatic mixture (2% cellulase Onozuka RS, Yakult Honsha, Tokyo; 0.3% pectolyase Y-23, Seishin Pharmaceutical Tokyo; 1.5% macerozyme R-200, Yakult Honsha, Tokyo;

and 1 mM EDTA pH 4.2). Spread either on glass slides for FISH or on tinted polyester membranes laid over the surface of the bottom of 35 mm plastic petri dishes for probe preparation. No staining is needed.

For laser dissection, direct-cloning and direct-labeling, use a standard laser dissection method described by Fukui *et al*. (1992) to cut out the disks of the polyester membrane on which around a hundred of barley nuclei are spread. Recover the disks into micro-centrifuge tubes (500 μl) with fine forceps using an inverted microscope.

Pretreat with a proteinase solution (1 mg/ml proteinase K, Wako Pure Chemical, Osaka, Japan, with 0.45% Tween 20 and Nonidet P-40, Sigma, in PCR buffer, Perkin Elmer Cetus, USA) at 55°C for 1 h.

Carry out a standard PCR method using a thermal cycler (Perkin Elmer Cetus, USA), according to the manufacturer's instructions. A programmed thermal cycle is 94°C (1 min), 55°C (2 min) and 72°C (2 min) and each cycle is repeated 30 times. Use a 100 μl aliquot reaction mixture containing 2.5 units of *Taq* polymerase, 0.2 mM DNA substrates with biotin-dUTP and 1 mM of a pair of primers in the PCR buffer. Use a pair of the primers with 20 bases for the direct-cloning and simultaneous direct-labelling of the 5S rDNA, which amplify the 301 bp coding region with the flanking spacer region. The sequences of the primers were:

5'—G A T C C C A T C A G A A C T C C G A A G—3' and

5'—C G G T G A T T T A G T G C T G G T A T—3'.

For direct-labelling with biotin-dUTP, use the combinations of 30% dTTP and 70% biotin-dUTP.

For fluorescent in situ *hybridization (FISH) and image analysis* treat slides with 100 μg/ml RNase A (Sigma) in 2× SSC at 37°C for 1 h. Denature the hybridization mixture (4 μg/ml biotin-labelled probe, 50% (v/v) formamide, 10% (w/v) dextran sulfate, 500 μg/ml salmon sperm DNA in 2× SSC) at 85°C for 10 min. Apply a 30 μl aliquot onto each slide previously denatured in a 70% formamide solution at 70°C for 1 min. After rinsing, apply 200 μg/ml fluorescein-avidin (Vector Lab, USA) at 37°C for 1 h. Counterstain with 1.25 μg/ml propidium iodide.

At least three photographs with the good *in situ* hybridization signals are taken using high sensitive colour positive films (Ektachrome ISO 400, Kodak). Enlarge the fluorescent signal recorded in the photographs by imaging methods using normalization, enhance contour and median filters.

5. Microdissection of B chromosomes and cloning (after Jamilena *et al*., 1995)

Microdissection: Fix root-tips of *Crepis capillaris* in 70% ethanol to prepare chromosome spreads. Macerate in 2% cellulase and 2% pectinase in 0.01M citrate buffer at 37°C for 1h and squash in a drop of 45% acetic acid on a coverglass. Stain chromosomes with 2% methylene blue and subject to microdissection. Isolate the B chromosomes with a glass needle using a Zeiss inverted microscope and an Eppendorf micromanipulator. Collect in a 0.5 ml tube containing 10 µl of water.

Chromosomal amplification and cloning: Amplify the isolated chromosomal DNA by DOP-PCR. Perform the reactions in a 50 µl final volume containing 60 ng of the degenerate primer 6-MW; 2.5 mM MgCl$_2$; 200 µM dNTPs, 1 × *Taq* DNA polymerase buffer (BRL) and 5U of *Taq* polymerase (BRL). Denature initially at 94°C for 10 min. Then run 5 cycles at 94°C for 1 min; 30°C for 1 min and 72°C for 1.5 min; followed by 30 cycles at 94°C for 1 min; 65°C for 1 min and 72°C for 1.5 min, with an extension step of 1 s per cycle and a final extension at 72°C for 10 min. Reamplify 10 µl of the first amplification product for 30 s at the higher annealing temperature.

Ligate one fiftieth of the PCR product from the second round of amplification to 100 ng of the plasmid PCT™ using the TA cloning kit from Invitrogen. Use one tenth of the ligation mixture to transform competent cells of *Escherichia coli* DH5α (BRL).

Use dot-blot hybridization to calculate the relative amount of B134 related sequences in both standard genome and B chromosomes. Denature defined amounts of B134 insert DNA and genomic DNA from OB and 3B plants, immobilise on nylon and hybridise to the labelled insert of B134. Estimate the relative amounts of this repeated sequence in the genome of OB and 3B plants by comparing densitometric scans of hybridization signals in genomic DNA and those obtained from B134 inserts.

6. Techniques adapted for isolation of metaphase chromosomes at different levels (after Schubert *et al.*, 1993).

Method (i): *Micromanipulation*:

Treat root tips for 3 h in 0.05% colchicine, fix for 15 min in 45% acetic acid and squash on coverglass by dry ice technique. Dehydrate in 70% and 96% ethanol, airdry and use.

Place the coverglass in a chamber (3 mm deep cut in a glass slide), filled with liquid paraffin with the squashed cells touching the paraffin oil. Place at its side another empty siliconized coverglass that is not touching the surface of the paraffin oil and is positioned on the slide holder to the microscope. Take up the chromosomes from the coverglass with tiny needles (with a 2 µm tip beat at an angle of 40°) of the micromanipulator guided by a mechanical system laterally into the chamber and transfer to the empty coverglass. Stick the back of the coverglass with the newly placed chromosomes onto a glass slide using Entellan. Wash the slides with chloroform and 96% ethanol to remove paraffin from chromosomes and airdry.

Method (ii): Isolation by preparing chromosome suspension:
Incubate seedlings of *Vicia faba* with 2 cm long main roots for 18 h at 24°C in aerated Hoagland solution containing 1.25 mM hydroxyurea to block cell cycle at S phase. Rinse in distilled water. Incubate excised root tips in fresh Hoagland solution for 6 h and then for 3 h in 0.05% colchicine to arrest cells at metaphase. Rinse in distilled water. Fix root-tips with 6% (v/v) formaldehyde in 15 mM Tris buffer, pH 7.5, for 30 min at 4°C. Wash twice for 20 min in Tris buffer at 4°C. Chop up the meristems of 30 root tips with a sharp scalpel in a petri dish containing 1 ml LBO1 buffer (15 mM Tris-HCl; 80 mM KCl; 20 mM NaCl; 2 mM disodium EDTA; 0.5 mM spermine; 0.1% Triton X-100; 15 mM mercaptoethanol, pH 7.5). Pass this suspension of released chromosomes and nuclei through a 50 μm nylon filter to remove tissue and cellular fragments. Syringe twice through a needle of 0.7 mm diameter. Layer 0.7 ml of the suspension in a glass tube on the top of 0.7 ml 40% sucrose. Centrifuge at 200 rpm for 15 min. Transfer supernatant carefully into sterile Eppendorf tubes. This suspension contains upto 500 chromosomes per μl. Drop immediately on ice cold slides in 5 μl portions. Air dry.

These slides can be used for the different banding and *in situ* procedures and scanning electron microscopy.

BIBLIOGRAPHY

PREFACE

DARLINGTON, C.D. AND LA COUR, L.F. (1968) *The handling of chromosomes.* George Allen and Unwin, London.

GLICK, B.R. AND THOMPSON, J.E. (1993) (eds.) *Methods of Plant molecular biology and biotechnology.* CRC Press, Boca Raton, FL. USA.

GOSDEN, J.R. (1994) (ed.) *Methods in molecular biology* **29**, Humana Press, Totowa, N.J., USA.

MALIGA, P., KLESSING, D.F., CASHMORE, A.R., GRUISSEME, W. AND VARNER, S.E. (1995) *Methods in plant molecular biology—a laboratory manual*, Cold Spring Harbor Lab. Press, USA.

SAMBROOK, J., FRITSCH, E.F. AND MANIATIS, T. (1989) *Molecular cloning— a laboratory manual,* 2nd ed. **1**, 2.

SHARMA, A.K. AND SHARMA, A. (1994) *Chromosome techniques—a manual*, Harwood Academic, Chur, Switzerland.

CHAPTER II.1

BELLING, J. (1926) *Biol. Bull.* **50**, 160.

CASPERSSON, T., FARBER, S., FOLEY, G.E., KUDYNOWSKI, J., MODEST, E.J., SIMONSEN, E. AND WAUGH, U. (1968) *Exptl. Cell Res.* **49**, 219.

CHAUDHURY, M., CHAKRAVARTY, D.P. AND SHARMA, A.K. (1962) *Stain Techn.* **37**, 95.

DARLINGTON, C.D. AND LA COUR, L.F. (1968) *The handling of chromosomes.* Allen and Unwin. London.

FEUGEN, R. AND ROSSENBECK, H. (1924) *Zelfs. Physiol. Chemie* **135**, 203.

HARRIS, B.J. AND BLACKMAN, G.E. (1954) *Nature* **173**, 642.

JOHANSEN, D.A. (1940) *Plant microtechnique*, McGraw Hill, New York.

LA COUR, L.F. (1941) *Stain Techn.* **16**, 169.

NEWTON, W.F.C. (1926) *J. Linn. Soc. (Bot.)* **47**, 339.

PELC, S.R. (1956) *Nature* **178**, 359.

SHARMA, A.K. AND BAL, A. (1953) *Stain Techn.* **28**, 255.

SHARMA, A.K. AND BHATTACHARJEE, D. (1952) *Stain Techn.* **22**, 20.

SHARMA, A.K. AND CHAUDRURI, M. (1962) *Nucleus* **5**, 137.

SHARMA, A.K. AND SHARMA, A. (1980) *Chromosome techniques—theory and practice.* 3rd ed. Butterworths, London.

TJIO, J.H. AND LEVAN, A. (1950) *Anal. Estac. Exptl. de Aula dei* **2**, 21.

WEISBLUM, B. AND DE HASETH, P. L. (1972) *Proc. Natl. Acad. Sci. US* **6**, 629.
ZEISEL, S. (1883) *Mh. Chem.* **4**, 162.

CHAPTER II.2

CARPENTER, B.G., BALDWIN, J.P., BRADBURY, E.M. AND IBEL, K. (1976) *Nucleic Acids Res.* **3**, 1739.
COSTAS, E. AND GOYANES, V.J. (1987) *Chromosoma* **95**, 435.
FUKUI, K. AND KAKEDA, K. (1990) *Genome* **33**, 450.
FUKUI, K. AND KAKEDA, K. (1994) *Jpn. J. Genet.* **69**, 537.
FUKUI, K. AND KAMISUGI, N. (1995) *Chromosome Research* **3**, 79.
GAY, H. AND ANDERSON, T.F. (1954) *Science* **120**, 1071.
GILLIES, C.B. (1983) *Kew Chromosome Conference* II, pp. 115. (ed.) BRANDHAM, P.E. AND BENNETT, M.D. George Allen and Unwin, London.
GOODPASTURE, C. AND BLOOM, S.E. (1975) *Chromosoma* **53**, 37.
HAAPALA, O. (1985) In *Advances in Chromosome and Cell Genetics* **1**, (ed.) SHARMA, A.K. AND SHARMA, A. Gordon Breach, London, pp. 173.
IIJIMA, K. AND FUKUI, K. (1991) *Bull. Nat. Inst. Agrobiol. Resource* **6**, 1.
KAMISUGI, Y., SAKAI, F., MINEZAWA, M., FUJISHITA, M. AND FUKUI, K. (1993) *Theor. Appl. Genet.* **85**, 825.
KORNBERG, R. (1980) *Nature* **292**, 579.
LASKEY, R.A. AND EARNSHAW, W.C. (1980) *Nature* **286**, 763.
LEWIS, C.D. AND LAEMMLI, U.K. (1982) *Cell* **29**, 171.
MARTIN, R., BUSCH, W., HERRMANN, R.G. AND WANNER, G. (1994) *Chromosome Research* **2**, 411.
MIRSKY, A.F. (1947) *Symp. Cold Spring Harbor Quant. Biol.* **12**, 143.
NAKAYAMA, S. AND FUKUI, K. (*1995*) *Jpn. J. Genet.* **70**, 267.
NOGUCHI, J. AND FUKUI, K. (1995) *J. Plant Research* **108**, 209.
STUBBLEFIELD, V. AND WRAY, W. (1971) *Chromosoma* **22**, 262.
VAN HOLDE, K.F. (1988) *Chromatin.* Springer Series in molecular biology Springer, 219.
WANNER, G., FORMANEK, H., MARTIN, R. AND HERRMANN, R.G. (1991) *Chromosoma* **100**, 103.
WANNER, G. AND FORMANEK, H. (1995) *Chromosome Research*, **3**, 368.

CHAPTER II.3

BENNETT, M.D. AND LEITCH, I.J. (1995) *Ann. Bot.* **76**, 113.
CHARLESWORTH, B., SNIEGOWSKI, P. AND STEPHEN, W. (1994) *Nature* **171**, 215.

CHOUDHURY, R.K., MUKHERJEE, S. AND SHARMA, A.K. (1986) In *Lathyrus and Lathyrism* (ed.) DELBOS, M. AND KUNTZ, M. *Proc. Lathyrus Colloq.* World Med. Res. Foundation, New York.

DE GROOT, B., DE LAAT, A.M.M., PUITE, K.J. AND SREE RAMULU, K. (1986) In *Gene Structure and Function in Higher Plants* (ed.) REDDY, G.M. AND COE, E.H. 167, Oxford—IBH, New Delhi.

FANTES, J.A., GREEN, A. AND SHARNEY (1994) In *Methods in Cell Biology* (ed.) GOSDEN, J.R. **29**, 205, Humana Press, Totowa, N.J., USA.

FLAVELL, R.B. (1980) *Ann. Rev. Plant Physiol.* **3**, 569.

GRAY, J.W. AND GRAM, L.S. (1990) In *Flow cytometry and sorting* (eds.) MELAMED, M.R. AND MENDELSSOHN, M.L. 503, Wiley-Liss, New York.

GUALBERTI, G., DOLEZEL, J. MACAS, J. AND LUCRETTI, S. (1996) *Theor. Appl. Genet.* **92**, 744.

HESLOP-HARRISON, J.S., SCHWARZACHER, T., ANAMTHAWAT-JONSSON, K., LEITCH, A.R., SHI, M. AND LEITCH, I.J. (1991) In *Technique—a Journal of Methods in Cell and Molecular Biology* 3, 109.

LEITCH, A.R., SCHWARZACHER, T., WANG, M.Z., MOORE, G. AND HESLOP-HARRISON, J.S. (1991) *Cytometry* Suppl. **5**, 39.

MUKHERJEE, S. AND SHARMA, A.K. (1993) *Cytobios* **75**, 33.

PATAU, K. (1952) *Chromosoma* **5**, 341.

POLLISTER, A. W., SWIFT, H. AND RASCH, E.M. (1969) In *Physical Techniques for Biological Research* **31**, 201, Academic Press, New York.

RAO, V.L.K. AND SHARMA, A.K. (1987) *Cytologia* **52**, 593.

REID, N. (1974) In *Practical Methods in Electron Microscopy* (ed.) GLAUERT, A.M., North Holland, Amsterdam.

SHARMA, A.K. (1983) In *Kew Chromosome Conference* II (ed.) BRANDHAM, P.E. and BENNETT, M.D., 35, George Allen and Unwin, London.

CHAPTER II.4

DANIELLI, J.F. (1947) *Symp. Soc. Exp. Biol.* **1**, 101.

ERRERA, M. (1951) *Biochim. Biophys. Acta* **7**, 605.

GOLDSTEIN, D.J. (1961) *Nature* **191**, 406.

GOMORI, G. (1939) *Proc. Soc. Exp. Biol. N.Y.* **42**, 23.

KURNICK, N.B. (1947) *Cold Spring Harb. Symp. Quant. Biol.* **12**, 141.

KURNICK, N.B. AND MIRSKY, A.E. (1950) *J. Gen. Physiol.* **33**, 265.

MANNHEIMER, L.H. AND SELIGMAN, A.M. (1948) *J. Nat. Cancer Inst.* **9**, 181.

MENTEN, M.L., JUNGE, J. AND GREEN, M.H. (1944) *J. Biol. Chem.* **153**, 471.

PEARSE, A.G.E. (1972) *Histochemistry—Theoretical and Applied*, Little Brown, Boston, Maryland.

SEMMENS, C.S. AND BHADURI, P.N. (1941) *Stain Techn.* **16**, 119.

SHARMA, A.K. AND MOOKERJEA, A. (1955) *Stain Techn.* **30**, 1.
TAKAMATSU, A. (1939) *Trans. Soc. Japan* **29**, 429.

CHAPTER II.5

CASPERSSON, T., HULTEN, M., LINDSTEN, J. AND ZECH, L. (1971) *Hereditas* **67**, 147.
CHATTOPADHYAY, D. AND SHARMA, A.K. (1988) *Stain Techn.* **63**, 283.
COMINGS, D.E. (1974) In *The Cell Nucleus* (ed.) BUSCH, H. **1**, New York, Academic.
DE CARVALHO, C.R. AND SARAIVA, L.S. (1993) *Heredity* **70**, 515.
DE LA TORRE, J. AND SUMNER, A.T. (1994)
DRETS, M.E. AND SHAW, M.W. (1971) *Proc. Natl. Acad. Sci. USA* **68**, 2073.
FUKUI, K. (1985) *The Cell* (Tokyo) **17**, 145.
FUKUI, K. (1986) *Theor. Applied Genet.* **7**, 227.
GERLACH, W.L. (1977) *Chromosoma* **62**, 49.
GILL, B.S., FRIEBE, B. AND ENDO, T.R. (1991) *Genome* **34**, 830.
HECHT, F., WYANDT, H.E. AND MCGENIS, R.F.H. (1974) In *The Cell Nucleus* (ed.) BUSCH, H. **2**, 32, Academic Press, New York.
KAKEDA, K., YAMAGATA, H., FUKUI, K., OHNO, M., FUKUI, K., WEI, Z.Z. AND ZHU, F.S. (1990) *Theor. Appl. Genet.* **80**, 265.
KAKEDA, K., FUKUI, K. AND YAMAGATA, H. (1991) *Theor. Appl. Genet.* **81**, 144.
KAMISUGI, Y. AND FUKUI, K. (1990) *Biotechniques* **8**, 290.
KAMISUGI, Y., IKEDA Y., OHNO, M., MINEZAWA, M. AND FUKUI, K. (1992) *Genome* **35**, 793.
KIHLMAN, B.A. AND KRONBORG, D. (1975) *Chromosoma* **51**, 1.
LATT, S.A. (1974) *Science* **185**, 74.
LAVANIA, U.C. AND SHARMA, A.K. (1979) *Stain Techn.* **54**, 261.
MARKS, G.F. AND SCHWEIZER, D. (1974) *Chromosoma* **44**, 405.
MATSUI, S.I. AND SASAKI, M.S. (1975) *Jpn. J. Genet.* **50**, 189.
MIRSKY, A.F. (1947) *Cold Spring Harb. Quant. Biol.* **12**, 143.
MODEST, E.J. AND SENGUPTA, S.K. (1973) *Nobel Symposium*, Academic Press, New York **23**, 327.
NAKAYAMA, S. AND FUKUI, K. (1995) *Jpn. J. Genet.* **70**, 267.
PARDUE, M.L. AND GALL, J.G. (1970) *Science* **168**, 1356.
SCHUBERT, I. (1990) *Caryologia* **43**, 117.
SEN, S. (1965) *Nucleus* **8**, 79.
SHANG, M.X., JACKSON, R.C. AND NGUYEN, H.T. (1988) *Genome* **30**, 956.
SHARMA, A.K. AND SHARMA, A. (1973) In *Encyclopedia of microscopy and microtechniquc* (ed.) GRAY, P. 77, Van Nostrand, Reinhold, New York.
SHARMA, A.K. (1975) *J. Ind. Bot. Soc.* **54**, 1.
SHARMA, A.K. (1978) *Proc. Ind. Acad. Sci.* **87B**, 161.

Sharma, A.K. and Sharma, A. (1980) *Chromosome Techniques—Theory and Practice.* Butterworth, London.

Sharma, A.K. and Sharma, A. (1994) *Chromosome Techniques: a manual.* Harwood Academic. Chur. Switzerland.

Stockert, J.C. and Lisanti, J.A. (1972) *Chromosoma* **37**, 117.

Sumner, A.T. (1994) In *Methods in Molecular Biology* **29**, (ed.) Gosden, J.R., p. 83, Humana Press, Totowa, N.J. USA.

Taylor, J.H., Woods, P.S. and Hughes, W.L. (1957) *Proc. Natl. Acad. Sci. (Wash.)* **43**, 122.

Templaar, M.J. *et al.* (1982) *Mutation Res.* **103**, 321.

Vosa, C.G. (1973) *Chromosoma* **43**, 269.

Vosa, C.G. (1977) In *Current Chromosome Research* 105 (ed.) James, K. and Brandham, P.E. Elsevier, Amsterdam.

Yanagisawa, T., Tano, S., Fukui, K. and Harada, K. (1993) *Jpn. J. Genet.* **68**, 119.

Yi, H. and Zhang, Z. (1992) *Plant Chromosome Research* (ed.) Tanaka, R., 153, Internat. Academic Publ., Beijing.

Zhang, Z. and Yang, X. (1986) *Acta Bot. Sinica* **28**, 595.

Zhang, S.L. *et al.* (1991) *Mutation Res.* **261**, 69.

Zhu, F. and Wei, J. (1987) *Plant Chromosome Research, Proc. Sino-Jpn. Symp. Plant Chromosomes* 135.

CHAPTER II.6

Brown, T.A. (1991) (ed.) *Essentials in Molecular Biology—a practical approach* 1 *and* 2 *IRL* Press, Oxford.

Busch, W., Herrmann, R.G., Houben. A. and Martin. R. (1996) *Plant Mol. Biol. Rep.* **14**, 149.

Davis, R.W., Thomas, M., Cameron, J., St. John, T.P. Scherer, S. and Padgett, R.A. (1980) *Methods in Enzymology* **65**, 404.

Dellaporta, S.L., Wood. J. and Hicks, J.B. (1983) *Plant Molecular Biology Res.* **1**, 19.

Dolezel, J., Binarova, P. and Lucretti, S. (1989) *Biologia Plantarum* **31**, 113.

Doyle, J. (1991) In *Molecular Techniques in Taxonomy* (ed.) Hewitt, G.M. *et al.*, Springer-Verlag, Berlin.

Griesbach, R.J., Malmberg, R.L. and Carlson, P.S. (1982) *J. Hered.* **73**, 151.

Hadlaczky, G. (1984) In *Cell Culture and Somatic Cell Genetics of Plants* **1**, 461, Academic Press, New York.

Kaufman, P.B., Wu, W., Kim, D. and Csekedil, L. (1995) (eds.) *Handbook of Molecular and Cellular Methods in Biology and Medicine*, 380, CRC Press, Boca Raton, FL.

MAULE, J. (1994) In *Methods in Molecular Biology* (ed.) GOSDEN, J.R. **29**, 221, Humana Press, Totowa. N.J., USA.

PASTERNAK, J.J. (1993) In *Methods in Plant Molecular Biology* (ed.) GLICK, B.R. AND THOMPSON, J.E., 29, CRC Press, Boca Raton, FL.

SCHUBERT, I., DOLEZEL, J., HOUBEN, A., SCHERTHEM, H. AND WANNER, G. (1993) *Chromosoma* **102**, 96.

SMITH, D.B. AND FLAVELL, R.B. (1974) *Biochem. Genet.* **12**, 243.

SONNEBICHLER, J., MACHIKAO, F. AND ZETL, I. (1977) In *Methods in Cell Biology* (ed.) PRESCOTT, D.M., Academic Press, New York.

WANNER, G., FORMANEK, H., MARTIN, R. AND HERRMANN, R.G. (1991) *Chromosoma* **100**, 103.

WU, T.Y. AND WU, R. (1987) *Nucleic Acids Res.* **15**, 5913.

WU, H.K., CHUNG, M.C., WU, T., NING, C.N. AND WU, R. (1991) *Chromosoma* **100**, 330.

XIE, Y. AND WU, R. (1989) *Plant Molecular Biol.* **13**, 53.

ZHAO, X., WU, T., XIE, Y. AND WU, R. (1989) *Theor. Appl. Genet.* **78**, 201.

CHAPTER II.7

ABBO, S., MILLER, T.E. AND KING, I.P. (1993a) *Genome* **36**, 815.

ABBO, S., DUNFORD, R.B., MILLER, T.E., READER, S.M. AND KING, I.P. (1993b) *PNAS*, USA **90**, 11821.

BEDBROOK, J.R., JONES, J., O'DELL, M., THOMPSON, R.J. AND FLAVELL, R.B. (1980) *Cell* **19**, 545.

BEJARANO, E.R., KHASHOGGI, A., WITTY, M. AND LICHTENSTEIN, C. (1996) *PNAS USA* **93**, 759.

BENNETT, M.D. (1996) In *Unifying Plant Genomes* (ed.) HESLOP-HARRISON, J.S. *Symp. Soc. Exp. Biol.* 45.

BENNETT, S.T., KENTON, A.Y. AND BENNETT, M.D. (1992) *Chromosoma* **101**, 420.

BENNETT, S.T., LEITCH, I.J. AND BENNETT, M.D. (1995) *Chromosome Research* **3**, 101.

DEJONG, J.H., GARRIGA-CALDERE, F. AND RAMANNA, M.S. *et al.* (1996) In IV *Kew Chromosome Conference* (eds.) BRANDHAM, P.F. AND BENNETT, M.D., 57.

FRANSZ, P.F., BLANCO, C.A., LIHARSKA, T.B., PEETERS, A.J.M., ZABEL, P. AND DEJONG, J. H. (1996) *The Plant Journal* **9**, 421.

FUCHS, J. AND SCHUBERT, I. (1995) *Chromosome Research* **3**, 94.

FUCHS, J., PICH, U., MEISER, A. AND SCHUBERT, I. (1994) *Chromosome Research* **2**, 25.

FUCHS, J., KLOOS, D. U., GANAL, M. W. AND SCHUBERT, I. (1996) *Chromosome Research* **4**, 277.

FUKUI, K., OHMIDO, N. AND KHUSH, G.S. (1994) *Theor. Appl. Genet.* **87**, 893.

GOSDEN, J.R. (1994) (ed.) *Methods in Molecular Biology* **29**, Humana Press. Totowa, N.J., USA.

GOSDEN, J.R. AND LAWSON, D. (1994) In *Methods in Molecular Biology* (ed.) GOSDEN, J.R. 323, Humana Press, Totowa, N.J., USA.

GUSTAFSSON, J.P., BUTLER, E. AND MCINTYRE, C.L. (1990) *Proc. natl. Acad. Sci., USA* **87**, 1899.

HESLOP-HARRISON, J. S., SCHWARZACHER, T., ANANTHAWAT-JOHNSON, K., LEITCH, A.R., SHI, M. AND LEITCH, I.J. (1991) *Technique* **3**, 109.

HESLOP-HARRISON, J.S. (1996) *Unifying plant genomes—comparison, colinearity and conservation*, Company of Biologists, Cambridge, 17.

HESLOP-HARRISON, J.S. AND SCHWARZACHER, T. (1996) In *Methods in genome analysis in plants* (ed.) JAUHAR, P.P., 163, CRC Press, Boca Raton, FL, USA.

HEYTING, C., DIETRICH, A.J.J., DEJONG, J.H. AND HARTSNIKER, E. (1994) In *Methods in Molecular Biology* **29**, (ed.) GOSDEN, J.R., Humana Press, Totowa, N.J., USA.

HOUBEN, A., BELYAEV, N.D., TURNER, B.M. AND SCHUBERT, I. (1996a) *Chromosome Research* **4**, 191.

HOUBEN, A., BELYAEV, N.D., LEACH, C.R. AND TIMMIS, J.M. (1997) *Chromosome Research* **5**, 235.

HOUBEN, A., BRANDES, A., PICH, U., MANTEUFFEL, R. AND SCHUBERT, I. (1996b) *Theor. Applied. Genet.* **93**, 477.

HUANG, P.L., HALLBROCK, K., AND SOMSSICH, I.E. (1988) *Mol. Gen. Genet.* **211**, 143.

JELLEN, E.N., GILL, B.S. AND COX, T.S. (1994) *Genome* **63**, 613.

JEPPESON, P. (1994) In *Methods in Molecular Biology* **29**, 253 (ed.) GOSDEN, J.R., Humana Press, Totowa, N.J., USA.

JIANG, J. AND GILL, B.S. (1994) *Chromosome Research* **2**, 59.

KENTON, A., PAROKONNY, A.S., GLEBA, Y.Y. AND BENNETT, M.D. (1993) *Molecular and General Genetics* **240**, 159.

KENTON, A., KHASHOGGI, A., PAROKONNY, A., BENNETT, M.D. AND LICHTENSTEIN, C. (1995) *Chromosome Research* **3**, 346.

KOCH, J. (1995) *Methods: A Comparison of methods in Enzymology* **9**, 122.

LAVANIA, U.C. (1998) Personal Communication.

LEITCH, A.R., MOSGOLLER, W., SCHWARZACHER, T., BENNETT, M.D. AND HESLOP-HARRISON, J.S. (1990) *J. Cell Science* **95**, 335.

LEITCH, I.J., LEITCH, A.R. AND HESLOP-HARRISON, J.S. (1991) *Genome* **34**, 329.

LEITCH, I.J., LEITCH, A.R., SCHWARZACHER, T., MALUSZYNSKA, J., ANAMTHAWAT-JONSSON, K., SHI, M., HARRISON, G. AND HESLOP-HARRISON, J.S. (1992) *Update* Brochure.

LEITCH, I.J., PAROKONNY, A. AND BENNETT, M.D. (1997) In *Chromosomes Today* (eds.) HENRIQUESHIL, N., PARKER, J.S. AND PUERTAS, M.J. **12**, 333, Chapman and Hall, London.

LEVI, N. AND MATTEI, G. (1995) In *Gene Probes* **2**, *A Practical Approach* (eds.) JAMES. B.D. AND HIGGINS, S.J., IRL Press, Oxford.

MEIER, T. AND FAHRENHOLZ, F. (1996) *A Laboratory Guide to Biotin Labelling in Biomolecalar Analysis—Biomethods*, Birkhauser-Verlag, Berlin.

MUKAI, Y., ENDO, T. R. AND GILL, B. S. (1990) *J. Hered.* **81**, 290.

MUKAI, Y. (1996) In *Methods of Genome Analysis in Plants* (ed.) JAUHAR, P.P., CRC Press. Boca Raton, FL.

MUKAI, Y. AND APPELS, R. (1996) *Chromosome Research* **4**, 401.

MUKAI, Y., NAKAHARA, Y. AND YAMAMOTO, M. (1993) *Genome* **36**, 489.

PALEVITZ, B.A. (1990) *Protoplasma* **157**, 120.

PAROKONNY, A.S., KENTON, A.Y., GLEBA, Y. AND BENNETT, M.D. (1992) *Plant J.* **2**, 695.

PINKEL, D., STRAUME, T. AND GRAY, J.W. (1986) *Proc. Natl. Acad. Sci. USA* **83**, 2934.

READER, S.M. ABBO, S., PURDIE, K.A., KING, I.P. AND MILLER, T.E. (1994) *Trends in Genetics* **10**, 265.

SCHWARZACHER, T., LEITCH, A. R., BENNETT, M. D. AND HESLOP-HARRISON, J. S. (1989) *Ann. Bot.* **64**, 315.

SCHWARZACHER, T. AND HESLOP-HARRISON, J.S. (1993) *Genome* **34**, 317.

SCHWARZACHER, T. AND HESLOP-HARRISON, J.S. (1993) In *Methods in Molecular Biology* **20**, chap. 26, (ed.) ISAAC, P.G., Humana Press, Totowa, N.J., USA.

SCHWARZACHER, T., HESLOP-HARRISON, J.S. AND LEITCH, A.R. (1994) In *Plant cell biology—a practical approach* (eds.) HARRIS, N. AND OPERKA, K.J., Oxford University Press, Oxford. 127.

SHEN, D., WANG, Z. AND WU, M. (1987) *Chromosoma* **95**, 311.

YANASIGAWA, T., TANO, S., FUKUI, K. AND HARADA, K. (1993) *Jpn. J. Genet.* **68**, 119.

SECTION III

ADAM BLORDON, A.F., SERIGNAC, M., DRON, M. AND BUNNEROT, L. (1995) In *Gene Probes—a practical approach*, (eds.) HAMES, B.D. AND HIGGINS, S.J., IRL Press, Oxford.

RIESEBERG, L. H., CHOI, H., CHAN, R. AND SPORE, C. (1993) *Heredity* **70**, 285.

CHAPTER III.1

HALL, L. (1995) In *Gene Probes 2—a practical approach* (eds.) HAMES, B.D. AND HIGGINS, S.J., IRL Press, Oxford.

INNIS, M.A., GELFUND, D.H., SNINSKY, J.J. AND WHITE, T. (1990) (eds.) *PCR Protocols, a guide to methods and applications*, Academic Press, New York.

KAUFMAN, P.B., WU, W., KIM, D. AND CSEKEDL, L. (1995) (eds.) *Handbook of Molecular and Cellular Methods in Biology and Medicine* 380, CRC Press, Boca Raton, FL.

MALIGA, P., KLESSING, D.F., CASHMORE, A.R., GRUISSEME, W. AND VARNER, S.E. (1995) *Methods in Plant Molecular Biology—a laboratory manual*, Cold Spring Harbor Lab. Press. USA.

ROWLAND, I.J. AND LEVI, A. (1994) *Theor. Applied Genet.* **87**, 563.

WEISING, K., NYBOM, H., KIRSTEN, W. AND MEYER, W. (1995) *DNA fingerprinting in Plants and Fungi*, CRC Press, Boca Raton, FL.

CHAPTER III.2

GLICK, B.R. AND THOMPSON, J.E. (1993) (eds.) *Methods of Plant Molecular Biology and Biotechnology*, CRC Press, Boca Raton, FL., USA.

SAMBROOK, J., FRITSCH, F.F. AND MANIATIS, T. (1989) *Molecular Cloning—a laboratory manual* **1**, Cold Spring Harbor Press, New York.

SLIGHTON, J.L., DRONG, R.F. AND CHEE, P.P. (1993) In *Methods in Plant Molecular Biology and Biotechnology* (ed.) GLICK, B.R. AND THOMPSON, J.E., 121, CRC Press, Boca Raton, FL.

CHAPTER III.3

CHANG, C., BOWMAN, J.L., DEJOHN, A.W., LANDER, E.S. AND MEYEROWITZ, E.M. (1988) *Proc. natl. Acad. Sci. USA* **81**, 1991.

CHAUDHURI, R.K. Personal communication.

DAVIES, K. F. (1990) (ed.) *Genome Analysis—a practical approach* IRL Press, Oxford.

EDWARDS, K.J., THOMPSON, H., EDWARDS, D., DE SAIZIEU, A., SPARKS, C., THOMPSON, J.A., GREENLAND, A.J., YERS, M. AND SCHUCK, W. (1992) *Plant Mol. Biol.* **19**, 299.

GIBSON, T., COULSON, A., SULSTON, J. AND LITTLE, P.F.R. (1987) *Gene* **53**, 275.

GRILL, E. AND SOMERVILLE, S. (1991) *Molecular Gen. Genet.* **226**, 484.

HWANG, I., KOHCHI, T., HAUGE, B.M. AND GOODMAN, H.M. (1991) *The Plant Journal* **1**, 367.

NAM, H.G., GIRAUDET, J., DEN BER, B., MOONAN, F., LOOS, W.D.B., HAUGE, B.M. AND GOODMAN, H.M. (1989) *Plant Cell* **1**, 699.

SAMBROOK, J., FRITSCH, E.F. AND MANIATIS, T. (1989) *Molecular cloning—a laboratory manual* **1**, Cold Spring Harbor Press, New York.

SCHMID, R., GNOPS, G., BANCROFT, I. AND DEAN, C. (1992) *Aust. J. Plant Physiol.* **19**, 341.

SULSTON, J., MALLET, F., STADEN, R., DURBIN, R., HORSNELL, T. AND COULSON, A. (1988) *Comput. Applic. Biosci.* **4**, 125.

CHAPTER III.4

CHAUDHURI, R.K. (1997) Personal communication.

DUNCAN, R.R. (1997) *Adv. Agron.* **58**, 201.

CHAPTER III.5

LIECKEELDT, E., MEYER, W. AND BORNER, T. (1993) *J. Basic Microbiol.* **33**, 413.

LIN, J.J., MA, J., AMBROSE, M. AND KUO, J. (1997) *Focus, Plant Biotechnology* **19**, 36.

WEISING, K., NYBOM, H., WOLFF, K. AND MEYER, W. (1994) (eds.) *DNA Fingerprinting in Plant and Fungi*, CRC Press, Boca Raton, FL.

CHAPTER III.6

BENNETT, M.D. (1996) In *Unifying Plant Genomes* (ed.) HESLOP-HARRISON, J.S. *Symp. Soc. Exp. Biol.* **2**, 45, Society for experimental biology.

DEVOS, K.M., MOORE, G. AND GALE, M.D. (1995) *Euphytica* **85**, 367.

FLAVELL, R.B., GALE, M.D., O'DELL, M., MURPHY, G. AND LUCAS, H. (1993) In *Chromosomes Today* **11**, 19, (eds) SUMNER, A.T. AND CHANDLEY, A.C., Chapman and Hall, London.

LAW, G.N. (1995) *Euphytica* **85**, 1.

MOORE, G., GALE, M.D., KURUTA, N. AND FLAVELL, R.B. (1993) *Biotechnol.* **11**, 584.

CHAPTER IV.1

BLAKESLEE, A.F. AND AVERY, A. (1937) *J. Hered.* **28**, 392.

GHOSH, S. AND SHARMA, A.K. (1968) *J. Cyt. Genet.* **3**, 54.

HARTWELL, L.H. AND WEINER, T.A. (1989) *Science* **242**, 62

HUSKINS, C.L. AND STEINITZ, L.M. (1948) *J. Hered.* **39**, 34.

LEVAN, A. (1949) *Hereditas* suppl. vol. 325.

MURRAY, A. AND HUNT, T. (1993) *The Cell Cycle: an introduction*, Oxford University Press, New York.

NISHIMOTO, T. UZAWA, S. AND SCHEGEL, S. (1992) *Current Opinions in Cell Biology* **4**, 174.

OEHLKERS, F. (1943) *Zl. A. V.* **81**, 313.

SCHUBERT, I., DOLEZEL, J., HOUBEN, A., SCHERTHEM, H. AND WANNER, G. (1993) *Chromosoma* **102**, 96.

SEN, S.(1970) *Res. Bull.* **2**, Dept. of Botany, Univ. of Calcutta.

SHARMA, A.K. (1996) In *Methods in cell science* **18**, 75, Kluwer Academic.

SHARMA, A. (1984) *Environmental Chemical Mutagenesis* Persp. Rep **6**, Golden Jub. INSA, 53.

SHARMA, A.K. (1978) *Proc. Ind. Acad. Sci.* **87B**, 161.

SHARMA, A. (1995) *II S. G. Sinha Mem. Lecture, Nat. Acad. Sci India Lett.* **18**, 117.

SHARMA, A. (1986) *J. Ind. Bot. Soc.* **64**, 9.

SHARMA, A.K. AND MOOKERJEA, A. (1954) *Bull. Bot. Soc. Bengal* **8**, 24.

SHARMA, A.K. AND ROY, M. (1956) *La Cellule* **58**, 109.

SHARMA, A.K AND SEN, S. (1954) *Genet. Iber.* **6**, 19.

SHARMA, A.K. AND SHARMA, A. (1960) *Internat. Rev. Cytol.* **10**, 101.

SHARMA, A.K. AND SHARMA, A. (1980) *Chromosome Techniques—Theory and Practice*, 3rd ed., Butterworths, London

SHARMA, A.K. AND SHARMA, A. (1994) *Chromosome techniques—a manual*, Harwood Academic, Chur, Switzerland.

STADLER, L.J. (1928) *Anat. Rec.* **41**, 97.

TORREY, J.C. (1960) *Physiol. Plant.* **20**, 265.

CHAPTER IV.2

GRANT, W.F. (1994) *Mutation Res.* **322**, 175

JI, Q. AND CHEN, Y. (1996) *Mutation Res.* **359**, 1.

FAIRBURN, D.W., OLIVE, P.L. AND O'NEILL K.L. (1995) *Mutation Res.* **339**, 37.

MA, T. (1982) *Mutation Res.* **99**, 257.

MA, T.H., XU, Z.D., XU, C.G., MC CONNELL, H., RAHAGO, E.V., ARREOLA, Z. AND ZHANG, H.G. (1995) *Mutation Res.* **334**, 185.

PARRY, J.M. (1985) In *Mutagenicity testing in environmental pollution control*, Ellis Horwood, Chicester.

SANDHU, S.S., DE SERRES, F.J., GOPALAN, H.N., GRANT, W.F., SVENSGAARD, J., VELEMINSKY, J. AND BECKING, G.C. (1994) *Mutation Res.* **310**, 257.

SHARMA, A.K. AND SHARMA, A. (1989) In *Management of Hazardous materials and Wastes* (ed.) MAJUMDAR, S.K., MILLER, E.W. AND SCHMALZ, R.F. 280, *Penn. Acad. Sci. Publ.*, Easton, Penn. USA.

SHARMA, A.K. (1994) In *Biological monitoring of the environment* (eds.) SALANKI, J., JEFFREY, D. AND HUGHES, M. 25, CAB International, Oxford.

SORSA, M., HEMMINKI, K. AND VAINIO, H. (1982) *Teratogenesis Carcinogenesis and Mutagenesis* **2**, 137.

CHAPTER IV.3

BURNHAM, C.R. (1954) *Maize genetics crop Newsletter* **28**, 59.

CHEN, P.D., FUJIMOTO, H. AND GILL, B.S. (1994) *Theor. Applied Genet.* **84**.

ENDO, T.R. (1990) *Jpn. J. Genet.* **5**, 135.

ENDRIZZI, J.E., TURCOTTE, E.L. AND KOHEL, R.J. (1985) *Adv. Genet.* **23**, 272.

FELDMAN, M. (1988) *Proc. Int. Wheat Genet. Symp.*, 23 (eds.) MILLER, T.F. AND KOEBNER, R.M., Cambridge.

JIANG, T., FRIEBE, B. AND GILL, B.S. (1994) *Euphytica* **73**, 199.

JOSHI, B.C. AND SINGH, D. (1979) *Proc. V Int. Wheat Genetics Symp.*, New Delhi, 342.

KOTA, R.S. AND DVORAK, J. (1985) *Canad. J. Genet. Cytol.* **27**, 549.

KUSPIRA, J. AND UNRAU, J. (1957) *Canad. J. Plant Sci.* **37**, 300.

LAURIE, D.A., O'DONOUGHNE, L.S. AND BENNETT, M.D. (1990) In *Gene manipulation in plant improvement* (ed.) GUSTAFSSON, J.P., 95, Plenum Press, New York.

LIN, C.J., ATKINSON, M.D., CHINOY, K., DEVOS, M. AND GALE, M.D. (1992) *Theor. Applied Genet.* **83**, 305.

MATZ, F. AND MAHN, A. (1994) *Plant Breeding* **113**, 125.

MENZEL, H.Y., DOUGHERTY, B.J. AND RICHMOND, K.E. (1991) In *Chromosome Engineering in Plants, Genetics, Breeding and Evolution* 471 (eds.) TSUCHIYA, T. AND GUPTA, P.K., Elsevier, Amsterdam.

RILEY, R., CHAPMAN, V. AND JOHNSON, R. (1968) *Genet. Res.* **12**, 199.

SEARS, E.R. (1954) *Miss. Agr. Exp. Stn. Res. Bull.* **572**, 1.

SEARS, E.R. (1956) *Brookhaven Symp. Biol* **9**, 1.

SEARS, E.R. (1972) *Stadler Genet. Symp.* **4**, 23.

SEARS, E.R. (1977) *Canad. J. Genet. Cytol.* **19**, 585.

SEARS, E.R. (1981) In *Wheat Science—Today and Tomorrow* (eds.) EVANS, T. AND PACKARD, N.J., Columbia Univ. Press, 75.

SHARMA, H.C. AND GILL, B.S. (1983) *Euphytica* **32**, 17.

SYBENGA, J. (1995) *Euphytica* **83**, 53.

TSUCHIYA, T. (1961) *Jpn. J. Genet.* **36**, 444.

WERNER, J.E., ENDO, T.R. AND GILL, B.S. (1992) *PNAS, USA* **89**, 14307.

ZELLER, F.J. (1993) *Proc. IV Wheat Genet. Symp.* Columbia, MO., 209.

ZHANG, X., LI, Z. AND CHEN, S. (1992) *Theor. Appl. Genet.* **83** 708.

CHAPTER IV.4

BOUVIER, L., FILLON, F.R. AND LESPINASSE, Y. (1994) *Plant Breeding* **113**, 343.

CARLSON, P.S., SMITH, H.H. AND DEARING, P.D. (1972) *PNAS, USA* **66**, 2292.

CHEN, Z.Z., SNYDER, S., FAN, Z.G. AND LOH, W.H. (1994) *Plant Breeding* **113**, 217.

HANSEN, A.L., PLEVER, C., PEDERSON, H.C., KEIMER, B. AND ANDERSON, S.B. (1994) *Plant Breeding* **112**, 89.

ICHIKAWA, H., TANNO-SUENAGE, L. AND IMAMWEA, J. (1988) *Plant Cell Tissue and Organ Culture* **12**, 201.

JONES, L.E., HILDEBRANDT, S.E., RIKER, A.J. AND WU, J.H. (1960) *Amer. J. Bot.* **47**, 468.

LEE, C.H. AND POWER, J.B. (1988) *Plant, Cell, Tissue and Organ Culture* **12**, 197.

MATZ, F. AND MAHN, A. (1994) *Plant Breeding* **113**, 125.

MURASHIGE, T. AND SKOOG, F. (1962) *Physiologia Plant.* **15**, 473.

MURRAY, A. AND HUNT, T. (1993) *The Cell Cycle—an introduction*, Oxford Univ. Press, New York.

NAVARRO-ALVAREZ, W., BAENZIGER, P.S., ESKRIDGE, K.M., HUGO, M. AND GUSTAFSON, V.D. (1994) *Plant Breeding* **112**, 192.

POWER, J.B., CUMMINS, S.E. AND COCKING, E.C. (1970) *Nature* **225**, 1016.

POWER, J.B. (1976) *Nature* **263**, 500.

SHARMA, A.K. AND SHARMA, A. (1994) *Chromosome Techniques—a manual,* Harwood Academic, Chur. Switzerland.

SCHIEDER, O. (1975) *Z PflanzenPhysiol.* **74**, 357.

WATTS, J.W. AND KING, J.M.M. (1984) *Bioscience Rep.* **4**, 335.

WILSON, S.B., KING, P.J. AND STREET, H.E. (1971) *J. Exp. Bot.* **21**, 177.

CHAPTER IV.5

GLIMELIUS, K. (1988) In *Progress in Plant Protoplast Research* (eds.) PUITE, K.J., DONS, J.J.M., HUIZING, H.J., KOOL, A.J., KOORNEEF, M. AND KRENS, F.A., Kluwer, Dordrecht.

PUITE, K.J. (1992) *Physiol. Plant* **85**, 403.

RAMULU, K.S., DIJHUIS, P., RUTGERS, E., BLAAS, J., VERBEEK, W.H.J., VERHOEVEN, A. AND COLIJN-HOOYMANS, C.M. (1995) *Euphytica* **85**, 255.

RAMULU, K.S., DIJHUIS, P., RUTGERS, E., BLAAS, J., KRENS, F.A., DONS, J.J.M., COLIJN-HOOYMANS, C.M. AND VERHOEVEN, H.A. (1996) *Genome* **39**, 921.

WAARA, S. AND GLIMELIUS, K. (1995) *Euphytica* **85**, 217.

CHAPTER IV.6

ASANO, Y. AND UGAKI, M. (1994) *Plant Cell Rep.* **13**, 243.

ARENCIBIA, A., MOLINIA, P.R., DELARIVA, G. AND SELMAN HOUSEIN, G. (1995) *Plant Cell Rep.* **14**, 305.

CHAUDHURY, A., MAHESWARI, S.C. AND TYAGI, A.K. (1995) *Plant Cell Rep.* **14**, 215.

DE LAAT, A.H.M., VERHOEVEN, H.A. AND SRIRAMULU, K.S. (1989) In *Biotechnology in Agriculture and Forestry* (ed.) BAJAJ, P.S. Springer, Heidelberg.

GREENBERG, B.M. AND GLICK, B.R. (1993) In *Methods in Plant Molecular Biology* (eds.) GLICK, B.R. AND THOMPSON, J.E., CRC Press. Boca Raton, FL. (USA).

GRIESBACH, R.J. (1987) *Biotechniques* **3**, 348.

HAMILTON, A.J., LYCELL, C.W. AND GRIERSON, D. (1990) *Nature* **346**, 284.

JEFFERSON, R.A., KAVANAGH, T.A. AND BEVAN, M. (1987) *Embo. J.* **6**, 391.

KAUFMAN, P.B., WU, W., KIM, D. AND CSEKEDIL, L. (1995) (eds.) *Handbook of molecular and cellular methods in biology and medicine* CRC Press, Boca Raton, FL.

KIELLSSON, G., SIMONSEN, V. AND AMMANN, K. (1997) (eds.) *Methods for risk assessment of transgenic plants*, Birkhauser-Verlag, Basel.

MALIGA, P., WESSIG, D.F., GRUSSEMED, W. AND BARKER, S.E. (1995) (eds.) *Methods in plant molecular biology—a laboratory course manual* Cold Spring Harbor Lab. Press. USA.

MARIANI, C., DE BEUCKELAAR, M., TRUETNES, J., LEEMANS, J. AND GOLDBURY, R.B. (1990) *Nature* **347**, 737.

POTRYKUS, I. (1991) *Annu. Rev. Plant Physiol. Plant Mol. Biol.* **42**, 205.

VAN DER KROL, A.R., MUR, L.A., DE LANGE, P., GERATS, A.G.M., MOL, J.N.M. AND SUITJE, A.R. (1990) *Mol. Gen. Genet.* **220**, 204.

VASIL, I.K. (1994) *Plant Mol. Biol.* **5**, 299.

YANG, N.S. AND CHRISTON, P. (1994) *Particle Bombardment Technology for Gene Transfer*. Oxford Univ. Press. UK.

SECTION V

BROWN, S.D.M. AND CAREY, A.H. (1994) In *Methods in Molecular Biology* **29**, 425 (ed.) GOSDEN, J.R., Humana Press, Totowa, N.J., USA.

FUKUI, K., MINEZAWA, M., KAMISUGI, Y., ISHIKAWA, M., OHMIDO, N., YANAJISAWA, T., FUJISHITA, M. AND SAKAI, F. (1992) *Theor. Appl. Genet.* **84**, 787.

BIBLIOGRAPHY

FUKUI, K., NOMIYA, A., NISHIGUCHI, M. AND FUJISHITA, M. (1995) *Zoological Studies* **34**, 35.

JAMILENA, M., GARRIDO-RAMOS, M., RUIZ REJON, M., RUIZ REJON, C. AND PARKER, J.S. (1995) *Chromosoma* **104**, 113.

KAMISUGI, Y. AND FUKUI, K. (1992) *Plant Chromosome Research* 103.

PONELIUS, N., STEIN, N. AND WEBER, G. (1997) *Nucleic Acids Res.* **25**, 3555.

SCHONDELMAIER, J., MARTIN, R., JAHOOR, A., HOUHEN, A., GRANER, A., KOOP, H.U., HERRMANN, R., AND JUNG, C. (1993) *Theor. Appl. Genet.* **86**, 629.

SCHUBERT, I., DOLEZEL, J., HOUBEN, A., SCHERTHAN, H. AND WANNER, G. (1993) *Chromosoma* **102**, 96.

AUTHOR INDEX

ABBREVIATIONS
h = hour, min = minute, s = second

Printed and bound by CPI Group (UK) Ltd, Croydon, CR0 4YY

23/10/2024

01778226-0008